Electron Spectroscopy:
Theory, Techniques and Applications

ELECTRON SPECTROSCOPY:
THEORY, TECHNIQUES AND APPLICATIONS

Chapters to be included in Volumes 2 and 3

Editors

C. R. BRUNDLE

Research Laboratory
Department K33–281
5600 Cottle Road
San Jose, CA 95193
U.S.A.

A. D. BAKER

Department of Chemistry
Queens College
Kissena Boulevard
Flushing, New York 11367
U.S.A.

Author	*Title*
C. S. Fadley	X-Ray Photoemission, Basic Concepts
G. K. Wertheim	Final State Structure in X-Ray Photoemission
J. J. Huang and J. W. Rabalais	UV Photoelectron Cross-sections and Angular Distributions
J. J. Huang and J. W. Rabalais	X-Ray Photoelectron Cross-sections and Angular Distributions
D. Dill and S. Manson	The Photoionization of Atoms: Cross-sections and Photoelectron Angular Distributions
J. A. R. Samson	Experimental Determination of Angular Distributions and Cross-sections in the Vacuum UV
W. Jonathan, J. Dyke, and A. Morris	Progress in Photoelectron Spectroscopy of Transient Species
J. H. D. Eland	Ion Fragmentation Mechanisms and their Relationship to Photoelectron Spectroscopy
C. R. Brundle	Electron Spectroscopy and Surface Studies
D. Briggs	Industrial and Analytical Aspects of X-Ray Photoemission
C. E. Kuyatt	Electron Energy Analyzer Design
D. S. Urch	X-ray Emission Spectroscopy
R. A. Bonham and H. F. Wellenstein	Electron Impact Spectroscopy: High Energy Aspects
R. Huebner and R. J. Celotta	Low Energy Electron Scattering from Atoms and Molecules
C. J. Todd	Auger Spectroscopy

Electron Spectroscopy: Theory, Techniques and Applications

Volume 1

Edited by

C. R. Brundle

*IBM Research
San Jose
California, U.S.A.*

and

A. D. Baker

*Queens College,
The City University of New York
Flushing, New York, U.S.A.*

1977

 Academic Press

London · New York · San Francisco

A Subsidiary of Harcourt Brace Jovanovich, Publishers

ACADEMIC PRESS INC. (LONDON) LTD
24–28 Oval Road, London NW1

U.S. Edition published by
ACADEMIC PRESS INC.
111 Fifth Avenue,
New York, New York 10003, U.S.A.

Library of Congress Catalog Card Number: 76-01691
ISBN 0-12-137801-2

Printed in Great Britain by
Whitstable Litho Ltd., Whitstable, Kent

Contributors to Volume 1

A. D. Baker, *Chemistry Department, Queens College, City University of New York, Flushing, New York, U.S.A.*

J. Berkowitz, *Physics Division, Argonne National Laboratory, 9700 South Case Avenue, Argonne, Illinois, U.S.A.*

C. R. Brundle, *IBM Research Laboratories, San Jose, California, U.S.A.*

R. L. DeKock, *Department of Chemistry, American University of Beirut, Beirut, Lebanon*

M. E. Gellender, *Department of Chemistry, Queens College, City University of New York, Flushing, New York, U.S.A.*

E. Heilbronner, *Physikalisch-Chemisches Institut der Universität Basel, 4056 Basel, Switzerland*

William L. Jolly, *Chemistry Department, University of California, and Inorganic Materials Research Division, Lawrence Berkeley Laboratory, Berkeley, California, U.S.A.*

J. P. Maier, *Physikalisch-Chemisches Institut der Universität Basel, 4056 Basel, Switzerland*

R. L. Martin, *Department of Chemistry, Lawrence Berkeley Laboratory, University of California, Berkeley, California, U.S.A.*

W. C. Price, *King's College, University of London, London, England*

D. A. Shirley, *Department of Chemistry, Lawrence Berkeley Laboratory, University of California, Berkeley, California, U.S.A.*

Foreword

D. W. TURNER

The last decade has seen a fusion of the different electron spectroscopic experiments to the point where a chemist recognizes that electron spectroscopy has taken its place as a distinct branch of spectroscopy capable of assisting in many tasks of structure and reactivity analysis.

The part played by photoelectron spectroscopy and especially by UV photoelectron spectroscopy has been on the one hand to give the chemist a picture of great directness and clarity regarding the orbital structure of even very complex molecules but, on the other hand, its inherently high resolution has allowed the physical limitations of electron spectroscopy as a whole to be thoroughly probed.

Photoelectron spectroscopy of molecules grew alongside and for a time quite separate from photoemission studies of surfaces (an unhappy term for future use since the confusion with photon emission is present), electron energy loss spectroscopy of solids and gases and x-ray photoelectron spectroscopy of solids. Behind these of course stretched a long history of electron scattering and Auger spectroscopy leading back to the original experiments of Rutherford, de Broglie and Franck and Hertz.

I must state frankly that for my part photoelectron spectroscopy was conceived in a mood of frustration not unmixed with laziness. Frustration with the lack of any exciting detail in the absorption spectra of the large molecules that I had been engaged, with the late E. A. Braude, on measuring; accompanied by a marked disinclination to endure for much longer the tedium of rigorous solvent purification for these vacuum uv solution studies. At that time photoionization threshold measurements had become popular and also photochemical applications of resonance lamps with the heavier rare gases were being reported. My initial intention was to use first the argon and then by extension the helium line to measure first ionization energies by photoelectron energy measurements, but my first test experiment showed with the helium source the presence of groups of lower energy electrons so that instead of only measuring a single parameter (I.P.) a spectrum was obtained. Hence molecular photoelectron spectroscopy.

Every research must have its most memorable moments and the high point for us at Imperial College was probably the first detection of vibrational fine structure in a nitrogen photoelectron spectrum. I recall poising at the shoulder of my student Al-Joboury, neither of us daring to move, even to breathe (the equipment with its vibrating reed electrometer and exposed leads was atrociously sensitive to movement), watching in rapture as the recorder pen wagged solemnly up and down. With the use of deflection analysers and electron-counting techniques our experiment soon became routine and many groups around the world have put this new branch of chemical spectroscopy to exciting uses. The range of application and certainty of interpretation are now breathtaking. It is no longer possible to do justice to every possible application in one book but the contributors to this series, who speak with the insight of long experience, range sufficiently widely for the work to become one of the most significant to appear on the topic.

It is extremely gratifying to see described so vividly here many possibilities which, though they could be guessed at from the first, have awaited expert hands for their full realization.

Fox Talbot in a beautifully modest disclaimer said of his own work in 1839, "I hope it will be borne in mind by those who take an interest in this subject that in what I have hitherto done I do not profess to have perfected an art, but to have *commenced* one, the limits of which it is not possible at present exactly to ascertain. I only claim to have based this new art on a secure foundation. It will be for more skilful hands than mine to rear the super-structure."

Here, more skilful hands have indeed reared superstructures in profusion!

Physical Chemistry Laboratory
Oxford, 1977   D. W. T.

Editorial Foreword to the Series

C. R. BRUNDLE and A. D. BAKER

Chemists first became aware of the potential of electron spectroscopy through the pioneering work of Kai Siegbahn and his colleagues at Uppsala, Sweden, in X-ray photoelectron spectroscopy, and of David Turner and his associates in UV photoelectron spectroscopy first at Imperial College and later at the University of Oxford in England. The impact of both branches of photoelectron spectroscopy in the chemical sciences within the past ten years has been pervasive; nowadays it is difficult to find a chemical journal that does not contain some articles related to photoelectron spectroscopy.

Studies of electron-impact types of electron spectroscopy, for example, scattering or energy loss studies, and Auger and Ramshauer effect experiments predated photoelectron spectroscopic measurements, and were mainly employed by physicists as major tools in probing the energy-absorbing properties of atoms and molecules. In recent years these techniques have taken on a new lease of life because of improved instrumental capabilities and in large part because of their direct application to the important and difficult area of surface science.

Believing that there is much to be gained by bringing the diverse types and the diverse applications of electron spectroscopy together, we have undertaken the task of organizing and editing these volumes. Among our motives is the strong feeling that advancement of scientific knowledge most often results from cross-fertilization of ideas emanating from individuals in areas of endeavour which, although seemingly diverse, are unified by some common but perhaps unrecognized aspect. We feel that in this regard there can be no doubt that the expertise of the more physics-oriented electron spectroscopists are of relevance to their chemical counterparts, and vice versa. This series of volumes is thus intended to be a central reference source for all forms of electron spectroscopy. It is inevitable, however, that since both editors are chemists by training, and have worked primarily in the areas of photoelectron spectroscopy, any bias should be toward chemical aspects of photoelectron spectroscopy.

Chapters of different styles and different intents are inevitable, but perhaps desirable in a series such as this. Comprehensive and rigorous reviews of given areas are to be found along with overview chapters covering particular applications or groups of techniques, and intended mainly for a more general readership. Thus we hope that both electron spectroscopy specialists and the scientific community in general will find something of interest among the contributed chapters.

Preface

Of the eight chapters in Volume 1 all but two are concerned exclusively with photoelectron spectroscopy and mainly with molecules rather than solid state or surface effects. Four chapters are devoted to UV photoelectron spectroscopy, three of these (Price, Berkowitz and DeKock) dealing with the delineation of electronic structure in small molecules, and one (Heilbronner and Maier) devoted to larger organic molecules. Two chapters deal with X-ray photoelectron spectroscopy. That of Martin and Shirley treats, in detail, the relationship between the initial (unionized) and final (ionized) states involved in the photoemission process. Jolly's contribution covers the relationship between core-level chemical shifts and charge distribution in inorganic systems and the use of the equivalent cores principle. The last chapter of the volume (Gellender and Baker) gives an overview of what has been achieved and what might be achieved in coincidence experiments where, for instance, measurements are made simultaneously on both the photoelectrons and the ions resulting from the photoemission process. The first chapter of the volume (Baker and Brundle) presents a general introduction to the different forms of electron spectroscopy together with examples of their applications.

We thank our many distinguished contributors for their time and effort. On a personal note we thank Professor W. C. Price, author of Chapter 4, for the advice, encouragement, and enthusiasm he imparted to us during the early years of our careers. Professor DeKock also merits our special thanks. He completed most of his manuscript during his stay at the American University in Beirut during the troubled period of civil strife in Lebanon.

January 1977

A. D. Baker
Queens College
New York, N.Y.

C. R. Brundle
IBM Research Laboratory
San Jose, California

Contents

1. An Introduction to Electron Spectroscopy

A. D. Baker and C. R. Brundle

2. Many-electron Theory of Photoemission

R. L. Martin and D. A. Shirley

3. The Application of X-ray Photoelectron Spectroscopy in Inorganic Chemistry

William L. Jolly

4. Ultraviolet Photoelectron Spectroscopy: Basic Concepts and the Spectra of Small Molecules

W. C. Price

5. Some Aspects of Organic Photoelectron Spectroscopy

E. Heilbronner and J. P. Maier

6. Ultraviolet Photoelectron Spectroscopy of Inorganic Molecules

R. L. DeKock

7. High Temperatures UPS Studies and Other Variations

J. Berkowitz

8. Two-parameter Coincidence Experiments

M. E. Gellender and A. D. Baker

1

An Introduction to
Electron Spectroscopy

A. D. BAKER

Queens College, Department of Chemistry,
City University of New York, Flushing, New York 11367
and

C. R. BRUNDLE

IBM Research Laboratory,
San Jose, California 95193

I. INTRODUCTION

Electron spectroscopy is the generic name given to a handful of individual techniques based upon the analysis of electron energies following a collision between an impacting particle or photon and an atom, molecule, or solid.

The individual techniques are listed in Table I, in which important previous reviews, or leading references in the absence of a review, are given.

The branches of electron spectroscopy were developed more or less independently, often by groups working in diverse areas, e.g. molecular spectroscopy as opposed to surface physics. The rapid growth in recent years of basic and application studies using electron spectroscopy can partly be attributed to a belated interaction between scientists working in quite different areas, but using electron spectroscopy, thus providing the means of exchanging ideas and techniques. A second factor is the advance in the technology and design of instrumentation. Whereas experiments in the past were frustrated for lack of high resolution energy analyzers, sensitive electron detection systems, or a sufficiently good vacuum for meaningful results, today there is an almost bewildering array of commercial technology in all these areas with which to implement bright (or not so bright) ideas. The only drawback is the price tag.

In this chapter some of the basic ideas and a few simple applications of the more important types of electron spectroscopy will be given as an introduction to the detailed discussions presented in the later chapters of this volume and the succeeding volumes. The areas in which applications have been found is remarkably wide, covering all cases in the gaseous or solid state (including surfaces) where elemental analysis or a knowledge of chemical bonding and electron structure is required. It complements other techniques over this wide range, but does not have the general applicability in any one area that, for instance, NMR offers in organic chemistry. A possible exception to this is in its applicability to surface studies. In addition to the many applications, it provides information of fundamental significance with respect to electron excitation mechanisms and the nature of electronically excited states.

As with most spectroscopic techniques, there are many different aspects to electron spectroscopy—instrumentation, basic theory of the phenomenon being studied, practical uses, theoretical and practical aspects of the situations to which it is applied, etc. Many of these considerations are interrelated—a full appreciation of the practicalities and limitations of one aspect is often impossible without an understanding of another aspect. Therefore there is something of a problem in presenting the basic ideas in a simple, logical, yet comprehensible form, while at the same time trying to bring out the important applications in many different fields. Our decision as to where to start and how much to cover in each area is therefore an arbitrary one, and is certainly

TABLE I

Types of Electron Spectroscopy

Name of technique	Abbrevia-tions	Basis of technique	References to leading reviews or papers
Photoelectron spectroscopy (ultraviolet excitation)	PES, UPS, or MPS	Electrons ejected from materials by monoenergetic ultraviolet photons are energy analyzed	(1)–(10)
Photoelectron spectroscopy (X-ray excitation)	ESCA or XPS	Electrons ejected from materials by monoenergetic X-ray photons are energy analyzed	(10)–(20)
Auger electron spectroscopy	AES	Auger electrons ejected from materials following initial ionization by electrons or photons (not necessarily monoenergetic) are energy analyzed	(21)–(25)
Ion neutralization spectroscopy	INS	Auger electrons ejected from surfaces following the impact of a noble gas ion are energy analyzed	(26)
Penning ionization spectroscopy	(PIS)	Metastable atoms are used to eject electrons from materials. The electrons are then energy analyzed	(27)–(29)
Electron impact energy loss spectroscopy	ELS	A monoenergetic electron beam is directed onto a sample, and the energies of the inelastically scattered electrons are measured	(30)–(36)
Autoionization electron spectroscopy		Similar to Auger electron spectroscopy. Electrons ejected in an autoionizing decay of superexcited states are measured. Electron or photon impact can be used to produce the super-excited state	(13), (37), (38)
Resonance electron capture / Electron transmission spectroscopy		The elastic scattering cross-section for electrons is measured as a function of the energy of the electron beam and the scattering angle	(39)–(42)

subjective and based on our own backgrounds, which are primarily chemical. We also rely heavily, for reasons of familiarity, on our own and collaborators' work for many of the example spectra presented.

Both authors of this chapter began their work in electron spectroscopy as graduate students of D. W. Turner in the fields of uv-photoelectron spectroscopy (UPS) and photoionization mass spectrometry. We have found this to be a good starting point to appreciate some of the other areas of electron spectroscopy with which we are now also involved. We have therefore chosen to begin our discussion with an initial consideration of uv- and X-ray photoelectron spectroscopy and to branch off from there. Later we return and consider applications of the photoelectron techniques and append, hopefully in a complementary fashion, some of the applications of the other electron spectroscopies.

II. PHOTOELECTRON SPECTROSCOPY: IONIZATION POTENTIALS

Once the quantum concept of electrons in atoms and molecules existing in orbitals with defined energies had been introduced in the mid-1920's, it became clear that a knowledge of these orbital energies would help greatly in understanding many phenomena in the physical sciences. Much effort has been devoted by chemists and physicists to developing quantum models and experimental techniques which provide estimates of these orbital energies. Experimentally the measurable parameter most closely related to the energy of an orbital is the *binding energy* or *ionization potential* of an electron occupying the orbital.

Prior to the introduction of photoelectron spectroscopy, various techniques were already available for measuring ionization potentials.[43-46] These involved the measurement of *ionization efficiency curves* (plots of an ion current versus energy of the impacting species, normally a photon or electron), or of the absorption spectrum in the vacuum ultraviolet region. For various reasons, both fundamental and practical, which will be mentioned later in this chapter, these methods were generally unable to provide much more than the binding energy of the most weakly bound electron (i.e. the one in the highest occupied atomic or molecular orbital) and sometimes one or two "inner" binding energies.

Photoelectron spectroscopy is in principle able to provide directly, as *primary* information, the binding energies of all the electrons in a species, from the most weakly bound to the most strongly bound. Such a body of data is clearly of fundamental importance for formulating theories of the behavior of atoms in bonding to form molecules or solids. In addition, comparison of trends in the photoelectron spectra of series of related families of substances has been found to be a rich source of what may be termed *secondary* information relating to the geometries of molecules, the operation of substituent electronic effects, the importance and electronic role of ligands in coordination compounds, the nature of species adsorbed onto surfaces, concepts of aromaticity, and many other factors which will be enumerated in the various chapters of these volumes.

It might be pointed out here that the other forms of electron spectroscopy to be considered in this chapter also provide primary data of a fundamental nature relating to the vacant orbital levels or energy bands in materials, and also produce secondary data of relevance in many fields, such as the understanding of impact and excitation phenomena in general, and in chemical analysis.

Photoelectron spectroscopy is a modern development of early experimentation on the Photoelectric Effect. It had been realized that photons of sufficiently short wavelengths would ionize atoms or molecules:

$$M + h\nu \rightarrow M^+ + e^-$$

Furthermore, it was soon shown that the kinetic energy of the ejected photoelectron depended only on the wavelength of the impacting photons, and not on their intensity. The Einstein Photoelectric Law, which is essentially an energy balance, relates the energy of the impacting photon, $h\nu$, with the ionization potential, I, of the target, and the kinetic energy, K.E., of the ejected photoelectron:

$$K.E. = h\nu - I \qquad (1)$$

Strictly speaking the law is an approximation even for the ionization of atoms, since not all the excess energy of the photon above the ionization potential (i.e. the quantity of energy $h\nu - I$) appears as kinetic energy in the photoelectron. A small quantity may appear, for example, as kinetic energy in the resulting ion, or, in the case of molecules, the resulting molecular ion may be formed with vibrational or rotational excitation energy.

The first of the above limitations of the Einstein Photoelectric Law, viz. the kinetic energy imparted to the ion, is usually inconsequential, since even in the most unfavorable case the principle of conservation of momentum requires that the ejected electron, being of small mass, carries off almost all of the energy as opposed to the relatively massive ion remaining behind which carries off very little.

The second limitation, the possibility that a photoionization act may produce an ion in an excited electronic, vibrational, or rotational state, is an important one in photoelectron spectroscopy. In general, the probability of a second photon collision occurring with a species which is in an excited state following a first collision, or of two photon collisions occurring simultaneously on one species, is very small. Thus only *one* photon interaction need be considered for any individual molecule in the sample ensemble. The result of the interaction may be different for the individual molecules in the ensemble however—an electron may be ejected from a variety of different molecular or atomic orbitals (Fig. 1). Ejection may occur from any orbital for which the associated binding energy is less than the energy of the incident photon, though the probability of the ionization will vary from orbital to orbital.

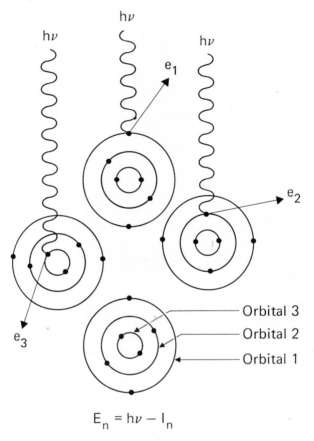

$$E_n = h\nu - I_n$$

Fig. 1. The three simple one-electron photoionization processes possible for an atom which has three atomic orbitals containing electrons.

Thus the net effect of photoionizing a large number of like molecules by photons of energy $h\nu$ is that some will lose an electron from one particular orbital, some from another particular orbital and so on. "Bunches" of electrons are therefore emitted, each electron originating from a different individual molecule and each bunch having a different energy according to the modified Einstein equation

$$E_n = h\nu - I_n \qquad (2)$$

where E_n represents the energies of the different bunches ($n = 1, 2, 3, ...$) and I_n represents the ionization potentials for an electron in the first (outermost) orbital, second (next outermost) orbital, etc.

A further "correction" to the Einstein equation is needed to cater for the incorporation of vibrational and rotational excitation energies, E_{vib} and E_{rot} into the ion produced. A photoelectron ejected from a molecule in an ionization act that gives rise to a vibrationally excited ion will obviously have less kinetic energy than one ejected from the same orbital in the same molecule by a photon of the same energy, but where the ion left behind is not vibrationally excited. Thus the appropriate form of the Einstein Photoelectric Law for the study of *molecules* by the photoelectron spectroscopy is:

$$E_n = h\nu - I_n - E_{vib} - E_{rot} \qquad (3)$$

Experimentally an energy analysis of the ejected electrons is carried out to measure E_n, and hence determine I_n, E_{vib}, and E_{rot}. The energy analyzers used are generally electrostatic or magnetic deflection types (see the chapter by Kuyatt, Volume II). The outgoing electrons enter the analyzer and execute different pathways depending upon their velocities and the value of the deflecting field. The deflecting field is smoothly changed so that progressively less energetic electrons are brought to a focus in turn on the analyzer's exit slit. An alternative mode of operation is to set the analyzer to pass one particular velocity and vary an accelerating potential applied to the electrons before they enter the analyzer. Electrons passing through the exit slit strike a detector and give rise to a signal which is ultimately displayed on chart paper. Thus the spectrum as recorded is a plot of number of electrons versus electron kinetic energy showing main bands corresponding to the bunches of electrons from all the different orbitals accessible to the photon energy. Fine structure relating to vibrational or rotational excitation energy imparted to the ion may also be resolved on the bands. Calibration of the kinetic energy scale of the spectrum enables ionization potentials to be read directly from the peak positions using Eq. (3) above. To summarize:

(a) Monochromatic photons are directed onto a sample.

(b) The energy spectrum of the ejected photoelectrons is recorded.

(c) This spectrum is expected to show, *to the simplest approximation*, a number of bands that correspond to the number of occupied molecular orbital levels. (There are various ways in which bands may be split so that *more* bands than orbitals are observed. They are discussed later and in detail in other chapters.)

(d) The positions of the bands in the spectra yield the binding energies or ionization potentials of the electrons in the sample.

(e) Each band also corresponds to the production of an ion in a particular excited electronic state (i.e. there is a "hole" in the outermost or one of the inner orbitals).

(f) Vibrational structure or other fine structure may be apparent within the bands. This corresponds to vibrational and/or rotational excitation of the ions produced in the ionizing acts.

The gross positions of the bands in spectra, the form of splitting, the relative band intensities and the detailed analysis of the fine structure are the sources of data derived from photoelectron spectra. These aspects will be covered briefly in a later section of this chapter, and in more detail in the individual chapters.

There is one further important point concerning the relationship of observed ionization potentials and the electronic structure of molecules that must be considered here. We have so far described this relationship in terms of a correspondence of the number of observed peaks and their energy ordering with the MO's of the molecule. In a calculation, the MO "energies" that are obtained by the usual SCF methods for a closed-shell electronic configuration are the canonical Hartree–Fock orbitals, so defined because they diagonalize the matrix of Lagrangian multipliers. The diagonal elements are called "orbital energies" because, with neglect of electronic rearrangement (relaxation) in the resultant ion (Koopmans' approximation,[47] also called the Frozen Orbital approximation, or a "bare" or "undressed" hole in solid state terminology), they represent the ionization energies of the individual electrons at the Hartree–Fock level. The canonical Hartree–Fock orbitals are usually a reasonable description of those orbitals (within the basis of all occupied MO's in the parent molecule) which best represent the electron charge density the molecule would lose for each electron, were that electron to be removed.

The above can simply be restated as meaning that both the *initial* (neutral) and *final* (ionized) states which are connected by the photoionization process are appropriately described, within the limits of Koopmans' approximation, by the *initial* state SCF MO's, so these MO's are particularly appropriate for describing the photoemission process. For instance, neutral CH_4 has the valence level SCF MO description $(2a_1)^2(1t_2)^6$ and ionization results in the two final states $(2a_1)^2(1t_2)^6$, $^2T_2$ and $(2a_1)^1(1t_2)^6$, $^2A_1$. Within the Frozen Orbital approximation the energies required to reach these final states, which are those observed in the photoelectron spectrum, are the SCF MO energies of the $(1t_2)^6$ orbital and $2a_1$ orbital, respectively. It is however well known that applying a unitary transformation to the total electronic wavefunction for a molecule, which alters the individual MO compositions, but leaves the total energy and electron density unchanged, results in an exactly equally valid orbital representation. In the case of CH_4 one obtains four sp^3 C—H bond orbitals. The orbitals and "orbital energies" in this transformed representation do not however, in general, diagonalize the matrix of Lagrangian multipliers and therefore do not, in general, represent observed ionization potentials. In other words, the four equivalent $(sp^3)^2$ C—H bond orbitals description is appropriate for neutral CH_4, but CH_4 minus a valence electron *cannot* be described in terms of three intact $(sp^3)^2$ C—H bond orbitals and one $(sp^3)^1$ C—H bond orbital. As any undergraduate theoretical chemistry text will reveal, the four degenerate final states obtained mix to yield the singly degenerate $^2A_1$ final state and the triply degenerate $^2T_2$ final state, exactly equivalent to the MO result. It is for these reasons—the straightforward connection between SCF MO's and ionization potentials via Koopmans' approximation and the usual lack of such a connection in a bond-orbital description—that one invariably finds the interpretation of the electronic structure of the neutral molecule, as studied by photoelectron spectroscopy, discussed in MO language. This was pointed out several years ago,[48] is occasionally neglected, and justifiably re-emphasized (see Chapters 2 and 4). The fact that an MO description is usually the easiest description to use for the photoemission process *says nothing* about whether the MO or bond-orbital electronic structure description of the neutral molecule is more appropriate. They are exactly equivalent and appropriateness is judged in terms of convenience. In practice the bond-orbital approach is usually the most convenient when considering chemical bonding, witness the very fact that we refer to CH_4 as having carbon—hydrogen bonds. When considering *reactivity* of molecules, examples can be found where MO descriptions are more convenient, e.g. the Woodward–Hoffman rules, but such situations turn out to be similar to the photoemission situation in that the reaction can be thought of as an electronic excitation process. More often the bond-orbital approach is usually the obvious choice since bonds are broken and reformed.

We have deliberately not considered the situation in photoemission where final state relaxation is so large that Koopmans' approximation cannot be justified. The question of whether or not it is justifiable must be an operational definition anyway. The present authors think of it being a justifiable approximation as long as features are observable in the photoelectron spectrum which can be related, in the way described above, to an SCF MO initial state description for the molecule concerned. The experimental IP will become less and less well represented by the SCF MO "energy" as the approximation worsens and additional features will appear strongly in the experimental spectrum owing to final state relaxation effects. Situations where Koopmans' approximation is so bad that there is no feature describable on the basis of the initial state SCF MO are rare. In such situations the initial state SCF MO description is completely unsuited to describing the final state and no one-electron process may be used to describe the photoelectron process observed. The effect of final state relaxation on photoelectron spectra is referred to again later in this chapter, and in detail in Chapter 2.

III. IONIZATION OR EXCITATION: EFFECT OF THE BOMBARDING SPECIES

Any radiation or particle of sufficient energy can in principle induce ionization or electronic excitation when striking a target. Thus different variants of electron spectroscopy can be envisioned in which the nature of the basic collision process is changed.

There are, however, fundamental differences in the natures of electron–molecule, photon–molecule, atom–molecule, and ion–molecule collisions that result in different ways of carrying out the various types of experiment, differing results, and in different potential applications for these different experiments.

A fundamental and important difference exists between excitation or ionization resulting from electron–molecule as opposed to photon–molecule collisions. A photon is annihilated in such an event whereas an electron is not; it is scattered with a lower energy than it had before the collision. The difference in energy, $E_p - E_s$, between the energy, E_p, of the primary electron and the energy, E_s, of the scattered electron is the energy imparted to the target molecule causing excitation or ionization. This is in marked contrast to the photon-impact case in which all the photon energy is imparted to the target molecule (except in the special case of Compton scattering).

In Section II we considered only ionization processes, resulting in the ejection of an electron from the sample, owing to impact by photons of a fixed frequency, $h\nu$. In electron impact studies, *excitation* from an occupied molecular energy level to an unoccupied level separated by an energy, ΔE, may occur and is necessarily accompanied by an energy loss, ΔE, in the energy of the primary electron (i.e. $E_p - \Delta E = E_s$). The technique of *electron impact energy loss spectroscopy* is based on this fact. An incident beam of mono-energetic electrons is passed through a sample, and the energy spectrum of the outgoing scattered electrons measured. Alternatively the energy of the incident beam could be varied, while the counting rate for scattered electrons of one particular energy is monitored. Either way an "energy-loss" spectrum results with a peak at each allowed energy, ΔE_n. The type of information obtainable from such a spectrum is similar to that supplied by the various types of optical absorption spectroscopy—infrared, visible, ultraviolet, etc., since the final excited molecular states are the same. However, since a photon must be annihilated by the process of excitation, excitation by photon impact is a *resonance process* which can only occur when $h\nu = \Delta E$, unlike the electron impact situation where $E_p > \Delta E$ is the requirement. Thus the photon absorption spectroscopies are necessarily performed by varying the photon energy and detecting absorption at those energies $h\nu_n = \Delta E_n$.

Ionization by electrons and ionization by photons are also rather different phenomena, the reason again being associated with the scattering, rather than annihilation of the impacting electrons. Ionization brought about by electron bombardment of energy, E_p, results in two electrons leaving the collision site, the scattered electron, with energy E_s, and the ejected electron, with energy E_e. The excess energy of the ionization process, $E_p - I_n$, is shared between the two electrons, unlike photon impact, where it is completely transferred to the ejected electron. Therefore, in electron impact energy loss spectroscopy, a weak continuum rather than a peak starts at the energy loss equal to the first ionization potential (I_1) since the two resultant electrons may take any values provided $E_p - I_1 = E_s + E_e$ (an analogous situation exists in vacuum-ultra violet absorption spectroscopy where band-to-band transitions give discrete absorption peaks, but ionizations produce continua).

Theor. Expt.	Theor. Expt.	Theor. Expt.	Theor. Expt.
(1) Negative ion formation by electron capture $A + e \rightarrow A^-$	(1) Photoionization $A + h\nu \rightarrow A^+ + e$	(1) Ionization by electrons $A + e \rightarrow A^+ + 2e$	(1) Double ionization by electrons $A + e \rightarrow A^{++} + 3e$
(2) Photoexcitation, $A + h\nu \rightarrow A^*$	(2) Excitation by electrons, $A + e \rightarrow A^* + e$	(2) Photoionization (double) $A + h\nu \rightarrow A^{2+} + 2e$	
	(3) Ion pair formation after excitation, $A + e \rightarrow A^* + e \rightarrow B^+ + C^-$		

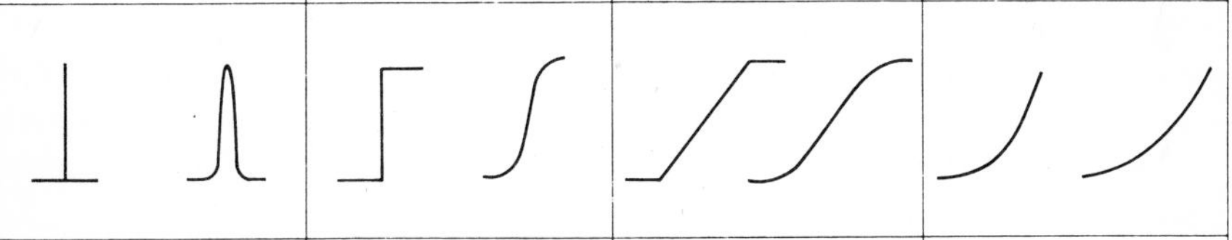

Fig. 2. Theoretical threshold laws and experimentally observed threshold behavior for the various excitation and ionization processes possible with photons and electrons.[46]

Figure 2 illustrates the different *threshold laws* for the electron and photon impact process. These threshold laws govern the rate at which an ion current changes as the energy of the impacting species increases just beyond the ionization threshold. The difference between the photon and electron impact threshold laws can be rationalized in terms of the electric field interactions of the species involved in and resulting from the collision. At the risk of over-simplification, it can be said that an impacting electron, being charged, remains in the vicinity of the collision complex (i.e. the ejected electron plus ion) where there is a good chance that it can neutralize the positive ion by being captured. Only when the electron has appreciable excess energy can it escape from the collision center allowing the ion current to grow. Morrison[49] has pointed out that the threshold law for ionization or excitation appears to be determined solely by the number of electrons leaving the collision site—e.g. double ionization by photons has the same threshold law as single ionization by electrons and single ionization by photons is comparable to excitation by electrons, as can be verified from the discussions given above. An obvious practical consequence of the differences in the threshold laws is that accurate ionization and appearance potentials (for positive ions) are more easily obtainable from photon impact ionization efficiency curves (because of the step-function threshold law) than from the corresponding electron impact curves.

A straight attempt to perform "photoelectron spectroscopy" by substituting a monoenergetic beam of electrons for hv does not work therefore. Nevertheless by using coincidence techniques to record the spectrum for only a fixed value of E_s, a discrete peak spectrum giving I_n values from measured E_e values becomes possible;[50-52] such an experiment ("electron–electron spectroscopy") is the electron impact analog of photoelectron spectroscopy (see Section VI). Coincidence instrumentation introduces added cost and complications to the basic experiment and greatly reduces signal intensities available. It does give more parameters (the electron and the angles between them) which carry information, and the energy of an electron beam is easily varied whereas the same is not true for photons.

Molecules bombarded with energetic atoms, ions, or other molecules are also subject to ionization and excitation. Electron spectroscopy is therefore a viable tool for studying these processes if electrons of discrete energy are released. Cermak and Herman originally suggested the measurement of the kinetic energies of electrons ejected following Penning ionization, a process in which metastable atoms or other neutral species ionize a target molecule.[53] They subsequently investigated the use of He 2^3S and 2^1S metastables as ionizing particles for electron spectroscopy.[54] Their technique is often called "Penning Ionization Spectroscopy". The helium metastables are produced by bombardment of helium with an electron beam. Since the energies released when these two excited states of He return to the ground state are 19·82 eV and 20·61 eV, respectively, this energy must be used up in the collision process causing the de-excitations, Penning Ionization Spectroscopy is phenomenologically equivalent to a UPS process involving photons of energy $hv = 19·92$ eV and 20·61 eV. Some differences occur in the spectrum compared to a UPS spectrum owing to the different mechanisms of ionization possible following a photon–atom collision and an excited atom–atom collision.[4, 28, 29, 55]

Ions have also been used as a bombarding source for electron spectroscopy. In Ion Neutralization Spectroscopy (INS), He^+ ions strike a solid sample, and Auger electrons (see next section) are ejected by the energy released. The energies of the ejected electrons are measured and since they originate only from the surface (the slow He^+ ions do not penetrate the surface), the technique is of great importance in surface studies, as discussed in Section VI.

IV. THE DIFFERENT CLASSES OF IONIZATION AND EXCITATION PHENOMENA

Irrespective of the difference in species causing ionization or excitation and the different mechanisms of these processes, the electrons emitted following these interactions may be classified into several different categories. The phenomena accompanying electron production (e.g. positive ion production and fragmentation, fluorescence) must also be considered.

A. Direct Ionization Electrons

Photoelectron Spectroscopy is an example of a technique in which an electron is ejected *directly* from the sample by the impacting photon, as are the ionization processes described so far for electron impact and He* impact.

B. Auger Electrons

Some of the other electron spectroscopic techniques discussed in this chapter are concerned with electrons released in a secondary step following an initial ionization or excitation. For example, Fig. 3 represents a gaseous molecule with a "hole" produced in an inner shell by ionization. The orbital energy level has been labeled, W, a general nomenclature used to refer to the level possessing the hole. The levels, X, Y, V are also general nomenclatures. X and Y refer to inner-shell MO energy levels and V refers to a valence-shell energy level. If the W hole is filled by an electron moving to it from a filled higher level, sufficient energy may be released to eject a second electron from the molecule (processes 1, 1′ or 2, 2′ in Fig. 3). This ejected electron is called an *Auger electron*.[56] The energy of the Auger electron is *independent* of the energy of the photon or electron which originally made the W hole. For instance, process 1, 1′ (termed a WXX Transition; process 2, 2′ is called a WVV Transition) produces an Auger electron of approximate energy:

$$\text{K.E.}_{\text{Auger}} = E_W - 2E_X$$

since, assuming no relaxation in the singly or doubly ionized states, energy

$E_W - E_X$ is released in transition 1, and E_X is used up in transition 1′. Thus, it is not necessary to use a monoenergetic beam of photons or electrons to excite Auger electron spectra. Experimentation may therefore be simpler, but exact interpretation of the resulting spectra is generally more difficult than

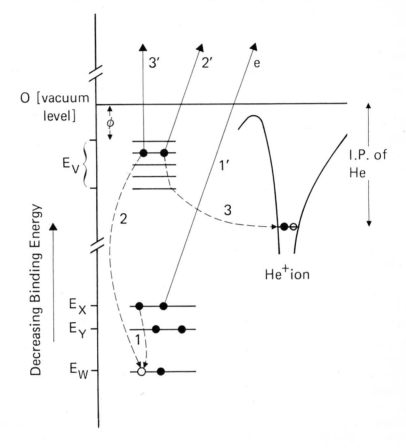

Fig. 3. Auger processes (1,1′; 2,2′) for a free molecule possessing a core-hole. 3,3′ represents the special type of Auger process possible when a gaseous ion approaches a solid surface (ion neutralization spectroscopy).

for photoelectron spectra, because of the three-electron nature of the whole process. Nevertheless, electron impact Auger spectroscopy is one of the most important tools in surface studies, and has specific applications in other areas which are discussed later in this chapter and in chapters by Todd and by Brundle in later volumes.

C. Autoionization

Autoionization is similar to Auger ionization in that it is a secondary process. It can occur following bombardment of a sample with sufficiently energetic photons, electrons, or other particles. It involves an initial excitation (not an ionization) step in which one, or more, electrons are promoted to give an excited atom or molecule. If the excitation energy exceeds the energy needed to eject (ionize) any of the other electrons present, the excited atom or molecule concerned will have sufficient energy to rearrange to an ion by expulsion of an electron (Fig. 4). The ejected electron is known as the "autoionization" electron. The difference, therefore, between the Auger and autoionization processes is that in the former the initial step is the ejection of an electron from an inner shell, rather than a promotion to a higher orbital. Furthermore, the final product of the Auger process, following the ejection of the Auger electron, is a doubly charged ion, whereas in autoionization the final product is a singly charged species. It is also quite possible for the two to appear in combinations when *ionization* of one electron and *promotion* of another occur at the initial step (a shake-up process—see Section VI).

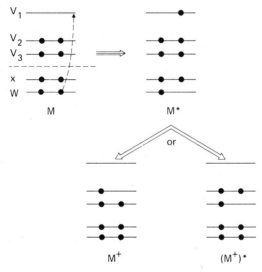

Fig. 4. Autoionization processes possible following an initial excitation process.

The nature of the autoionization process implies that it will occur in photon impact studies only as a resonance process, since the incident photon, which is annihilated in the process, must have an energy exactly equal to the difference in energy between the ground state atom or molecule and the "superexcited" state produced (M^* in Fig. 4). Autoionization is therefore common in photon impact ionization efficiency curves (since $h\nu$ is continuously varied) showing up as resonance peaks superimposed on the expected "steps" for the various ionization potentials (see Fig. 5 and the threshold laws of Fig. 2). Partly because of the preponderance of features due to autoionization, the step features due to inner ionization potentials are often obscured in ionization efficiency curves, making it a less than satisfactory way of obtaining these IP's.

Autoionization will only be detected in conventional photoelectron spectroscopy if the photon energy used happens to coincide with an auto-ionization level. In general, freedom of spectra from complications due to autoionization is one of the major reasons that photoelectron spectroscopy is the best technique for measuring inner ionization potentials.

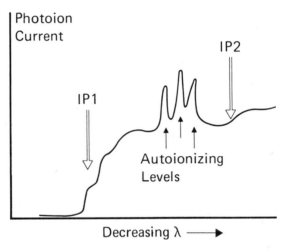

Fig. 5. Schematic of photoionization efficiency curve showing resonance autoionization structure.

Selective study of autoionization by electron spectroscopy is possible, however. If electrons are used as the bombarding species (or alternatively a continuum photon source) and the kinetic energy spectrum of the emitted electrons measured, discrete peaks appear due to autoionization electrons as shown in Fig. 6. For photon impact, some peaks might also appear that are attributable to Auger electrons (generally at higher energies). For electron impact, additional discrete features appear owing to inelastically scattered electrons (i.e. the energy loss spectrum described earlier), or to the auto-detachment of electrons from transient negative ions formed by initial electron attachment.[39-42] An electron impact emission spectrum is thus obviously rather complex, containing resonances associated with energy loss, Auger, and autoionization processes, plus continua associated with direct ionizations. By working under optimized experimental conditions, various workers have developed reasonably selective methods for the study of the autoionization electron process.[37, 38, 58] The results are of interest because they provide information about orbitals which are not populated in the normal unexcited molecule.

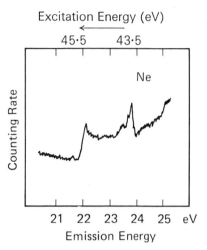

Fig. 6. Autoionization peaks observed in the electron impact emission spectrum of Ne.[57]

D. *Fate of the Ion Produced*

Fluorescence competes with Auger electron emission as a decay process following the creation of an inner-shell hole. Indeed, fluorescence may be the only available route for the de-excitation of some species produced in electron spectroscopy.

Considering first the case of processes following the creation of a core-shell hole, Auger electron ejection is the major de-excitation route for compounds containing light atoms (atomic number < 12), and fluorescent emission is the major route for heavier atoms (Table II). The fluorescence emitted will of course be in the x-ray region. X-ray fluorescence is an important analytical tool in its own right; it is of potential interest in electron spectroscopy as a complementary process.

TABLE II

Probability, P, of Auger Electron Emission versus X-ray
Emission as a Function of Atomic Number, Z

Z	10	15	20	25	30	35	40	45	50	55	60	65	70
P	0·99	0·95	0·86	0·71	0·55	0·43	0·31	0·22	0·16	0·13	0·11	0·09	0·07

Fluorescence of a longer wavelength from ions with valence level vacancies is also of interest as a complementary way of obtaining information related to that provided by photoelectron spectroscopy of the valence levels. Thus, it is possible to obtain information about excited states of ions by observing the fluorescent radiation accompanying their de-excitation (Fig. 7) as well as by studying uv-photoelectron spectra. Examples of fluorescent yield curves can be found in various publications of Cook *et al.*, and others.[60-63] A fluorescent yield curve for CO_2^+ is shown in Fig. 8. Recently Turner *et al.*[64] have begun

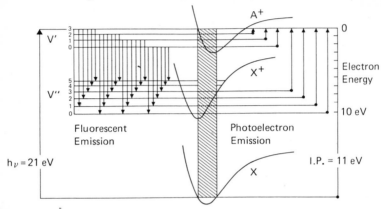

Fig. 7. Fluorescent radiation decay processes possible following valence level ionization to an excited ionic state.

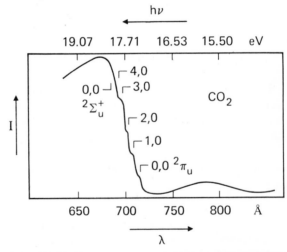

Fig. 8. Fluorescent decay yield, I, for CO_2^+ as a function of incident photon energy, $h\nu$. The steps marked 0, 0; 1, 0, etc., represent fluorescent intensity from the 0, 1, etc., vibrational levels of the $^2\Pi_u$ state of CO_2^+ to the 0 vibrational level of the $^2\Sigma_u^+$ CO_2^+ ground state.

investigation on fluorescence/photoelectron coincidence experiments, although they conclude that few molecular ions of any complexity display intense enough fluorescence to be generally useful.

Various workers have tried to relate observed features in photoelectron spectra with the fragmentation of the ions produced.[65] This is possible either by correlations based on band shapes and known appearance potentials of fragment ions, or better through coincidence experiments in which both the fragment ions and photoelectrons are examined. The latter is in principle a very powerful technique (see the chapter by Baker and Gellender in this volume and that by Eland in Volume III).

V. INSTRUMENT DESIGN AND OPERATION

There are a great variety of electron spectrometers in existence. In Table III are listed the commercially available instruments at the time of writing. Design varies widely because of the different objectives of the users.

TABLE III

Commercially Available Electron Spectrometers

Company	Type	Analyzer	Characteristics
Associated Electrical Industries (AEI) Ltd., UK	XPS (UPS and Auger options)	HMA	Solids, surface compatible, monochromator option
DuPont Corporation, USA	XPS		Solids, not suitable for surface work, rapid loading, relatively cheap for an XPS instrument
Hewlett–Packard, USA	XPS	HMA	Monochromator standard, not fully surface compatible, rapid sample loading, expensive
Leybold–Hereaus, Germany	XPS (UPS and Auger options	HMA	Specialist surface instrument, but gas version available
McPherson Corporation, USA	XPS (UPS option)	HMA	Solids, surface compatible, handles irregular-shaped samples, fully automated
McPherson Corporation, USA	ELS	HMA	Gas-phase instrument
Perkin–Elmer, UK	UPS	127° sector	Gas-phase instrument. Facility for vaporizing solids
Physical Electronics Industries, USA	AES (XPS option)	CMA	Specialist surface physics instrument. Options such as LEED and SIMS available

TABLE III—continued

Company	Type	Analyzer	Characteristics
Varian Associates, USA	XPS	HMA	First commercial XPS instrument. Poor vacuum and sample-handling facilities. No longer in production
Varian Associates, USA	AES	CMA	Specialist surface instrument
Varian Associates, USA	AES/LEED	Retarding grids	Specialist surface instrument
V.G. Scientific, UK	UPS	HMA	Gas-phase or specialist surface instruments available
V.G. Scientific, UK	UPS	HMA	Angular resolved photoemission specialist instrument
V.G. Scientific, UK	XPS	HMA	Specialist surface instrument. Gas-phase version available
V.G. Scientific, UK	XPS/UPS/AES	HMA	Specialist surface combination instrument. Options such as LEED and SIMS available
V.G. Scientific, UK	AES	HCMA	Specialist surface instrument

HMA—hemispherical mirror analyzer.
CMA—cylindrical mirror analyzer.
HCMA—hemicylindrical mirror analyzer.

The fundamental components of an electron spectrometer are (a) a source to provide the impacting particles, (b) a means of analyzing the energies of electrons after impact, (c) an electron detecting and recording system, and (d) a vacuum chamber housing—(b) is discussed in detail in the chapter by Kuyatt in Volume II and so will only be briefly mentioned here.

A. Sources

1. *Photon Sources.* The HeI resonance line at 21·2 eV energy may be obtained by either a dc- or microwave-induced discharge in flowing helium (Fig. 9). It was the first photon energy used in gaseous UPS work, is still the most popular in that area, and is now also routinely used in solid-state and surface photoemission work. One problem with this source can be the

presence of a few per cent intensity of other photon energy lines in the discharge and care has to be taken to either suppress these or to recognize their contribution to the photoelectron spectrum recorded. Sometimes it is useful to

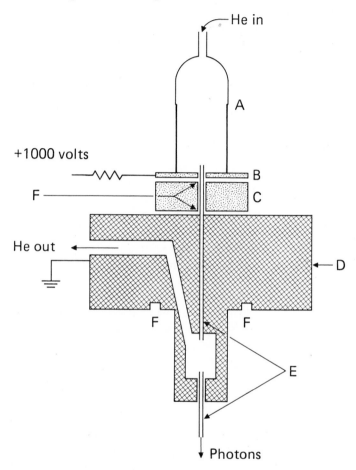

Fig. 9. Cross-section through a DC discharge lamp used for production of the HeI resonance line.[1] A is a copper–glass seal with a glass capillary He inlet; B is the copper anode; C is a teflon spacer; D is the aluminum cathode; E are quartz capillaries forming the discharge region between anode and cathode, and a photon-collimating capillary. F indicates the points at which O-ring vacuum seals are made. The cathode mates to the main body of the spectrometer at F.

enhance the high energy lines (HeIIα at 40·8 eV, HeIIβ at 48·4 eV) by changing design parameters[66] and then using these lines to produce photoelectron spectra, thus allowing orbital ionizations deeper than 21·2 eV to be observed. The same lamps can usually be used with other discharge gases

(Ne, Ar, Kr, H_2) to produce other photon energy lines,[67] though there are usually at least two intense photon energy components in these discharges, limiting their usefulness without monochromatization.

An alternative procedure in the vacuum-uv photon energy range is to use many-line or continuum-sources in conjunction with a vacuum monochromator to provide monochromatic radiation between 3 eV and 80 eV range. The introduction of the monochromator causes loss in available photon intensity and there are not available suitable many-line and continuum sources to cover smoothly the whole 3–80 eV range. Synchroton radiation sources in conjunction with monochromators have been suggested as potential sources for a long time[67] and they are now becoming more generally available. They can, in principle, supply radiation up to several hundred eV energy[68] though photon flux falls off dramatically at higher energies.

For any of the vacuum-uv sources it is generally necessary to use windowless systems, unless the photon energy being used is below 11 eV, in which case LiF windows can be used. Windowless systems set rather stringent requirements on design concerning differential pumping arrangements to keep the discharge gas (at about 0·1–1·0 Torr pressure) out of the ionization and electron analysis regions. Partial pressure of up to 10^{-6} Torr of discharge gases are tolerable in gas phase work, but for surface work (see Section VI) they must be kept below 10^{-9} Torr.

X-ray photoelectron work has mostly been performed using the $MgK\alpha$ (1253·6 eV) or $AlK\alpha$ (1486·6 eV) lines, using conventional X-ray tubes. To ionize very deep core-levels higher energy lines are required ($CuK\alpha$ at 8048 eV). Recently successful attempts at using very low X-ray energies have been made.[69] Thin aluminum or beryllium windows are usually used to separate the X-ray source region and the rest of the spectrometer, which allows separate vacuum pumping arrangements, an important consideration in ultra-high vacuum surface work applications.

A major difference between the vacuum ultraviolet (VUV) and X-ray sources is the source line widths which of course limit the ultimate photoelectron spectral resolution available. The VUV line widths are less than 0·01 eV, whereas the X-ray line width varies from about 1 eV at the lower energies to several eV for $CuK\alpha$. Two commercial instruments, Hewlett–Packard (HP) and Associated Electrical Industries (AEI), provide monochromators with their X-ray sources (AEI as a separate option; HP as standard) which reduces the source line width contribution to about 0·2 eV. The X-ray monochromator also has the advantage of removing the X-ray Bremsstrahlung and low intensity satellite lines ($K\beta$ lines) present in the unmonochromatized radiation. An incidental advantage is that the X-ray falling on the sample is reduced, thereby reducing the probability of X-ray-induced chemical change.

All the VUV and X-ray photon energy lines which have found practical use are listed in Table IV.

TABLE IV

Photon Energy Lines used in Photoelectron Spectroscopy

Line	Energy (eV)
H Lyman α	10·1986
ArIα	11·6233
	11·8278
NeIα	16·6704
	16·8476
HeIα	21·2175
HeIβ	23·0865
HeIIα	40·8136
HeIIβ	48·3702
YMξ	132·3
MgKα	1253·6
AlKα	1486·6
CuKα	8048

2. *Electron Sources.* The electron source required is dictated by the type of electron impact spectroscopy being performed. For Auger spectroscopy a monoenergetic source is not required, Auger electron energies being independent of the ionizing species energy, so focused beams from a heated filament and lens system suffice. Energy Loss Spectroscopy is often carried out in conjunction with Auger and, of course, the impacting beam should be monoenergetic for good resolution. The energy spread of the filament beam can be reduced to about 0·25 eV, which is adequate for many applications. For high-resolution energy-loss work an electron energy monochromator is incorporated between the filament supply and the sample region (Fig. 10). The requirements of an energy analyzer used as a monochromator are that a sufficiently large beam current be passed and that the beam can be focused to a small area. The hemispherical deflection analyzer, in conjunction with a lens system (Fig. 1b), is generally used for this requirement (discussed in detail by Kuyatt in Volume II).

3. *Excited Metastable Atom Sources.* Figure 11 shows a source for producing excited atoms. Helium atoms enter region A through a multi-channel tube, T; and are excited to higher electronic states by a beam of electrons, E. Only those states living longer than 10^{-5} sec (the 2^3S and 2^1S spectroscopic levels of energies 19·81 eV and 20·61 eV) manage to leave through orifice O. An electric field between P_1 and P_2 prevents ions from

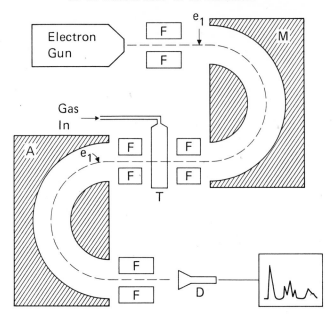

Fig. 10. Schematic representation of a commercial energy-loss spectrometer for gaseous studies (courtesy of McPherson Corporation). M is a hemispherical electron energy analyzer used as an electron monochromator. T is the gas target chamber. A is a second energy analyzer for detecting the scattered electrons. D is a channeltron detector. F are the various lens elements used to focus, accelerate, and retard the electron beams in this slitless system.

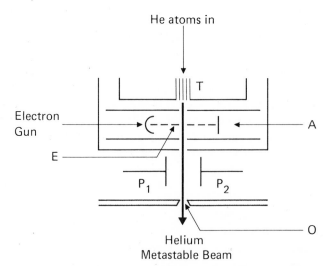

Fig. 11. Source for the production of He metastable atoms (see text).

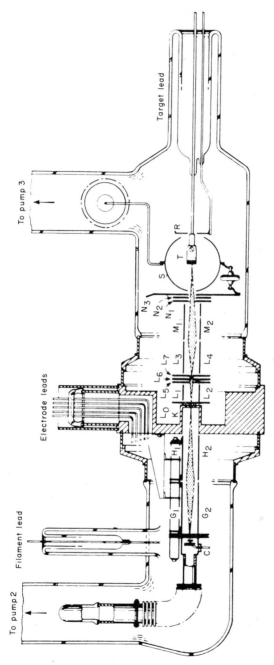

Fig. 12. Ion-beam apparatus for INS studies.[26] The ionizing electron beam passes through chamber C, and the He⁺ ions formed are extracted and focused by lenses G, H and slit K. (The ion beam is represented by the shaded area.) The lens system L, M focuses the electrons onto the target T, causing the ejection of Auger electrons (see text) whose energies are analyzed at the electron collector. Adjustment of the potentials on LM allows variations of the impacting He⁺ kinetic energy. Pumps 2 and 3 provide differential pumping between source and target regions through slit K.

leaving, and ways of suppressing one of the metastables to provide a single energy particle stream have been devised. Neon metastable atoms have been used also.

4. *Ion Beam Sources.* He^+ beam production is based upon a conventional mass spectrometry ion source. The He^+ beam, produced by electron impact, is collimated by a magnetic field and passed through electrostatic lenses which focus it and reduce the kinetic energy. Ne^+ and Ar^+ beams have also been used. Figure 12 shows such an ion beam source incorporated into a system for taking INS data.[26]

B. Electron Energy Analyzers

Three main methods of energy analysis have found favor in electron spectroscopy. The least popular, the Wien filter, involves the use of combined electrostatic and magnetic fields. It has been used for analysis of high electron energies,[70] and will not be described further here.

In the second method (Fig. 13a), the electrons to be analyzed are retarded by a voltage between grids G_1 and G_2. Only those with enough energy will

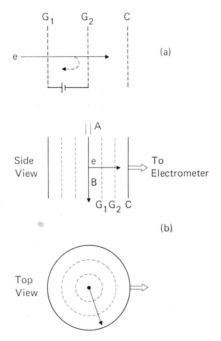

Fig. 13. Retarding grid (cylindrical) electron energy analyzer. G_1 and G_2 are concentric cylindrical grids. A represents the photon collimating capillary of a discharge lamp. C is the cylindrical collector electrode. B is the ionization region.

penetrate G_2 and reach collector plate C. For example, if the electrons passing through G_1 have energies of 10, 12, and 15 eV, sweeping the retarding voltage between G_1 and G_2 would give a spectrum as in Fig. 14a. A rise in detected current occurs as each group of energetically different electrons

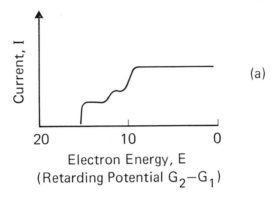

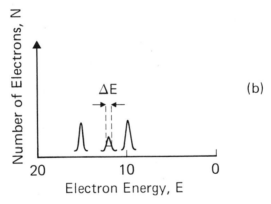

Fig. 14. (a) Integrated electron energy distribution spectrum obtained using retarding field grid analyzers. (b) Direct electron energy distribution ·obtained using deflection analyzers.

manages to penetrate G_2 and reach C. There are many variations of retarding potential analyzers. The one of Fig. 13 is a cylindrical system; also common is a system of spherical geometry.

The third method of analysis utilizes deflection of the electron beam by a magnetic or electrostatic field. A 127° electrostatic cylindrical analyzer forms part of the electron spectrometer illustrated in Fig. 15.[71] Electrons pass through a slit into the analyzer and are deflected round it by the voltages V_1 and V_2 on the analyzer plates. By varying the voltages, electrons of different energy can be brought to focus at the exit slit in turn. The relationship between the voltages and the energy, E, of the electrons focused is given by

$$V_2 - V_1 = 2ER \log(R_1/R_2)$$

where R_1 and R_2 are the radii of the plates and R is the mean radius. The spectrum produced is the differential of a retarding potential spectrum, i.e. peaks are obtained instead of steps (Fig. 14b). Many other types of deflection analyzers exist. An example of a double focusing type is the hemispherical system. This consists of two hemispherical deflection plates instead of the two 127° cylindrical sectors of Fig. 15. In this analyzer, focusing occurs irrespective of entry angle into the analyzer. Small holes, suitable for beams of small cross-sectional area, may therefore be used instead of slits.

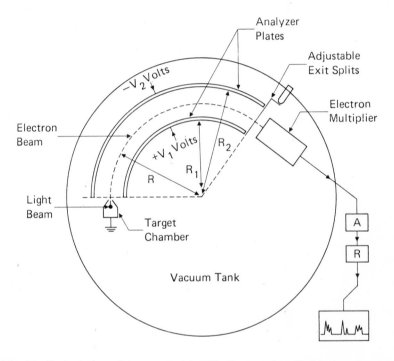

Fig. 15. Basic design of a commercial 127° electrostatic cylindrical analyzer UPS spectrometer for gaseous studies.[71] A is a pulse amplifier and R a ratemeter.

There has also been considerable interest in the cylindrical mirror analyzer because of its second-order focusing property.[72] Comparisons of different analyzer types in terms of performance, versatility, etc., are difficult because of the radically different configurations employed. However, several generalizations can be made. Retarding analyzers suffer from inherent problems of relatively poor signal-to-noise ratio in comparison to deflection analyzers because of the integrated nature of the spectrum. A spherical retarding field analyzer can collect over all emission angles however, whereas deflection analyzers collect over a small solid angle in a fixed direction. These considerations become important when considering angular distributions of emitted electrons (covered in detail in chapters in Volume III). Deflection analyzers which can rotate round a sample region, which itself can turn, have been devised for sophisticated angular measurements where wide variation of polar and azimuthal angles are required. For all the focussing deflection analyzers, factors such as transmission, resolution, source area, and "brightness" of source are inter-related and dictate the performance obtained.[73, 74]

C. Detecting and Recording Systems

The electron current leaving the analyzer can be measured either by an electrometer or by electron-counting procedures involving an electron multiplier detector and pulse amplification. In practice, electron multiplier detection has largely superseded electrometer detection. There are several variations available within this detection scheme however. Phosphor-coated sealed photon multiplier tubes are sometimes used to separate the working vacuum from the contaminant sensitive multiplier surfaces. The so-called channel plate detectors (a two-dimensional array of individual channeltron devices) may be used in conjunction with a phosphor screen and Vidicon scanner.[75] One thus records the entire spectrum over a large energy range simultaneously, rather like using a photographic plate. A large increase in the rate of data-accumulation is obtained which is particularly useful when low count-rates are encountered. Channel plate detection and display systems will probably also become important in angular distribution measurements, the ultimate device being a spherical channel plate.

D. Vacuum Housing

Vacuum requirements vary with the type of experimental information required. For gas-phase studies only a moderate 10^{-5} Torr vacuum capability is required. The limiting factors are that some electron multipliers cannot be operated above the 10^{-4} Torr, and that pressures must be kept low enough to avoid electron molecule collisions after the original ionization process.

For solids the requirements are more stringent, because the surface nature of the electron spectroscopic techniques requires that if the electron spectrum being taken is to be characteristic of the sample, the sample surface must be kept clean. At 10^{-6} Torr pressure a monolayer of material can be adsorbed at a surface in 1 sec. Vacuum systems capable of maintaining pressures in the 10^{-10} Torr range are therefore generally required.

E. The Complete Spectrometer

The individual components can be arranged in a variety of fashions depending on the objective of the experiment. Figure 10 illustrates the arrangement in a commercial electron impact energy loss spectrometer for gaseous studies and Fig. 15 shows the basic design for the simplest commercial uv-photoelectron spectrometer for gaseous studies. In both these systems the gas sample is bled into the target chamber and pumped away through the electron exit slit. For solid-state and surface studies the target chamber is replaced by a specimen holder which is usually rather sophisticated to allow for lateral motion and rotation, as well as providing heating and cooling facilities. Figure 16 shows a block diagram of a commercial XPS/UPS/Auger

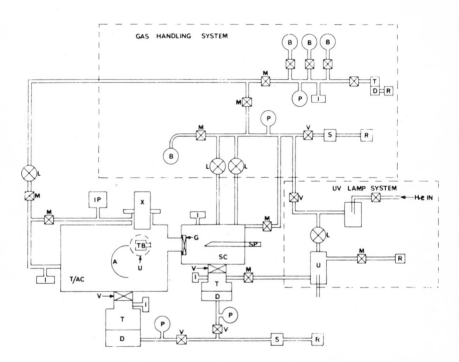

Fig. 16. Block diagram of a commerically available combination spectrometer (XPS/UPS/Auger) specifically for surface studies.[76] All seals between the ultra-high vacuum (10^{-10} Torr) and atmosphere are copper gasket or gold O-ring. T/AC is the target/analyzer chamber; SC, sample preparation chamber; X, X-ray source; U, uv source (the dotted circle indicates the position of the lamp-mating flange, at the rear of the system); A, hemispherical analyzer; SP, sample manipulator; T, liquid nitrogen trap; D, diffusion pump; S, molecular sieve trap; R, rotary pump; I, ion gauge; IP, ion source ion pump; P, Pirani gauge; M, metal valve; V, Viton-sealed valve; L, metal leak valve; G, gate valve; B, gas bulbs. The Auger electron gun is mounted on the same axis as the uv lamp, but from the front of the system.

combination spectrometer for surface studies.[76] All three types of spectra can be recorded sequentially by simply rotating the sample towards the x-ray, vacuum-uv, or electron beam sources.

VI. INTERPRETATION AND APPLICATIONS OF ELECTRON SPECTROSCOPY

Sample spectra from the different types of electron spectroscopy are first presented to illustrate the characteristic features present, and then the applications of the main branches are discussed. As in the earlier sections, photoelectron spectroscopy forms the heart of the discussion and the other branches of electron spectroscopy are treated in terms of the additional information that can be obtained.

A. Typical Examples of Electron Spectra

1. *Photoelectron Spectroscopy.* Figure 17 shows the UPS spectrum of carbon monoxide excited by 21·2 eV photons.[1] Each of the three bands, formed by photoejection from the three MO's, 5σ, 1π, and 4σ, has associated structure representing the vibrational levels in the three resultant ionized states. (Rotational structure cannot be resolved in photoelectron spectroscopy.) The *adiabatic* IP's (zeroth vibrational level) are 14·02 eV, 16·54 eV, and 19·69 eV. The *vertical* IP's (most intense vibrational peaks) are 14·02 eV, 16·91 eV, and 19·69 eV. On the right hand of Fig. 18 the energy level diagram for the transitions involved is shown. The Franck–Condon principle states that ionization is rapid compared to the time of a vibration, meaning that during ionization the internuclear distance remains fixed. At the risk of oversimplification, transitions can be represented by vertical lines in Fig. 17. Maximum ionization cross-section occurs when both the initial and final states have a maximum in their vibrational wave function. There will be zero cross-section between positions where a wave function is zero, i.e. all the internuclear distances not falling between the two lines, A, the Franck–Condon principle accounts for the experimental peaks having the observed intensities. Since line B connects the maximum wave function of the ground state with a maximum at the zeroth vibrational level of the first ionized state, the zeroth vibrational peak in the first IP band is the most intense. For the second ionized state, B connects with a wave function maximum at the $v = 2$ level and so this peak is strongest in the second IP band.

Figure 18 shows a typical photoemission spectrum of a metal, nickel, taken with 21·2 eV photons.[77] The band nature of the energy levels in a metal results in the continuous spectrum observed, with no discrete IP's corresponding to discrete MO's and no vibrational structure. Electrons which have been scattered many times by collisions on passing through the solid before

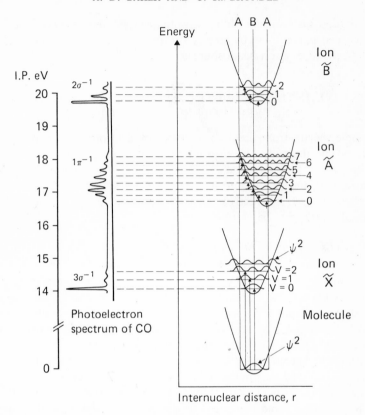

Fig. 17. UPS spectrum of CO, excited by HeI radiation.[1] The ionization potential (or binding energy) scale is given at the left (21·23 V − K.E. of electron) and the potential energy wells and vibrational energy levels for the neutral molecule (ground state) and the three ionized states at the right.

escaping into the vacuum lose most of their energy, and so contribute a large part of the spectrum at low electron energies. This is often termed the secondary electron background.

Core levels of both gases and solids are discrete and therefore both give individual peaks rather than bands in a photoelectron spectrum. Most core levels have B.E.'s greater than the photon energies available in UPS and therefore must be observed using XPS. Figure 19 shows the C(1s) core-level XPS spectrum of ethyltrifluoroacetate, taken using AlKα radiation.[12] The C(1s) binding energies of the carbon atoms are all slightly different because their chemical environment is slightly different (see later) and so four peaks are observed. In a spectrum taken without an X-ray monochromator (top panel, Fig. 19) the resolution is insufficient to resolve fully the four component peaks, but with a monochromator (bottom panel, Fig. 19) the peaks are completely separated. It can be seen that to a first approximation they are all of equal intensity because the $(1s)$ ionization cross-sections for the different carbon atoms are the same, the cross-section being an atomic property.

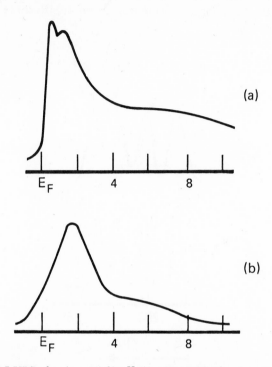

Fig. 18. (a) HeI UPS of a clean Ni film.[77] (b) AlKα XPS of a clean Ni film.[78]

XPS spectra from the valence-level regions of solids have lower resolution than UPS and there may be other differences caused by the nature of the photoemission process in solids (see later sections). The XPS of the Ni valence region[78] is shown in Fig. 18b for comparison to the UPS spectrum. For gas-phase XPS the valence region spectrum shows discrete IP's, as for UPS. Figure 20 shows the complete XPS spectrum for CO.[13] The 0–21·2 eV region may be compared to the UPS spectrum taken with the HeI resonance line (Fig. 17). The major differences are the restriction on resolution in XPS, washing out the vibrational structure, and the different intensity ratios of the peaks, due to different ionization cross-sections at the two photon energies used.

Figure 21 gives an overall picture of the types of spectra from the two branches of photoelectron spectroscopy for gases and solids. In both UPS and XPS studies of solids, only electrons from the surface or near the surface are detected (see subsequent sections and the chapter by Brundle in Volume III). This means that in addition to the aforementioned resolution and ionization cross-section differences between XPS and UPS, the two techniques may sample different thicknesses of the surface region because different electron energies are involved and electron penetration distances are a function of their energy.

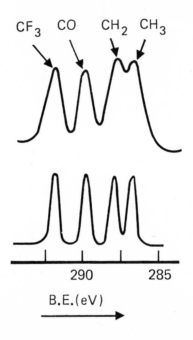

Fig. 19. XPS spectrum of ethyl trifluoracetate showing the C(1s) region. Upper trace non-monochromatized X-ray source,[12] lower panel monochromatized X-ray source.

2. *Penning Ionization Spectroscopy*. Figure 22 shows the spectrum of CO taken using He atoms in the 2^3S and 2^1S excited states as the source.[54] The energies of the ejected electrons are given by the equations:

$$K.E._n = 20 \cdot 61 - I_n \quad \text{(for He } 2^1S \text{ impact, peaks "a" in Fig. 22)}$$

$$K.E._n = 19 \cdot 86 - I_n \quad \text{(for He } 2^3S \text{ impact, peaks "b" in Fig. 22)}$$

Each Ionization Potential, I_n, therefore appears twice in the spectrum (cf. the UPS spectrum of Fig. 17). The poor resolution is partly the consequence of using a poor analyzer. It can be improved and also the He 2^3S metastables can be suppressed, removing the doubled nature of the spectrum and leaving a spectrum very similar to the UPS spectrum of Fig. 17.

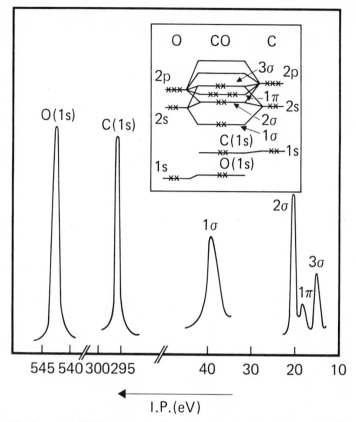

Fig. 20. Complete XPS spectrum of gaseous CO showing the core-levels and the valence level MO's made up by the bonding between C and O atomic valence orbitals (see inset).[13]

3. *Auger Spectroscopy.* Gaseous Auger studies of small molecules have been made under high-resolution conditions. The kinetic energy of the ejected Auger electron is determined by a combination of two transitions (see preceding discussion) and since a large number of transitions are possible the total Auger spectrum can be very complex, particularly if vibrational structure is resolved. Figure 23 shows that part of the electron impact Auger spectrum of CO and CO_2 involving the filling of a C(1s) hole as the final transition.[79] In a different energy region the transitions involving the O(1s) hole-state are observed. Very similar spectra would result from X-ray impact since the Auger process is independent of the way in which the original hole is formed. For solids all vibrational structure is smeared out so solids can sometimes be studied effectively under low resolution and, until recently, retarding-field grid analyzers were usually used for this work. Figure 24 shows a typical

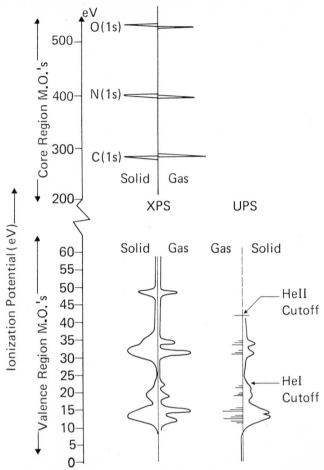

Fig. 21. UPS and XPS schematic spectra of a carbon-, nitrogen-, and oxygen-containing species in the gaseous and solid state.

example, a first derivative Auger spectrum from a steel sample.[80] Taking a derivative spectrum is a standard method used to enhance the small peaks and to suppress the sharply rising scattered electron background on which these peaks sit. Figure 25 is the schematic complete electron spectrum (not differentiated) which might be obtained by impact of 1000 eV electrons. The small Auger peaks fall typically in the kinetic energy range 50–700 eV. WXY transitions have a lower energy than WVV, since the core levels X and Y are deeper than the valence levels V.

4. *Ion Neutralization Spectroscopy* (*INS*). The processes observed in INS are equivalent to WVV Auger processes except that the first step involves an

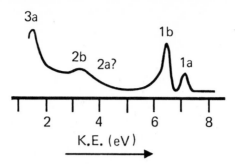

Fig. 22. Penning Ionization Spectrum of CO using He 2^1S and He 2^3S metastables.[54] The three ionization potentials of CO (see Fig. 17) appear as peaks 1, 2, and 3 for He 2^1S (peaks a) and He 2^3S (peak b) induced ionization.

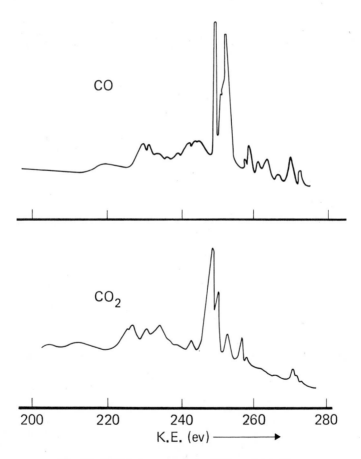

Fig. 23. C KLL Auger spectra of CO and CO_2.[79]

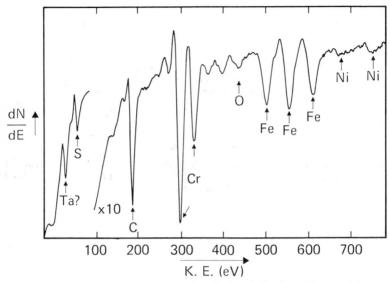

Fig. 24. Typical Auger spectrum, recorded as the first derivative with respect to energy, dN/dE, of a steel surface revealing the presence of Ni, Cr, O, S, C and possibly Ta, in addition to Fe.[80]

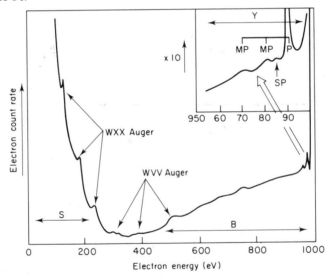

Fig. 25. Schematic electron impact spectrum for a solid showing the elastically scattered primary beam (1000 eV); plasmon excitation inelastic scattering from that beam (region Y); the region B in which energy-loss peaks caused by inelastic collisions resulting in band-to-band transitions may occur, the small Auger electron peaks; and the multiply scattered secondary electron region S. In typical practice the spectrum would be differentiated to enhance the small Auger peaks and suppress the sharply varying background.

electron falling from the valence level to the vacant level in He^+, instead of to a hole (Fig. 3). Figure 26a is the INS spectrum of a clean Ni(100) crystal.[81] One might expect a WVV Auger peak to be similar in *shape*, though the *absolute* energy of the electrons would be different owing to the different energy of the initial hole in the two processes. In practice the shapes are different also because of differences in the transition matrices and in surface sensitivity (see later). The valence bandwidth and shape can be approximated by taking the derivative of the spectrum or more correctly by a mathematical "unfolding" process, to remove the effect of the two-electron nature of the process. The unfolded INS Ni(100) spectrum is also shown in Fig. 26a, where the valence level bands show up clearly.

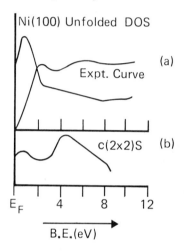

Fig. 26. INS spectra of (a) a clean Ni(100) surface and (b) a c(2 × 2) overlayer surface formed by the adsorption of sulphur.[81] Only the unfolded data are shown for the adsorbed situation. The dotted trace shows a UPS spectrum of sulphur on Ni.[77]

5. *Electron Impact Energy Loss Spectroscopy.* Figure 27 shows part of the electron spectrum obtained from impact of 34 eV electrons on helium.[31] Each peak represents promotion of an electron from the filled atoms 1s orbital to an empty one. The two 1s states used for Penning ionization, 2^3S and 2^1S, are observed at energies of 19·82 eV and 20·16 eV. In the solid state this type of transition occurs also. Promotions from valence levels to unfilled levels are known as band-to-band or interband transitions. Similar promotions occur from deeper levels. The Auger spectrum of Fig. 25 shows several peaks interpretable as band-to-band energy loss features (region B). There are other types of energy loss features also in Fig. 25. The energy region which is close to the elastically scattered 1000 eV peak is shown in expanded form in the figure. It contains peaks which are caused by collective excitations (plasmons)

of the metal "electron gas", i.e. they are energy loss processes from the impacting 1000 eV beam, the amount of energy loss being given by the plasmon energy, P. Multiple plasmon excitations or surface plasmon excitations may also be observed. Since in Fig. 25 both Auger and energy loss features occur, it is necessary to establish which peaks are Auger and which energy loss to obtain a satisfactory interpretation of the spectrum. Besides the fact that the two types of peaks tend to fall in different energy regions, the distinction is fairly easy experimentally. If the energy of the impacting beam is changed, the energy of the scattered beam must change by a similar amount for a given loss process, i.e. the separation between the energy loss features and the elastically scattered peak remains constant. Auger electrons, being independent of the impacting energy, maintain their kinetic energies irrespective of changes in the energy of the incident beam.

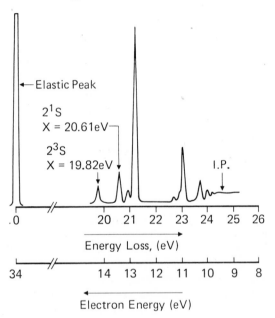

Fig. 27. Energy-loss spectrum resulting from the impact of 34 eV electrons with helium.[31]

B. Photoelectron Spectroscopy

1. *Determination of Electronic Structures and Related Electronic Effects.* Applications in this area form the largest segment of basic research within photoelectron spectroscopy and within electron spectroscopy as a whole. It forms the subject matter for many of the later chapters and the basic ideas involved in making the connection between observed spectra and electronic

structure are comprehensively treated elsewhere in these volumes. A sufficient outline of some of those ideas is given here so that the reader may appreciate the type of information available and the problems existing in making the interpretations.

A photoelectron spectrum is the best experimental approximation to a direct display of molecular orbital energies, though as pointed out earlier a molecular orbital description is not the only "correct" way of describing the electronic structure of a molecule. UPS spectra display the features resulting from ionization of the outer (valence) MO's and XPS spectra all the MO's down to the inner-shell core levels (see Fig. 21). For many years theoreticians have performed calculations yielding MO energies. The same calculations yield more tangible quantities such as bond lengths, dipole moments, and heats of formation, but photoelectron spectroscopy allows for the first time an experimental check on the primary quantities involved, orbital energies. Conversely, accurate calculations are helpful in assigning experimental IP's to the correct MO's. Though the ordering and energies of the valence MO's are important since they help decide the chemistry of a molecule, more information than simply this is contained in the photoelectron spectrum. In Section VI-A it was shown in Fig. 17 how the Franck–Condon principle determines the shapes of the photoelectron bands. Removal of an electron contributing little to the bonding in a molecule (i.e. an electron in a nonbonding MO such as the oxygen lone pair of H_2O) makes little difference to the bond length. The potential well of the ion will therefore be almost exactly above that of the molecule (case $\tilde{X}$ in Fig. 17). If a strongly bonding electron were removed, the equilibrium bond length, r_e, would increase as the bond is weakened, and the potential well of the ion would be shifted to the right in Fig. 17 (case $\tilde{A}$). Removal of an antibonding electron would increase the bond strength, reducing r_e, and shift the potential well to the left. Thus we are able to distinguish between nonbonding electrons and bonding or antibonding electrons. For diatomic molecules, at least, a short vibrational series with maximum intensity at the zeroth level implies a nonbonding electron; a long vibrational series a bonding or antibonding electron. Further information about bonding character may be obtained from the vibrational frequency, ω, observed, as is discussed in the chapters on UPS of gases.

For solids the UPS features are related to the solid-state band structure in the valence region, though there is not necessarily a direct one-to-one correspondence between the experimental spectrum and the density of states (DOS). A major interest has been in elucidating the nature of the photo-emission process in a solid and thereby relating the experimentally observed spectrum to electronic structure effects.[82] If we consider the hypothetical band-structure ε,k diagram of a solid given in Fig. 28, the photoionization process may be considered as raising an electron occupying an allowed energy level, ε, in k space, by energy $h\nu$. However, unlike the situation in the free molecule the electron is still within the crystal at this point and must end up occupying another allowed energy band. For the quantum restriction of k-conservation, transitions can only occur at those values of k where there is a gap of exactly $h\nu$ between initial and final states. Having reached the final state the electron can now be transported to the surface and ejected. Changing $h\nu$ will in general alter the transitions allowed (e.g. changing from $h\nu_1$ to $h\nu_2$ in Fig. 28) and result in different experimental spectra which are representative of a joint density of states—(initial * final) at the different photon energies, unlike the free-molecule situation where a continuum of final states (free electron) is available to connect with the initial state no matter what the value of $h\nu$ (Fig. 28c). At high enough $h\nu$ (generally above 40 or 50 eV), final state structure is so dense for the solid case that it approximates continuum-like behavior and so the experimental spectrum becomes representative of an initial DOS, modified by whatever cross-section effects may be present. Since the technique is surface sensitive there is always the possibility that the experimental spectrum represents a surface electronic structure which is different from that of the bulk (see later).

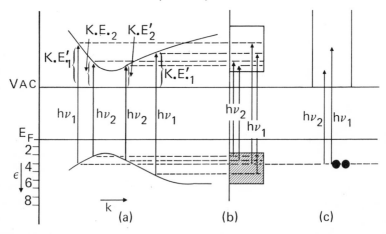

Fig. 28. Hypothetical εk band-structure diagram (a), and DOS (b), of a solid, illustrating the effect of k conservation and final state structure on the photoemission process. (c) represents the equivalent photoemission process for a free molecule.

For XPS, electronic structure determination can be tackled either by looking directly at the valence levels or by examining the response of the core levels to changes in the valence electron distribution. For the former the advantage for gaseous studies over UPS has been to be able to observe those valence levels beyond the HeI or HeII cut-off values. The main disadvantages are low resolution and low ionization cross-section. An important complementary feature of XPS valence level studies to UPS results from the drastic changes in relative ionization cross-sections that occur for different atomic orbitals as a function of $h\nu$. Thus the C $2s/2p$ ratio increases by more than an order of magnitude in going from HeII to MgKα radiation energy. These changes, which have become qualitatively well understood experimentally and theoretically and are discussed in detail in chapters in Volumes II and III, are extremely useful in helping assign orbital character to experimental peaks, both in the gas phase and in the solid state. For the solid state, XPS valence level studies are complementary to UPS in a further way. The photon energy is always high enough for the modifying effect of final-state band structure to approximate a continuum-like behavior and become unimportant, so the experimental spectrum is a cross-section modified (in a generally known way) version of the initial DOS. It should be borne in mind, however, for both UPS and XPS that if crystalline samples are concerned strong angular anisotropies may occur. These may also relate to electronic structure effects[83] and may be utilized in electronic structure interpretation, but since most spectrometers operate with a fixed small acceptance angle at the present time one has to be careful of anisotropy effects.

The second method of tackling electronic structure by XPS, examination of core-level behavior is what originally led to the name ESCA, Electron Spectroscopy for Chemical Analysis.[12, 13] For both gases and solids inner IP's are characteristic of the individual atoms, since the electrons concerned are tightly bound to those atoms. Thus the energies of the $1s$ electrons in the elements carbon, nitrogen, and oxygen, which are found by XPS to be about 284 eV, 399 eV, and 530 eV, do not change much when these atoms form part of molecules. Thus one can tell from the IP's observed which elements are present. In methane (gaseous) the carbon $1s$ energy is 290·8 eV, which in CH_3F becomes 293·6 eV.[48] In $FeSO_4$ (solid) the oxygen 1s energy is 532·5 eV, whereas in $Fe_2(SO_4)_3$ it is 531·8 eV.[84] This slight variation in energy of a particular inner-core level is related to the molecular environment of the atom concerned. It may be used to help determine the atom's environment, i.e. provide information on the bonding, and to distinguish between atoms of an element in non-equivalent positions such as the carbon atoms in $CF_3COCH_2CH_3$ (Fig. 19).

The inner-core electron energies change with molecular environment because of the electrostatic interaction between the core electrons and the valence electrons. In the simplest description of a chemical shift the core electrons can be considered to be classically screened by the shell of valence electrons at greater radius from the nuclei. Any charge change in that shell will be reflected in a change in the amount of energy required to remove a core electron through the valence shell to infinity, i.e. the B.E. changes by a "chemical shift". The "chemical shifts" may be calculated theoretically. Figure 29 illustrates the measure of success obtainable from some rather early calculations.[85] The calculated absolute energies shown in Fig. 29 are in-accurate, but the straight-line correlation shows that the chemical shifts are predicted rather well. Correlations can also be made between observed chemical shifts and oxidation states, oxidation numbers, or charge on an atom, allowing measurement of such properties in unknown cases. A very useful tabulation of XPS core-level binding energies can be found in the "Handbook of Spectroscopy".[86]

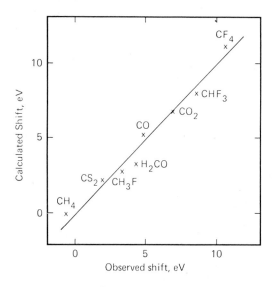

Fig. 29. Plot of the observed C(1s) chemical shift in a number of molecules against the calculated shift.[85]

Another concept which has received attention is the "Principle of Equi-valent Cores" proposed by Jolly (see Chapter 3), in which thermodynamic data are correlated with XPS chemical shifts. The whole concept is based on the approximation that when a core electron is removed from a molecule or ion, the valence electrons relax as if the nuclear charge had increased by one unit. Thus, atomic cores (i.e. nuclei plus core shells) having the same net charge can be considered to be chemically equivalent, e.g. ejection of a core electron from the N 1s orbital of ammonia will leave a species that is chemically equivalent to H_3O^+. By using a type of Born–Haber cycle, Jolly has developed equations relating changes in chemical shifts with heats of formation of species such as H_3O^+; often the thermodynamic parameters he is able to deduce would be difficult to determine by alternate means.

Variations in both valence shell and core ionization potentials within structurally related series of compounds have been correlated with substituent effects, such as electronegativities and Hammett constants, and with features observed in related spectroscopic techniques such as uv and NMR spectroscopies. Frequently features observed in the photoelectron spectra enable substituent and related electronic and steric effects to be looked at in a new light.

It was stated at the beginning of this section that the photoelectron B.E.'s were the best experimental *approximation* to orbital energies, ε_j. It should be realized however that they are only approximations and that the photoemission process does not measure ε_j, but the difference in energy between the initial state of the total n electron system before ionization and the $n-1$ electron final state after ionization. Even in the absence of multiple final-state possibilities caused by spin–orbit or spin–spin interactions of the remaining unpaired electron, B.E.$_j$ would only equal ε_j if, when an electron is removed from the jth orbital, all the other energy levels remain undisturbed (the approximation of Koopmans[47]). In fact they do not remain undisturbed, instead there is a relaxation of valence electrons towards the positive hole created by the photoionization process which reduces the energy of the final state by E_j^R, the *Relaxation Energy*. Thus B.E.$_j = \varepsilon_j - E_j^R$. In the gas phase E_j^R is often, for calculation and descriptive purposes, split into two components, the atomic relaxation representing the relaxation of the valence levels of the atom possessing the core hole, and the extra-atomic relaxation representing relaxation from the rest of the system. Any arguments concerning initial-state electronic structure differences between related species based only on equating B.E.$_j$'s to ε_j's therefore relies for its validity on E_j^R being similar in the different species. If they are not similar then some theoretical treatment of E_j^R or experimental measure must be included in the comparisons.

So far we have considered what we might term the primary information sources from photoelectron spectron, measured B.E.'s (valence and core), the effect of varying ionization cross-sections, and vibrational structure in the valence region. A secondary source of information exists in the positions, intensities, and numbers of "extra" peaks associated with the main peaks in the spectra. These "extra" peaks have their origins in effects such as spin–orbit coupling, "multiplet or spin–spin" splitting, and "shake-up" phenomena giving rise to satellite peaks.

The effects of these phenomena are discussed in detail in other chapters and so are only very briefly introduced here.

Spin–orbit coupling takes place if an electron ejection leads to an ion being produced with a hole in any orbital other than an s-orbital. Thus, even at the simplest level, ionization of an argon atom through ejection of a $3p$ electron can lead to two possible ion states, either $^2P_{\frac{1}{2}}$ or $^2P_{\frac{3}{2}}$. These have different energies, and thus two peaks, not one, are obtained in the region of the UPS photoelectron spectrum corresponding to the Ar $3p$ electrons. Information on electron delocalization can be obtained from the differences. Thus the spin–orbit splitting observed for the ionized states resulting from I $5p$ electron injection in HI is 0·66 eV; that for C_2H_5I is only 0·60 eV, indicating greater delocalization in the latter case. However, for ethyl iodide, the observed splitting between the peaks also relates in part to the interaction of the pseudo-π-orbital associated with the ethyl group and one of the iodine lone pairs. A clear exposition on the competing factors has been given by Brogli and Heilbronner.[87]

Multiplet splitting is the result of spin interaction between an unpaired electron resulting from the photoionization process and other unpaired electron(s) present in the system. A typical situation resulting in the multiplet splitting type phenomenon in X-ray photoelectron spectra is that of ejection of an inner electron from a compound of a transition element; a specific example being the case of $3s$ electron ejection from a Mn^{II} compound. Mn^{II} contains five unpaired d-electrons; thus ejection of a $3s$ electron can lead to either a pentuplet or septet state:

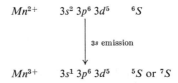

$$Mn^{2+} \qquad 3s^2\ 3p^6\ 3d^5 \qquad {}^6S$$

$3s$ emission

$$Mn^{3+} \qquad 3s^1\ 3p^6\ 3d^5 \qquad {}^5S \text{ or } {}^7S$$

depending on whether the resulting unpaired $3s$ electron is antiparallel or parallel with the five unpaired $3d$ electrons. Since two states can be formed following the ejection of a $3s$ electron, two peaks might be expected in the Mn$3s$ ESCA spectrum. Two peaks are often observed in such cases, and the point of real interest is that the separation between the two peaks varies depending upon the environment of the atom concerned. For example, Fig. 30 shows the different multiplet splittings observed in the XPS spectra for some chromium compounds. The data included in the figure are taken from the work of Carver et al.,[88] who were among the first to show that multiplet-splitting could provide chemically useful information. An advantage of looking at differences in multiplet splittings rather than at differences in chemical shifts is that, in the former case, calibration in terms of absolute binding energies (a problem that has proved difficult in XPS) is not critical since the difference between the two peaks is being measured. Multiplet splittings have been the subject of considerable theoretical effort also and the degree of quantitative understanding of the phenomena is reasonably good in those systems which have been treated in detail.[17, 89] Basically in a case where two final states are possible, e.g. ${}^7S$ and ${}^5S$ for ${}^3S$ ionization from Mn^{II} or

Fe^{III}, the total energies of the two ions. (i.e. ^7S and ^5S), calculated by the spin-unrestricted Hartree–Fock free-ion method, are expected to differ only in the exchange integral, $G^2(3s, 3d)$,

$$G^2(3s, 3d) = C^2(0, 0, 2, 0) \int_0^\infty \int_0^\infty R_{3s}(r_1) R_{3d}(r_2) R_{3s}(r_2) R_{3d}(r_1) \left(\frac{r_<^2}{r_>^3}\right) dr_1 dr_2$$

where R_{3s}, R_{3d} are the radial wave functions for the respective $3s$ and $3d$ orbitals, and $C^2(0, 0, 2, 0)$ is a Clebsch–Gordon coefficient.

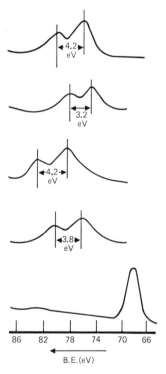

Fig. 30. XPS Cr $3s$ spectra for a series of Cr^{III} compounds (Cr_2O_3, Cr_3S_2, CrF_3, $CrCl_3$, and $K_3Cr(CN)_6$ from top to bottom) showing the multiplet splitting of the $3s$ level.[88]

Qualitatively it can be said that the greater is the exchange integral, the greater will be the splitting in the spectrum, and therefore anything which tends to decrease the d-electron density at the metal (e.g. charge delocalization to ligands) will decrease the splitting. Thus a study of multiplet splitting effects in a series of compounds may be related to such parameter as the number of unpaired valence d-electrons present and the percentage covalency of the compounds.

The so-called "shake-up" structure observed in XPS (and sometimes UPS) is an area of much current theoretical[90] and experimental effort.[17, 91] Experimentally, satellite peaks on the high B.E. side of the core-level XPS peaks have been observed frequently in compounds of the transition elements, in rare gases, alkali metal halides, and in several organic compounds. They may be viewed as the excitation of a valence electron to an unoccupied level simultaneously with the core-electron ejection (see Chapter 2). The separation of the shake-up satellites from the main core-level peak therefore corresponds to the valence level excitation energies concerned. "Shake-off" processes can also occur where the second electron is ejected rather than excited (i.e. a double ionization process). The partition of kinetic energy between the two ejected electrons results in steps followed by a continuum instead of the peaks found for shake-up.

Shake-up and Shake-off Structure is a manifestation of the relaxation phenomena mentioned earlier. The perturbation of the Coulombic potential felt by the valence electrons when a core-level electron is ejected causes their collapse towards the positive hole, creating the relaxation energy and allowing the possibility of excitation of the valence electrons.

In principle the Sudden Approximation Theory (instantaneous change in the Coulomb potential), and the Sum Rules relating the spectral weighting of shake-up and shake-off intensity to the relaxation energy, E_j^R, may be used to calculate the probabilities and energies of the shake-up and shake-off structure.[17] The sum rules apply a lever-arm principle to the experimental spectrum with the hypothetical unrelaxed core B.E. position (i.e. Koopmans' Theorem frozen orbital value) acting as a pivot. E_j^R is the difference between this hypothetical value and the actual observed main core line at lower B.E. The intensity of the main peak multiplied by E_j^R must be balanced by the (intensity × energy separation) of all the shake-up and shake-off structure to the high B.E. side of the pivot position. Since there is always a finite value of E_j^R it follows that there must *always* be shake-up and shake-off structure, even if the intensity is weak. The relationship between relaxation and satellite structure is discussed in Chapter 2, and elsewhere.[17, 83]

Frequently rather large changes in satellite intensities and positions are observed with change in geometry or ligand. Thus, the satellite peaks associated with the main 2p peaks in the spectra of the octahedral $CoCl_2$ and the tetrahedral $[Et_4N]_2[CoCl_4]$ are of different intensities.[92] This can possibly be explained by the different numbers of transitions in the two cases that are allowed by the appropriate charge-transfer selection rules. For the Mn^{II} halides, observed differences (Fig. 31) may be related to the charge densities on the metals and ligands.[93] Thus x-ray photoelectron spectra can be usefully used to provide information on both geometries and electron distributions in metal-ion complexes.

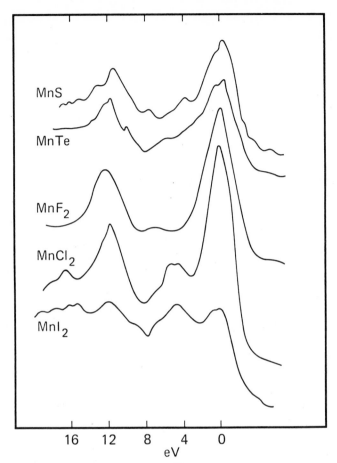

Fig. 31. Mn $2p_{\frac{3}{2}}$ XPS region for some Mn^{II} halides showing the variation in relative intensities and positions of the shake-up structure.[93]

2. *Applications in Analytical Chemistry.* In principle both UPS and XPS can be applied to a number of analytical problems, since both are capable of detecting very small amounts of sample either in the gas phase or at surfaces, and both give information about the structure of the sample. Quantitative analysis is possible in some cases because the intensity of a given peak can be expected to be proportional to the concentration or absolute amount of the species giving rise to the peak.

Since the energies of inner electrons are characteristic of the atom concerned, XPS, by measuring these energies, is able to provide an elemental analysis of a sample, both qualitatively (in terms of the elements present) and quantitatively (from peak areas). The strength of the technique qualitatively is illustrated by, for example, the easy detection of the single cobalt atom in vitamin B12[12] and by the detection of an abundance of impurities in the originally much publicized, but now derided, "polywater".[94]

Quantitative measurements are not so straightforward, however, owing to (1) the variation of ionization cross-section with both atomic and orbital number (discussed in detail in Volumes II and III); (2) the problem that not all of the intensity goes into a single main peak (shake-up and shake-off phenomena); and (3) for solids the analysis relates to the surface rather than the bulk. To allow for (1) one can either rely on theoretically calculated ionization cross-sections, or use suitable calibration procedures. The latter has the advantage that for any given spectrometer, instrumental factors such as the variation of transmission with K.E. are automatically included. Relative sensitivities for detection of different elements by XPS have been determined experimentally for solids many times and tables compiled. Unfortunately it is rarely possible for one person to use another's tables satisfactorily because of unknown instrumental parameters and the generally unknown condition of the sample surface.

In terms of elemental analysis, UPS can at best provide a qualitative guide to the presence within molecules of certain atoms (e.g. halogen, O, S, N) that give rise to distinctive lone-pair peaks. In general UPS peaks, since they represent valence MO's, are characteristic of the whole molecule rather than an individual atom. UPS can also provide limited information in instances where it is necessary to have a method for monitoring one or more of a mixture of structurally similar compounds which have dissimilar UPS spectra but similar IR, NMR, etc. spectra. A UPS GC link-up in these circumstances is a viable proposition.[95]

Analyses of mixture by UPS and XPS without prior separation into components have been attempted.[2, 13, 96] Such analyses can be relatively simple if each of the components gives rise to a distinctive peak, or set of peaks, not overlapped by peaks due to other components. In addition one is usually restricted in UPS (for analytical purposes) to the discrete peaks found in gas-phase spectra. Stripping the known spectrum of one known pure component from a spectrum of a mixture can reveal the spectrum of another component or components, which may then be identified.[97]

A novel method for trace analysis using XPS has been investigated—the method consisting essentially of using glass fiber "mats" with chelating groups on their surfaces.[98] Here the surface nature of XPS is being exploited for a bulk analysis. The surface of the glass fiber is treated in such a way that a chelating moiety is incorporated onto the surface; the trace element to be determined is then extracted from aqueous (or other) solution onto the fiber mat, thereby improving the sensitivity of XPS for that element manyfold by concentrating it onto the surface. Detection limits of about 10 ppb were observed for lead, calcium, thallium, and mercury. Brinen and McClure have also used XPS for trace analysis by electrodepositing metals on mercury-coated platinum electrodes, but no estimate of detection limits was made.[99]

3. *Solid State and Surface Chemistry and Physics.* Both XPS and UPS have become established techniques for studying surfaces and surface reactions[100–103] (see also the chapter by Brundle in Volume III). In general, electrons originating deep in a solid cannot escape without being inelastically scattered by interactions with the solid. The result is a loss of intensity from the electron peak at its original energy and the production of peaks and continua at lower electron energies, representing the energy losses and secondary ionizations of the scattering processes. The depth from which an electron can escape with unimpaired energy is governed by its *mean free path length*, L_e, the latter quantity being dependent on the K.E. of the electron concerned. Experimental determinations of L_e in XPS and UPS exist for a number of metals and for a number of different electron energies; values lie between a few ångströms and about 100 Å (Fig. 32).[100] Thus it can be seen that UPS spectra might consist of electrons with L_e values as low as 5 Å whereas an AlKα XPS spectrum of the same valence levels will consist of electrons with L_e some 25–30 Å. The XPS spectrum is "cleaner", however (Fig. 32), because the multiple scattered electrons cascade down towards zero

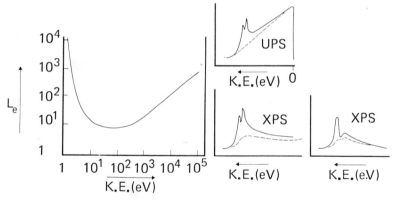

Fig. 32. (a) Relationship between inelastic mean-free path length, L_e, and electron kinetic energy. (b) Typical contribution of scattered electron background to a valence level spectrum recorded by XPS and UPS. (c) Typical background contributions to XPS core-level spectrum.

energy and are therefore removed from the high K.E. region of an XPS valence spectrum, whereas those near zero-energy electrons in UPS form a high-intensity background on which the UPS spectrum sits.

A material which has its surface completely free of adsorbate material and has unchanging physical structure from the bulk right up to the vacuum interface will still have a different electronic structure at the surface because of the very existence of the interface. The difference may be just a narrowing of the DOS owing to the reduced coordination of surface atoms, or strong additional surface states may be present. In either case such effects are more easily observed in UPS than in XPS because of the generally smaller L_e values involved.

If atoms or molecules are adsorbed at the solid surface, the electronic structure of the surface region of the substituent will be further modified and, more importantly, new structure will be observed in the UPS relating to the modified "orbitals" of the adsorbate. Identification of these orbitals usually allows one to distinguish between nondissociative and dissociative adsorption and often to identify exactly which orbital or orbitals are involved in the bonding to the surface.[100] An example is given in Fig. 33 for the adsorption of CO on Mo under two different conditions.[104] At room temperature no MO's reminiscent of molecular CO are observed in the UPS, and in fact the adsorbate-induced features closely resemble those obtained by superimposing spectra of carbon atom and oxygen atom adsorption. At 80 K, CO molecular

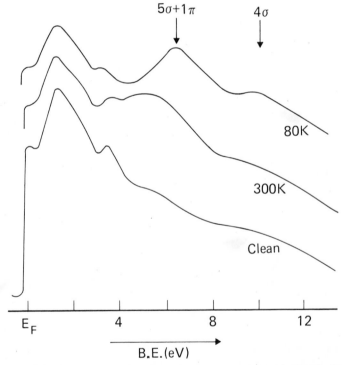

Fig. 33. HeI photoelectron spectra for a clean Mo surface; CO absorbed at 80 K; and CO absorbed at 300 K.[104]

orbitals are observed but the 5σ carbon lone-pair orbital has been significantly perturbed relative to the 1π and 4σ levels, leading to the conclusion that the 5σ is heavily involved in the bonding. On warming to room temperature the CO molecule dissociates giving the C plus O atom spectrum. Attempts are just beginning to account theoretically for the magnitude of molecular orbital B.E. shifts on adsorption in terms of bond formation to the surface,[105–107] and to extract such parameters as heats of adsorption and predict surface reactions.[108, 109]

XPS core-level intensities of the adsorbate material first of all allow a quantitative analysis of the elements present. Besides providing an estimate of the total amount of adsorption, this can be invaluable in establishing whether adsorbate stoichiometry is maintained or whether dissociation plus desorption or lattice penetration occurs.[77, 100] The B.E. energies can often provide confirmation of the UPS interpretations, e.g. for the CO on Mo case at room temperature the C(1s) and O(1s) core-level values are the same as for atomic absorption; but for 80 K absorption they are not.[104] After extensive surface reaction (e.g. corrosion) UPS loses its diagnostic effectiveness because of the band nature of the valence region. XPS is often still able to identify the species formed. In attempting to characterize the bonding and electronic structure involved it is as important to use the information available from the "secondary effects" as it is in nonsurface work. The formation of CoO rather than Co_3O_4 on initial oxidation of a Co surface can easily be diagnosed by looking not just for a chemical shift effect but also taking account of multiplet splittings, spin–orbit splittings, and shake-up structure, all of which are dependent on electronic structure.[110] The $2p_{\frac{3}{2}}$ and $2p_{\frac{1}{2}}$ regions of CoO and Co_3O_4 are shown in Fig. 34 to illustrate the differences in all the above effects.

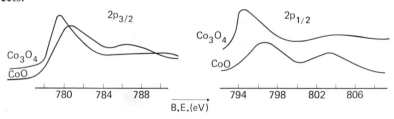

Fig. 34. The XPS Co $2p_{\frac{3}{2}}$, $2p_{\frac{1}{2}}$ region for CoO and Co_3O_4, showing the combined effects of a chemical shift, multiplet, and shake-up structure differences.[110]

Angular resolved photoemission measurements are also becoming important in solid state and surface work.[83] They fall into two broad categories. Those which have been performed with the object of providing a depth distribution of the elements and compounds present in the surface region, and those which provide information on the symmetry of the orbitals being photoionized. Unfortunately the two effects on angular measurements cannot always be easily separated, which can make interpretation difficult for either. The depth-distribution measurements are in principle straightforward, merely involving variation of the angle of ejection with respect to the surface (polar angle) at which electrons are accepted for energy analysis. Going to small angles increases the path length through the solid that an electron must take before escaping into the vacuum, thereby increasing the probability of inelastic scattering out of the XPS peak. The effective L_e value is thus decreased. The XPS peak accordingly contains a greater contribution from the surface layer at small angles than at high angles. In practice it is hard to get really quantitative depth-distribution measurements and allowance has to be made for surface roughness,[111-113] but if done with care and analyzed correctly the technique is probably the best nondestructive method (cf. ion-etching techniques) available for the first 20 Å or so of surface.

If the sample concerned is a single crystal and ordered surface structures are formed during absorption, then the above analysis runs into problems because of the angular effects introduced by the periodicity. These may be complicated since there will be both initial state effects (angular anisotropy depending on the symmetry and orientation of the orbital concerned) and final state effects (scattering effects). How important are the relative contributions of each is as yet unclear,[83] but high-quality experimental data are becoming available which clearly illustrates the potential in detecting and characterizing directed orbitals at surfaces. Perhaps the best known of these at the present time is that on the layer compound $TaSe_2$ where an azimuthal angular distribution plot showed a threefold symmetry of photoemission from the Ta d-orbitals.[114] In a later study a different type of $TaSe_2$ was examined, where alternate layers of $TaSe_2$ are rotated through 180°.[115] The symmetry then observed was close to sixfold—in fact the superimposition of two threefold patterns, the one resulting from the second layer being less intense than the first because of electron attenuation. The authors concluded that symmetry analysis of UPS data offers a powerful way of detecting situations where the top layer of surface atoms is arranged with different rotational symmetry compared to the bulk.

4. *General Examples in Chemistry.* In this short section we mention some of the more general uses of UPS and XPS in chemistry. Some of these could have been included under Section B-1 since they relate to electronic structure effects, but are included here because the observed effect is a means to a different end, or because the interpretations are couched in more traditional chemical parameters than molecular orbital descriptions. Organic chemistry is based upon the wide-ranging abilities of carbon to bond to H, O, N, S, the halogens, and other carbon atoms. Since photoelectron spectroscopy easily

identifies these elements and provides information on bonding properties, it is a powerful technique for studying organic problems. Changing the reactivity, chemistry, and biological activity of molecules by introducing different chemical "groups" occupies a significant fraction of organic chemists' time (e.g. morphine to codeine by substitution of $-CH_3$ for $-H$), and such substitution effects can be followed by UPS. The first IP of NH_3 (10·15 eV) relates to the removal of a "lone-pair" electron on the nitrogen. Substitution of a methyl group to give methylamine reduces the basicity of the molecule (the ability of the lone pair to form a bond with another molecule) and is accompanied by a corresponding reduction in the lone pair IP (8·97 eV). Changes in the oxygen $2p$ lone-pair IP of a series of alkyl alcohols, $R-OH$, measure the electron-donating properties of the alkyl groups, R (Inductive effect, $+I$). Aromatic compounds provide examples where both resonance effects (R) and Inductive effects of substituents can be followed by UPS.[1] Donation or withdrawal of electrons by the substituent changes the atomic charge distribution which means the effects are also followable by chemical shifts in XPS.

Structure determinations are possible. The UPS of $CH_2=CH-CH=CH_2$ and $CF_2=CF-CF=CF_2$ reveal the π MO's as the lowest IP's. In organic chemistry terms these MO's are formed by double bond conjugation, significant only for planar molecules. The conjugation observed for C_4F_6 is only one-third that of C_4H_6 (Fig. 35), indicating that C_4F_6 is nonplanar.[116]

Because UPS is so effective in showing up interactions between π-orbitals or between π and lone pair orbitals, it has proved to be a most useful technique for probing the mechanisms of these interactions—"through bond" or "through space." Direct interaction of two orbitals through space will in general produce two new orbitals with energies symmetrically distributed about the energies of the interacting orbitals, and in general, the antisymmetric combination will have a higher energy (lower binding energy) than the symmetric combination (this is known as the "natural" sequence). This type of "through space" interaction can be expected to be of prime importance when the two interacting orbitals are close together in space and have appropriate symmetries and energies. It is exemplified by the interaction of the π-orbitals in butadiene, discussed above, or between the sulphur $3p$ lone pairs in disulphides:

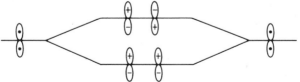

S: R - S - S - R   S:

The concept of through-bond interaction between lone pairs was first introduced by Hoffmann *et al.*[117] It involves an interaction through the intermediacy of C—C σ-orbitals of appropriate symmetry and energy. In general, "through bond" interactions take place when all the interacting orbitals are properly aligned with high-energy C—C σ-orbitals of the same symmetry—this can produce either the "natural" sequence (antisymmetric

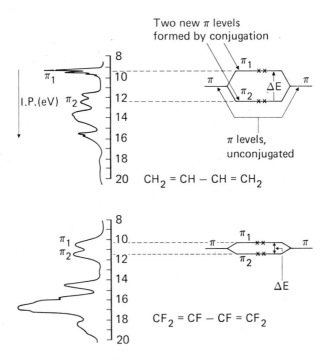

Fig. 35. HeI photoelectron spectra of C_4H_6 and C_4F_6 showing the reduced π_1, π_2 separation in C_4F_6.[116]

above symmetric) or the "inverse" sequence (symmetric above antisymmetric). By studying appropriate series of compounds, it is possible to decide which of the interaction types through bond or through space) is predominant. This is discussed in detail in Chapter 5 of this volume, and some examples are also discussed in a review by Baker *et al.*[118]

The identification of nonequivalent atoms in XPS may be of significance to structural analysis without any electronic structure analysis of the chemical shifts being involved. XPS showed *two* sulphur inner-shell peaks for $R_2S_2O_2$, for example, indicating the structure (I) not (II):[12]

$$
\begin{array}{ccc}
\text{O} & & \\
\downarrow & & \\
\text{R—S—S—R} & & \text{R—S—S—R} \\
\uparrow & & \uparrow \quad \uparrow \\
\text{O} & & \text{O} \quad \text{O} \\
\text{(I)} & & \text{(II)}
\end{array}
$$

In the study of carbonium ions, a C(1s) XPS peak indicative of the positively charged carbon atom can be expected at higher binding energy than the peak for uncharged carbon atoms. Difficulties in generating clean surface layers of the reactive carbonium ions on the probe of an XPS instrument is, of course, a serious practical problem, but nevertheless sufficiently good spectra have been claimed to distinguish, for example, between "classical" and "non-classical" ions. In XPS, the time-scale of the measurements is very short so the technique can in principle distinguish between a non-classical structure (I) and equilibrating classical structure (II) for the norbonyl cation.

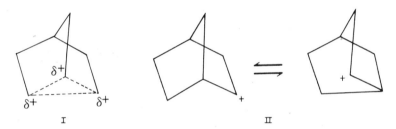

It was observed that the cation showed only one broad C(1s) band indicating considerable charge delocalization, in agreement with ^{13}C NMR and Raman data and pointing to the non-classical structure.[119, 120]†

The study of polymers, or rather polymer surfaces, has received much attention[19, 121] and is a good example of the role XPS has played in bringing different disciplines together (in this case, polymer and theoretical chemistry, and surface physics and technology).

In transition metal chemistry, identification of nonequivalent atoms is often not as straightforward as in nontransition chemistry because of the complexities introduced by the drastic differences in *d*-level electronic structure often occurring on a change of oxidation state. Changes from high-spin states to low-spin states, for instance, cause large changes in multiplet splitting and shake-up structure (e.g. the CoO and Co_3O_4 example given

† It has recently been argued, however,[168] that the interpretation of the XPS data has been oversimplified.

earlier). This of course provides extra data to base identifications on, but it may take more expertise to make a correct analysis, whereas the approach of one peak–one type of atom in simple situations can be applied by a relative novice.

It is likely that XPS will have applications in studying biological systems in the future where heavy metal ions are bonded to active sites, e.g. magnesium in chlorophyll, iron in hemoglobin. Already some studies on copper and iron proteins have appeared in the literature.[122, 123]

Finally, the chemical physics of the ionized states produced by the photo-emission process has importance in its own right. Ionized molecules are important in the upper atmosphere, both biologically (e.g. the ozone layer) and militarily (e.g. ions formed in rocket exhausts are important in missile guidance and detection), and in extraterrestrial atmospheres. These ions can be identified from emission spectra (followable from the earth's surface) provided enough data exist on the excited species. UV and vacuum-uv absorption and emission spectroscopy are important analytical and structural tools in chemistry. The basic quantities measured by these techniques are the energies for transitions from occupied orbitals to empty ones, which of course can be found also by electron energy-loss measurements. The interpretation of these data is made easier if one already knows the energies of the filled orbitals from UPS and XPS data.

C. Penning Ionization Spectroscopy

Compared to the extensive amount of work going on in the areas of XPS and UPS, the number of research groups working on PIS is very limited and there are no commercial spectrometers for PIS available. Potential applications fall into two areas, (a) the determination of molecular electronic structure in a like manner to UPS, and (b) the study of the basic mechanisms of ionization by excited atom impact, particularly with reference to differences from photon impact.

For electronic structure studies there is little information which is not obtainable more accurately by the more usual technique of UPS.[27, 28, 124–128] There are no cases where Penning Ionization electron spectroscopy has detected MO energy levels which have not been detected by PES, though sometimes the relative intensities of bands in a spectrum are different.[129] The technique can be used "in reverse" to measure the unknown energies of excited atomic states by using these as source atoms for a target molecule of known IP's. From the energies of the ejected electrons, the basic energy balance equations will then reveal the unknown energy of the excited atom. Long-lived states of N_2^* excited by electron impact were detected in this fashion.[130]

For (b) such studies are of practical worth because ionizing reactions between excited atoms and other atoms or molecules are important in gaseous discharges, photolysis, radiolysis, and plasma effects. In PIS when the excited source atom, A*, collides with the target molecule, XY, a short-lived collision complex, (AXY)*, is formed before the ejection of an electron occurs. The "normal" result of this ejection (termed the Penning process) is to give $A + XY^+ + e$, the energy of e being determined by the appropriate energy balance equation. However, there are alternative processes depending on the lifetime of the collision complex and its energetics. It is these alternatives which may complicate the spectrum, compared to the equivalent UPS spectrum, and make the determination of the electronic structure of XY by PIS difficult. A discussion of the mechanisms of the alternative processes is beyond the scope of this chapter, but a brief description of their effects upon the electron spectrum is given. The interested reader is referred elsewhere for detailed discussions.[28, 131, 132]

First the population of the vibrational levels in the ionized states may be influenced by the formation of the complex (AXY)*. In the cases originally studied however, this does not seem to have occurred and the Penning ionizations could be regarded as direct Franck–Condon transitions between X—Y and XY^+, since vibrational intensities in PIS (when resolved) were very similar to those in UPS. Recently there has been a report[129] that significant differences in the vibrational intensities compared to UPS are observed for a number of molecules, suggesting appreciable modification of the potential energy diagram of the target molecule by the impacting metastable He atom.

A common occurrence in PIS is that the IP's deduced using a straightforward energy balance equation are displaced by ΔE (positive or negative) from their true positions as found by UPS. This can be due to two processes. In the first ΔE represents translational energy which can either be given to the final ion XY^+ (ΔE would then be negative, i.e. the measured IP would be lowered), or which is present in (AXY)* and transferred to the energy of the ejected electron (ΔE positive).[131] This occurs in the case of

$$He^* + H_2 \rightarrow [HeH_2]^* \rightarrow H_2^+ + e + He$$

where $\Delta E \simeq 0 \cdot 1$ eV. In the second process an *Associative Ionization* occurs, i.e.

$$(AXY)^* \rightarrow AXY^+ + e \quad \text{or} \quad AX^+ + Y + e$$

Depending on the shape and position of the potential well of AXY^+ relative to (AXY)*, ΔE can be positive or negative. Associative ionization to give HeH_2^+, HeH^+, and $HeHg^+$ has been suggested.[131, 133] In a Penning Ionization spectrum the above processes may occur together, with a range of ΔE's, plus the "normal" Penning process, such that one obtains a spread in measured IP instead of a sharp peak.

D. Applications of Ion Neutralization Spectroscopy

The INS process applied to solids is perhaps the most surface sensitive of the electron spectroscopic techniques in terms of the depth of surface examined. Unlike XPS, UPS, or AES it does not rely for its sensitivity on a low L_e value for the ejected Auger electron. The impinging slow beam of He^+ ions does not penetrate the surface and so the ionization process actually takes place outside the surface by means of an electron-tunnelling mechanism of electrons from the surface region. This is the major feature distinguishing the technique from electron impact AES looking at WVV Auger transitions (Fig. 3). Both are two-electron spectroscopies and therefore give a self-convolution of the one-electron transition density. An example was given in Fig. 26a, where the self-convolution of the one-electron transition density from the d-bands of a Ni(100) surface is recorded. Inverting the self-convolution integral to obtain the one-electron transition density was originally a difficult mathematical problem, which has been fully described by Hagstrum and co-workers, the originators and almost sole practitioners of the INS technique.[26] Once acceptable procedures had been established it became a routine computational operation. The unfolded one-electron transition density for the Ni(100) sample (Fig. 26a) clearly shows the initial-state d-band DOS in the surface region.[81] This will differ from that obtained by applying a similar unfolding procedure to a WVV Auger transition for two reasons. First the DOS sampled by the WVV Auger process will have contributions from several layers thick, whereas the INS actually only samples the tails of the wave functions protruding into the vacuum above the surface. Second, the experimental WVV band is likely to be strongly modified by transition matrix effects since the probability of filling the W core-hole from one region of the valence band will be different from that from another region if the electronic character is different (i.e. s-, p-, d-type regions).

Perhaps of more importance than INS work on clean surfaces is that concerned with adsorption at surfaces. Surface orbital resonances are observed in the unfolded data, which may be compared directly with equivalent UPS data. An example, an ordered $c(2 \times 2)$ structure induced by sulphur absorption on Ni(100) is shown in Fig. 26b,[81] together with an HeI UPS trace for S absorption on an Ni film.[77] The S($2p$) related structure is observed in both spectra but its intensity relative to that from the underlying Ni d-bands is much greater in the INS case, indicating the superior surface sensitivity of INS. One can learn quite a lot about the depth distribution of adsorbate material by comparing INS and UPS data on the same system. Hagstrum[26] has reported cases where adsorbate resonances are observable in UPS but not in INS, and interpreted this as implying that the adsorbate is present in the sub-surface regions, but not on the surface.

E. Applications of Auger Spectroscopy

1. *Solids.* Electron impact AES has become almost a routine technique for element detection on surfaces[32] owing to three factors: (a) the high probability of the W shell-hole being filled by an Auger process; (b) in a WXX transition *both* levels defining the energy of the Auger electron levels are atomic in nature, and in a WVV transition *one* is, therefore the energy of an Auger electron is highly characteristic of an individual atom, cf. XPS; and (c) Auger electrons useful in element identification fall within the 20–1000 eV range, i.e. the small L_e range (see Section VB-3), making it a surface-effective technique. A study of all factors involved (the same factors as for UPS and XPS) indicates that AES is usually limited to about 50 Å depth, and often much less, and that low-energy electrons from high atomic number materials have the lowest escape depths.[23]

Figure 24, the complete AES spectrum from a stainless steel sample referred to earlier,[80] represents a typical example of AES use in element identification at a surface. Besides the expected Fe and Cr, there are also present Ni, S, O, C, and Ta. The chromium transitions are very intense, indicating a much higher surface proportion of Cr in the surface regions than in the bulk, giving the sample the stainless quality. Probably 90% of all Auger data accumulation is of this type—qualitative identification of elements present at a surface for technological control reasons. The technique has become so prominent in technological applications because of the high spatial resolutions obtained by the use of a finely focused electron beam as the probe and because of its development specifically for surface applications by the commercial instrument companies (cf. XPS where the original intention of the instrument companies was a more *general* analytical tool). Many applications in thin-film technology (including integrated circuits, photo-sensitive devices, and magnetic films) have been found.[134] Typical technological studies involve evaluation of the cleaning procedures used in film fabrication, the study of surface and grain boundary segregation, corrosion problems, and semiconductor surface doping. Alloying problems have been extensively studied in the metallurgical industries, particularly with reference to corrosion and segregation problems.

Quantitative elemental analysis by AES has been very limited, though one could be excused for believing otherwise from the number of published papers with the word "quantitative" in them. One usually finds that "quantitative" refers to an estimate of surface coverage rather than any quantitative determination of the elemental composition present. In the latter case the problems are much more severe than for XPS. One needs to have either theoretical estimates of relative Auger transition sensitivities for different atoms or calibrated values, as for XPS. The theoretical values are not available and calibrations are difficult because it is not possible to measure an Auger peak area in the derivative, $dN(E)/dE$, recording mode. Using peak-to-peak heights as an estimate of intensity (as is often done) is only valid if an element Auger peak shape (in the $N(E)$ curve) does not differ from compound to compound—a situation known to be untrue in many areas. The second difficulty is the often very destructive nature of the electron beam on the surface being examined, leading to dissociation, desorption, segregation, and bulk dissolution. Chang has recently attempted to put AES on a quantitative footing by providing experimental element relative Auger sensitivities.[135] He points out the problem of peak shape changes when working in the $dN(E)/dE$ mode, and does not apply the results to any adsorbate/substrate systems where electron beam damage would occur.

The origin of XPS chemical shifts was briefly discussed earlier and it was shown how information on the bonding of atoms could be obtained. Since inner-shell levels are involved in the Auger process, chemical shift behavior will occur here also, though it will be more complex because the difference of individual core-level shifts are involved. There are two types of chemical shift. For a WXX process all levels involved are inner-shell, and therefore the AES shift will be related to the individual shifts of all the core-levels involved. The simple equation

$$\text{K.E.}_{\text{Auger}} = E_W - 2E_X$$

indicates that a shift of ΔE in levels W and X would result in a shift of $-\Delta E$ in K.E.$_{\text{Auger}}$, i.e. WXX Auger chemical shifts should be similar to the XPS X level shift. The equation is actually a very poor approximation however, and frequently Auger chemical shifts are much larger than XPS values.[136] The reason for the non-correspondence of the shifts is connected with the chemical dependence of the large amount of additional relaxation that can occur in the doubly ionized Auger final state. For instance Cu^I can be distinguished quite easily from Cu in the Auger spectrum (electron or x-ray induced) whereas the XPS $2p_{\frac{3}{2}}$ B.E.'s of the two oxidation states are almost identical. The relationship between Auger chemical shifts and core-level chemical shifts has been discussed by Shirley and co-workers[137] and by Wagner.[138]

The second type of Auger "shift" occurs when transitions involving the valence levels are involved (WVV or WXV). The inclusion of valence-level binding energies, which behave quite differently from the core-levels from compound to compound (since they are characteristic of the ensemble rather than being an atomic property), means that any measured Auger "shift" from compound to compound will have no simple correlations with XPS core-level shifts, or with atomic charge, oxidation state, etc. As indicated in the previous section on INS the shape of a WVV Auger peak (unfolded) is related to the DOS of the valence levels, modified by transition matrix terms and representing the depth region appropriate to the L_e value of the Auger electron concerned. Attempts have been made to extract DOS information for metals from WVV Auger transitions. The only example approaching anything like success is the free electron-like metal, Al.[139, 140] For Ag, Cu, and Zn the experimental spectra contains fine-structure reminiscent of atomic spectra,[139, 141] an effect which may be due to localization caused by the core-hole initial state.

Shape changes within the *WVV* band envelope (i.e. reflecting the valence level changes occurring) are used as a means of empirically following chemical changes at surfaces. The *WVV* Si transition stretches from about 65 to 95 eV and contains sharp fine-structure which is significantly modified on going from a silicon oxide surface.[142, 143] The shape of the C *KLL* Auger transition varies with the chemical form of the carbon. Thus molecular CO adsorption, carbide formation, and graphitic carbon deposit can all be distinguished when present on an Mo surface.[144]

2. *Gases*. More detailed structure occurs in the spectra of gaseous molecules because of the discrete rather than band nature of the valence levels. In addition, since the energy of an Auger electron is independent of the photon or electron energy creating the initial hole-state, its width will not be affected by the energy width of the exciting X-ray or electron beam. Auger peaks may therefore be narrower than XPS core-level peaks in some cases, and even show vibrational fine-structure when valence-level transitions are involved. In consequence the effect of chemical environment on an Auger transition can be studied in more detail. Figure 23 showed the carbon *KLL* Auger transition (*WVV* Type) for CO and for CO_2.[79] The general energies (approximately 250 eV) are characteristic of carbon, but the detailed structure is characteristic of the individual molecules. Thus we have an elemental *and* a molecular analysis, though it is not yet clear whether gas-phase Auger spectroscopy can be a viable analytical technique. If these had been solid-state spectra of a carbon compound, one broad band (but with subsidiary maxima and minima) stretching over approximately 40 eV would have been observed.

Many of the peaks in Fig. 23 are not "normal" Auger peaks. Just as in XPS, the ejection of a *W* electron may be accompanied by the simultaneous promotion or ionization of a valence electron, leading to satellite Auger peaks. These can be either at higher or lower energies than the normal Auger lines, depending on the details of the energy levels involved. Detailed studies of satellites have been made with the object of identifying all the initial and final ionized states involved in an Auger process. Gaseous Auger spectra may also be used to obtain double ionization potentials, i.e. the minimum energy required to remove *two* electrons from a molecule.[79] This information is usually obtained from mass spectrometry by measuring the appearance potential of the doubly charged ion, but such data are less reliable because the ion, M^{2+}, must remain stable for 10^{-5} s to be detected.

F. Electron Impact Energy Loss Spectroscopy

1. *Gases.* Energies of electronically excited states of molecules can be mapped out from their energy-loss spectra. An example for helium was given in Fig. 27. This is also the object of visible, uv, and vacuum-uv absorption spectroscopy—i.e. optical spectroscopy, and it is worth comparing the two techniques. Optical spectroscopy has higher resolution, especially at the low-energy end of the spectrum, but is restricted in the energy range that may be covered because of the lack of suitable continuum light sources above 12 eV. There are selection rules operating in optical spectroscopy which result in not all possible transitions being allowed. The selection rules are less rigorous for electron impact, and by varying E, and the scattering angle, it is possible to observe transitions not observed in optical spectroscopy. Energy-loss spectroscopy has the advantage that losses from a few meV (vibrational excitation) to hundreds of eV can be studied in one apparatus, and in one spectrum. To cover the same range in optical spectroscopy,[31] separate spectrometers are required for the infrared (vibrational excitation), visible, uv, and vacuum-uv regions.

The major users of electron impact energy loss spectroscopy to date, physicists, have concentrated on atoms, and diatomic and triatomic molecules. From the variation of peak intensities with scattering angle and incident energy the type of transition (electron-dipole allowed, electron-dipole forbidden, electron-quadrupole allowed, etc.) can be determined. Tabulations of oscillator strengths of transitions and vibrational peak intensities have been made.[31] Transient species between electron and target molecule (negative ions) can sometimes be detected.[145] Work is now, however, beginning to accumulate on the use of energy-loss spectroscopy to complement optical and UPS studies in determining the electronic structures of fairly complex molecules. Kupperman *et al.*, for example, have recently studied allene at 20 eV, 40 eV, and 60 eV impacting energies, and at scattering angles of 6° to 80°.[146] They assigned two transitions with maxima at 4·28 eV and 4·89 eV as singlet $\rightarrow$ triplet excitations, and argue that the magnitude of the splitting between these transitions (0·61 eV) is a measure of the "interaction" between the two perpendicular π molecular orbitals. This compares with a splitting of 1·69 eV in the conjugated diene *s-trans* 1,3-butadiene. Tam and Brion[147] and Huebner *et al.*[148] have also applied energy-loss spectroscopy to organic molecules, using impacting energies up to 100 eV. Among molecules studied were aldehydes and ketones, alcohols, and unsaturated compounds.

Brion and colleagues have been largely responsible for the extension of energy-loss spectroscopy to the studies of K-shell excitations in various molecules.[149] In their experiments, the energy of the impacting electrons is much greater than that used for studies of valence shell excitations (2·5 keV for the K-shell excitations compared to 100 eV or less for the valence-shell work). They see peaks attributable to the excitation of K-shell electrons to Rydberg orbitals, which converge to the K-shell ionization limit. Among molecules studied have been CO_2, N_2O, and CO.

Van der Wiel and Brion[150] have reported that the conditions of photo-electron spectroscopy may be simulated using fast (3·5 keV) electron impact, with coincidence detection of scattered and ejected electrons. Brion and his colleagues[150, 151] have made several investigations using this technique. They showed that the absolute total photoabsorption cross-section could be determined using optical sum rules. They have, for example, studied partial oscillator strengths for the ionization of water, carbon monoxide, and ammonia. Although the resolution of their "photoelectron" spectra does not compare with the standard UPS method, their technique does permit the measurement of partial "photo" ionization cross-sections over an energy range that is difficult to cover in UPS. For water, they find that the oscillator strength for ionization from the $1b_2$ orbital decreases rapidly near to threshold, strongly indicating that the $(1b_2)^{-1}$ state of H_2O^+ leads to dissociative ionization, a conclusion in agreement with the measurements of Dibeler *et al.*[152] giving 18·05 eV as the appearance potential of OH^+ from water, and with an earlier analysis of the photoelectron spectrum of water given by Turner *et al.* This example is indicative of the sort of information that the technique can contribute in this area.

An interesting new type of process observable by electron impact spectroscopy, the formation of transient negative ions at specific impacting energies by resonance electron capture, has recently been studied by various groups. Schulz and Sanche[145] were responsible for the pioneering work. The principle of the technique is as follows. An electron beam of progressively increasing energy is passed through a gas, and the transmitted electron current monitored. At certain values of the impacting electron energy, electrons can be captured by gas molecules, thereby forming metastable negative ions. These negative ions then dissociate after a short period of time, re-emitting the initially absorbed electron, with its initial kinetic energy. Though "nothing overall has happened", the re-emitted electrons come off over a wide angular range, leading to a change in the measured "straight-through" transmitted current. Beautiful spectra are obtained showing many different and previously uncharacterized negative ion species for various molecules. Vibrational fine structure associated with these negative ion states has also been resolved. Studies on atoms, diatomics, triatomics,[145] benzene and other aromatics,[153] carbon monoxide,[154] and the halogen acids[155] have been published.

Very practical uses of energy-loss spectroscopy have been suggested. McPherson markets a high-resolution spectrometer (resolution ca. 10 meV), based on a National Bureau of Standards design, as an analytical tool for studying gaseous mixtures. A spectrum of air plus ca. 5% CO is shown in Fig. 36.[156] Peaks corresponding to the individual component gases are easily identified from the spectra of the pure gases. McPherson suggests quantitative applications (peak heights are proportional to concentration) in air-pollution studies for detecting small amounts of such contaminants as SO_2, NO, CO, CH_4. The system would also be suitable for post-analysis of samples produced by gas chromatography. The same applies to UPS as mentioned earlier. In fact, since UPS produces simpler spectra than ELS, the former would be more suitable, but the absolute amounts of material detectable are probably lower in ELS because of higher count rates. The restriction in both techniques is that complex molecules have very complex spectra (many filled MO's) limiting the useful analytical applicability.

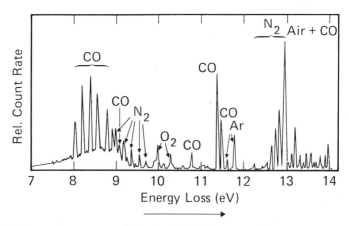

Fig. 36. Energy-loss spectrum of air plus 5% CO (courtesy of McPherson Corporation).

2. *Solids.* The transitions in solids analogous to those of gases, promotion of electrons from filled energy bands to empty ones, have been used to help interpret the electronic structures of solids and the energy levels of the unfilled bands, important properties in, for example, semiconductor techno-logy. The fine structure observed on the electron-impact threshold edges for core-level excitation processes in solids has been a source of interest to theoretical physicists for a long time. So-called "edge-singularities" have sometimes been ascribed to many-body effects to the delight of many-body theorists. Something of a reappraisal of the situation seems now to be under

way since it appears that, although some manifestation of many-body effects must appear in the experimental data, much of the observed fine structure can be explained on the basis of strictly one-electron band structure effects. Of interest also are the plasmon energy loss features of Fig. 25. Plasmon processes are sometimes responsible for satellite peaks accompanying Auger peaks[157] since any Auger electrons escaping through the solid causing plasmon oscillations suffer plasmon energy losses, resulting in small satellite peaks at lower kinetic energy than the main Auger peak. Discretion is needed to avoid confusing these with genuine Auger structure. Plasmon losses can also occur in photoemission studies,[158] causing peaks which could be mis-interpreted as part of the electronic band structure. Since plasmon energies are dependent on the solid structure and are in principle calculable, it is feasible to use them for chemical studies, e.g. the composition and properties of a range of alloys may be studied by comparing the plasmon spectra.[159]

There have been many studies of energy-loss spectroscopy applied to surface chemistry and physics, related to plasmon excitations, band-to-band excitations, phonon spectra, and surface vibrational spectra. Surface plasmons, which may be regarded as two-dimensional collective oscillators, may be excited. Theoretically these have $1/\sqrt{2}$ of the bulk plasmon energy for an ideal surface of a free-electron material, and $1/\sqrt{3}$ for a surface containing spherical grains. For titanium,[160] chromium,[160] and aluminum[161] the surface and bulk plasmon losses have been distinguished on the clean surfaces and it has been shown that the surface plasmon mode is sensitive to adsorption.

Most of the band-to-band excitation studies (i.e. the electronic excitation processes) have been performed on semiconductor surfaces, particularly silicon. Very beautiful and detailed work on different single crystal surfaces of silicon has been carried out by Rowe and Ibach where many electronic transitions have been assigned, some involving electronic surface states.[162, 163] The results are complemented by UPS measurements and from the effects on both types of spectra of chemisorption, Rowe et al.[163, 164] have been able to draw conclusions concerning the chemical bonding at clean and covered silicon surfaces, particularly with respect to the nature of the so-called "dangling bonds."[165]

Ibach has performed very high-resolution energy loss studies on Si looking closely at the phonon losses for the clean surface (meV energies). Similar resolution applied to adsorbate systems would allow detailed study of the vibrational spectra on the adsorbate (the 0–2 eV energy loss range) as an alternative to IR methods of making such measurements. Early work was carried out by Propst and Piper,[166] who followed the interaction of H_2O on W as a function of time. They observed the appearance of the W—O stretching frequency followed at a later stage by an O—H frequency, suggesting the formation of an oxide surface followed by adsorption of H_2O

or OH on the new surface. Little subsequent work in this area was performed for several years, but the recent results of Ibach[167] indicate its great potential.

VII. SUMMARY

It seems clear to us that the individual branches of electron spectroscopy will go on expanding for some years to come with an increasing interaction across disciplinary boundaries. The behavior of electrons constitutes a great deal of physics and even more of chemistry, so it is not surprising that measurements of the energies and probabilities of electron excitation processes should be very important from a basic research point of view. It is because this aspect has inevitably become coupled, or intermeshed, with the chemical analysis and the surface sensitivity aspects of the subject that interactions between scientists with different disciplines and objectives will increase. This can only be to the benefit of both the personnel and disciplines concerned. Both the authors have, merely as a consequence of using electron spectroscopy, steadily increased their contact with subject matter far beyond the original narrow confines of their own work. This contrasts with the often held view of a "spectroscopist" practising his particular spectroscopy and becoming ever more isolated in his subject.

REFERENCES

1. D. W. Turner, C. Baker, A. D. Baker and C. R. Brundle, "Molecular Photo-electron Spectroscopy" (Wiley, London, 1970).
2. A. D. Baker and D. Betteridge, "Photoelectron Spectroscopy—Chemical and Analytical Aspects" (Pergamon, Oxford, 1972).
3. J. H. D. Eland, "Photoelectron Spectroscopy" (Halsted Press, New York, 1974; Butterworths, London, 1974).
4. A. D. Baker, C. R. Brundle, and M. Thompson, *Chem. Soc. Rev.* 355 (1972).
5. C. R. Brundle and M. B. Robin, in: "Determination of Organic Structures by Physical Methods", Vol. III, F. Nachod and G. Zuckerman, eds. (Academic Press, New York, 1971).
6. A. Hamnett and A. F. Orchard, in: "Electronic Structure and Magnetism of Inorganic Compounds", Vols. II and III, P. Day, ed., Specialist Periodical Report of the Chemical Society (London).
7. H. Bock and P. D. Mollere, *J. Chem. Ed.* **51**, 506 (1974).
8. D. Betteridge and M. Williams, *Analyt. Chem.* **46**, 125R (1974); A. D. Baker, M. Brisk, and D. Liotta, *Analyt. Chem.* **48**, 281R (1976).
9. W. C. Price, in: "Adv. Atomic and Molecular Physics", D. R. Bates and B. Bederson, eds. (Academic Press, New York, 1974), Vol. 10.
10. T. A. Carlson, *Physics Today*, **25**, 30 (1972).
11. C. Nordling, *Angew. Chem. Int. Ed.* **11**, 83 (1972).
12. K. Siegbahn, C. Nordling, R. Fahlman, R. Nordberg, K. Hamrin, J. Hedman, G. Johansson, T. Bergmark, S.-E. Karlsson, I. Lindgren, and B. Lindberg, "ESCA: Atomic, Molecular, and Solid State Structure Studied by Means of

Electron Spectroscopy," Nova Acta Regiae Soc. Sci., Upsaliensis, Ser. IV, Vol. 20 (1967).

13. K. Siegbahn, C. Nordling, G. Johansson, P. F. Heden, K. Hamrin, U. Gelius, T. Bergmark, L. O. Werme, R. Manne, and Y. Baer, "ESCA Applied to Free Molecules" (North-Holland, Amsterdam, 1969).
14. D. Hercules and J. C. Carver, *Analyt. Chem.* **46**, 133R (1974).
15. D. A. Shirley, *Adv. Chem. Phys.* **23**, 85 (1973).
16. W. Bremser, *Fortschr. Chem. Forsch.* **1**, 36 (1973).
17. C. S. Fadley, in: "Electron Emission Spectroscopy", Dekeyser, Fiermans, Van der Kalen, and Vennik, eds. (D. Reidel, 1973).
18. C. J. Allan and K. Siegbahn, "MTP International Review of Science", "Physical Chemistry", Series One, Vol. 12, p. 1 (1973).
19. D. T. Clark, *Ann. Repts. Chem. Soc. (London)*, **69**, 66 (1972).
20. H. G. Fitzky, D. Wendisch, and R. Holm, *Angew. Chem. Int. Ed.* **11**, 979 (1972).
21. W. Melhorn, "Proc. Intl. Conf. Inner Shell Ionization Phenomena—Future Applications" (Publ. 1973, U.S. Atomic Energy Comm., Oak Ridge, Tenn.), p. 437 (1972).
22. J. C. Tracy, as in ref. 17.
23. C. C. Chang, in: "Characterization of Solid Surfaces", P. F. Kane and G. B. Larrabee, eds. (Plenum Press, New York, 1974).
24. L. A. Harris, *Analyt. Chem.* **40**, 24A (1968).
25. A. Joshi, L. E. Davis, and R. W. Palmberg, in: "Methods of Surface Analysis", A. W. Czandera, ed. (Elsevier, New York, 1975).
26. H. Hagstrum, *J. Vac. Sci. Tech.* **12**, 1 (1975).
27. V. Cermak, *J. Chem. Phys.* **44**, 3781 (1966).
28. H. Hotop, in: "Advances in Mass Spectrometry" (Institute of Petroleum, London, 1971), Vol. 5, p. 116.
29. M. J. Shaw, *Contemporary Physics*, **15**, 445 (1974).
30. R. S. Berry, *Ann. Rev. Phys. Chem.* **20**, 357 (1969).
31. S. Trajmar, J. K. Rice, and A. Kupperman, *Adv. in Chem. Phys.* **18**, 15 (1970).
32. C. E. Brion, C. A. McDowell, and W. B. Stewart, *J. Electr. Spectr.* **1**, 113 (1972).
33. G. R. Wight and C. E. Brion, *J. Electr. Spectr.* **4**, 313, 327, 335, 347 (1974), and references quoted therein.
34. W. Tam and C. E. Brion, *J. Electr. Spectr.* **4**, 139 (1974), and references quoted therein.
35. C. E. Brion, in: "MTP International Review of Science", Physical Chemistry, Series One (Butterworths, London, 1972), Vol. 5, p. 55.
36. O. Klemperer and J. P. G. Shepherd, *Adv. Phys.* **12**, 355 (1963).
37. J. Comer and F. H. Read, *J. Electr. Spectr.* **1**, 3 (1972/73).
38. N. Oda, F. Nishimura, and S. Tahira, *Phys. Rev. Letters*, **24**, 42 (1970).
39. J. P. Zeisel, I. Nenner, and G. J. Schulz, *J. Chem. Phys.* **63**, 1943 (1975).
40. I. Nenner and G. J. Schulz, *J. Chem. Phys.* **62**, 1747 (1975).
41. L. Sanche and G. J. Schulz, *Phys. Rev.* **A6**, 69 (1972).
42. L. Sanche and G. J. Schulz, *J. Chem. Phys.* **58**, 479 (1973).
43. A. Streitweiser, *Prog. Phys. Org. Chem.* **1**, 1 (1963).
44. D. W. Turner, *Adv. Phys. Org. Chem.* **4**, 31 (1966).
45. L. G. Christophorou, "Atomic and Molecular Radiation Physics" (Wiley–Interscience, London, 1971).

46. J. Collin, "Proc. NATO Adv. Study Inst., Glasgow", p. 201 (1964).
47. T. Koopmans, *Physica*, 1, 104 (1933).
48. C. R. Brundle, M. B. Robin, and H. Basch, *J. Chem. Phys.* 53, 2196 (1970).
49. J. D. Morrison, "Institut International de Chemie, 12th Conseil de Chemie, Bruxelles", p. 397 (1962).
50. C. Backx, M. Klewer, and M. J. van der Wiel, *Chem. Phys. Letters*, 20, 100 (1973).
51. M. J. van der Wiel and C. E. Brion, *J. Electr. Spectr.* 1, 309, 439, 443 (1973).
52. G. R. Branton and C. E. Brion, *J. Electr. Spectr.* 3, 129 (1974); M. J. van der Wiel and C. E. Brion, *J. Electr. Spectr.* 1, 439 (1973).
53. V. Cermak and Z. Herman, *Nature*, 199, 588 (1963).
54. V. Cermak and Z. Herman, *Coll. Czech. Chem. Comm.* 30, 169 (1965).
55. V. Cermak, *J. Chem. Phys.* 44, 3774 (1966).
56. P. Auger, *J. Phys. Radium*, 6, 205 (1925).
57. J. Comer and F. H. Read, *J. Electr. Spectr.* 1, 3 (1972).
58. V. Cermak, M. Smutek, and J. Sranek, *J. Electr. Spectr.* 2, 1 (1973).
59. E. Bolduc and P. Marmet, *Canad. J. Phys.* 51, 2108 (1973).
60. G. R. Cook, P. H. Metzger, and M. Ogawa, *J. Chem. Phys.* 44, 2935 (1966).
61. G. R. Cook, P. H. Metzger, and M. Ogawa, *Canad. J. Chem.* 43, 1706 (1965).
62. P. H. Metzger, G. R. Cook, and M. Ogawa, *Canad. J. Chem.* 45, 203 (1967).
63. L. C. Lee, R. W. Carlson, D. L. Judge, and M. Ogawa, *J. Chem. Phys.* 63, 3987 (1975).
64. D. L. Ames, J. P. Maier, F. Watt, and D. W. Turner, *Faraday Soc. Discussions*, No. 54, 277 (1972).
65. An example of this approach is shown in the work of C. R. Brundle and D. W. Turner, *Intl. J. Mass Spectroscopy and Ion Physics*, 2, 195 (1969).
66. F. Burger and J. P. Maier, *J. Electr. Spectr.* 5, 783 (1974).
67. J. A. R. Samson, "Techniques of Vacuum Ultraviolet Spectroscopy" (Wiley, New York, 1967).
68. U.S. Department of Commerce National Technical Information Service Report BNL 50381 "Research Applications of Synchroton Radiation". Proceedings of a Study-symposium held at Brookhaven National Laboratory, September 1972.
69. M. O. Krause, *Phys. Letters*, 5, 341 (1971); *Chem. Phys. Letters*, 10, 65 (1971).
70. H. Boersch, J. Geiger, and W. Stickel, *Z. Physik*, 180, 415 (1964).
71. Courtesy of Perkin–Elmer U.K. Ltd. See also ref. 1.
72. H. Hafner, J. A. Simpson, and C. E. Kryatt, *Rev. Sci. Instr.* 39, 33 (1968); P. W. Palmberg, G. K. Bohn, and J. C. Tracy, *Appl. Phys. Letter*, 15, 254 (1969).
73. C. Kuyatt, chapter to appear in Vol. 2 of this series.
74. M. Gellender and A. D. Baker, *J. Electr. Spectr.* 4, 249 (1974).
75. E.g. Hewlett–Packard 5950A ESCA spectrometer.
76. C. R. Brundle, M. W. Roberts, D. Latham, and K. Yates, *J. Electr. Spectr.* 3, 241 (1974).
77. C. R. Brundle and A. F. Carley, *Faraday Discussions*, 60, 51 (1975).
78. C. S. Fadley and D. A. Shirley, *Phys. Rev. Letters*, 21, 980 (1968).
79. W. E. Moddeman, T. A. Carlson, M. O. Krause, B. P. Pullen, W. E. Bull, and G. K. Schweitzer, *J. Chem. Phys.* 55, 231 (1971).
80. R. W. Joyner, personal communication.
81. H. D. Hagstrum and G. E. Becker, *J. Chem. Phys.* 54, 1015 (1971).

82. W. E. Spicer, in: "Survey of Phenomena in Ionized Gases" (Intern. Atomic Energy Agency, Vienna, 1968), p. 271.
83. J. W. Gadzuk in "Electronic Structure and Reactivity of Metal Surfaces", E. G. Deroudne and A. A. Lucas, eds. (Plenum Press, New York, 1976).
84. B. J. Lindberg, K. Hamrin, G. Johansson, U. Gelius, A. Fahlman, C. Nordling, and K. Siegbahn, *Physica Scripta*, **1**, 286 (1970).
85. K. Siegbahn, Uppsala University Institute of Physics report No. 714, July 1974. More recent plots of this type can be found in reference 13.
86. "Handbook of Spectroscopy", Vol. I, J. W. Robinson, ed., (Chemical Rubber Company, Cleveland, Ohio, 1974).
87. F. Brogli and E. Heilbronner, *Helv. Chim. Acta*, **54**, 1423 (1971).
88. J. C. Carver, T. A. Carlson, L. C. Cain, and G. K. Schweitzer, in: "Electron Spectroscopy", D. Shirley, ed. (North-Holland, Amsterdam, 1972), p. 803.
89. C. S. Fadley, D. A. Shirley, A. J. Freeman, P. S. Bagus, and J. V. Mallow, *Phys. Rev. Letters*, **23**, 1397 (1969); A. J. Freeman, P. S. Bagus, and J. V. Mallow, *Int. J. Mag.* **4**, 35 (1973); P. S. Bagus, A. J. Freeman, and F. Saski, *Phys. Rev. Letters*, **18**, 850 (1973).
90. L. J. Aarons, M. F. Guest, and I. H. Hillier, *J. Chem. Soc. Farad. Trans.* **68**, 1866 (1972); H. Basch, *J. Electr. Spectr.* **5**, 463 (1975), and references given therein.
91. M. A. Brisk and A. D. Baker, *J. Electr. Spectr.* **7**, 197 (1975).
92. M. A. Brisk and A. D. Baker, unpublished results; see also J. Escard *et al.*, *Inorg. Chem.* **13**, 695 (1974); L. J. Matienzo *et al.*, *Inorg. Chem.* **12**, 2762 (1973).
93. M. A. Brisk and A. D. Baker, *J. Electr. Spectr.* **6**, 81 (1975).
94. R. W. Davis, "Electron Spectroscopy", Proc. Intl. Conf. 1971, D. A. Shirley, ed. (North-Holland, Amsterdam, 1972), p. 903.
95. D. Betteridge, A. D. Baker, P. Bye, S. K. Hasanuddin, N. R. Kemp, D. I. Rees, M. A. Stevens, M. Thompson, and B. J. Wright, *Z. Anal. Chem.* **263**, 286 (1973).
96. J. Dainteth and D. W. Turner, in: "Electron Spectroscopy", D. A. Shirley, ed. (North-Holland, Amsterdam, 1972).
97. D. Betteridge, M. Thompson, A. D. Baker, and N. R. Kemp, *Analyt. Chem.* **44**, 2005 (1972).
98. D. M. Hercules, L. E. Cox, S. Onisick, G. D. Nichols, and J. C. Carver, *Analyt. Chem.* **45**, 1973 (1973).
99. J. S. Brinen and J. E. McClure, *Anal. Lett.* **5**, 737 (1972).
100. C. R. Brundle, in: "Surface and Defect Properties of Solids", Vol. 1, J. M. Thomas and M. W. Roberts, eds., Specialist Periodical Report of the Chemical Society (London) 1972; *J. Vac. Sci. Tech.* **11**, 212 (1975); *J. Electr. Spectr.* **5**, 291 (1974); *Surface Science*, **48**, 99 (1975); in: "Electronic Structure and Reactivity of Metal Surfaces", as ref. 83.
101. W. E. Plummer, in: "Topics in Applied Physics", Vol. 4, R. Gomer, ed. (Springer-Verlag, New York, 1975).
102. A. Bradshaw, L. Cederbaum, and W. Domke, in: "Photoelectron Spectroscopy" (Springer-Verlag, 1975).
103. D. Menzel, *J. Vac. Sci. Tech.* **12**, 313 (1975).
104. S. J. Atkinson, C. R. Brundle, and M. W. Roberts, *Chem. Phys. Letters*, **24**, 175 (1974).
105. P. S. Bagus and K. Hermann, *Solid State Comm.* to be published.
106. I. P. Batra and P. S. Bagus, *Solid State Comm.* **16**, 1097 (1975).
107. I. P. Batra and C. R. Brundle, *Surface Science*, **57**, 12 (1976).

108. J. E. Demuth and D. E. Eastman, *Phys. Rev. Letters*, **32**, 1123 (1974).
109. K. Y. Yu, W. E. Spicer, I. Lindau, P. Pianetta, and S. F. Lin, *J. Vac. Sci. Tech.* **13**, 227 (1976).
110. T. J. Chuang, C. R. Brundle, and D. W. Rice, *Surface Science*, to be publtshed.
111. C. S. Fadley, R. J. Baird, W. Siekhaus, T. Novakov, and S. A. L. Bergström, *J. Electr. Spectr.* **4**, 93 (1974).
112. C. S. Fadley, *J. Electr. Spectr.* **5**, 725 (1974).
113. C. S. Fadley, *Faraday Discussions*, **60**, 18 (1975).
114. M. M. Traum, N. V. Smith, and F. J. DiSalvo, *Phys. Rev. Letters*, **32**, 124 (1974).
115. N. V. Smith and M. M. Traum, *Surface Science*, **45**, 745 (1974).
116. C. R. Brundle, M. B. Robin, and N. A. Kuebler, *J. Am. Chem. Soc.* **94**, 1451 (1972).
117. R. Hoffman, A. Imamura, and G. Zeiss, *J. Am. Chem. Soc.* **89**, 8215 (1967).
118. A. D. Baker, M. Brisk, and D. C. Liotta, *Analyt. Chem.* **48**, 281R (1976).
119. G. A. Olah, G. D. Mateescu, and J. L. Riemenschneider, *J. Am. Chem. Soc.* **94**, 2529 (1974).
120. G. D. Mateescu, J. L. Reimenschneider, J. J. Svoboda, and G. A. Olah, *J. Am. Chem. Soc.* **94**, 7191 (1972).
121. D. T. Clark, *Faraday Discussions*, **60**, 183 (1975).
122. E. I. Solomon, P. J. Clendening, H. B. Gray, and F. J. Gruntharer, *J. Am. Chem. Soc.* **97**, 3878 (1975).
123. Y. Isaacson, M. Brisk, Z. Majuk, M. Gellender, and A. D. Baker, *J. Am. Chem. Soc.* **97**, 6603 (1975).
124. V. Cermak, *Advances in Mass Spectrometry*, Vol. 4, p. 697 (1968).
125. V. Cermak, *Coll. Czech. Chem. Comm.* **33**, 2739 (1968).
126. V. Cermak and A. J. Yencha, *J. Electr. Spectr.* **8**, 109 (1976).
127. D. S. C. Yee, A. Hamnett, and C. E. Brion, *J. Electr. Spectr.* **8**, 291 (1976).
128. D. S. C. Yee and C. E. Brion, *J. Electr. Spectr.* **8**, 377 (1976).
129. D. S. C. Yee and C. E. Brion, *J. Electr. Spectr.* **8**, 313 (1976).
130. V. Cermak, *Chem. Phys. Letters*, **4**, 515 (1970).
131. H. Hotop and A. Niehaus, *Z. Physik.* **228**, 68 (1969).
132. H. Hotop and A. Hiehaus, *Int. J. Mass. Spec. Ion. Phys.* **5**, 415 (1970).
133. V. Cermak and Z. Herman, *Chem. Phys. Letters*, **2**, 359 (1968).
134. R. E. Weber, *Solid State Tech.* p. 49 (December 1970).
135. C. C. Chang, *Surface Science*, **48**, 9 (1975).
136. C. D. Wagner and P. Biloen, *Surface Science*, **35**, 82 (1973).
137. S. P. Kowalczyk, R. A. Pollak, F. R. McFeely, L. Ley, and D. A. Shirley, *Phys. Rev.* **B8**, 2387 (1973).
138. C. D. Wagner, *Analyt. Chem.* **47**, 1201 (1975).
139. C. J. Powell, *Phys. Rev. Letters*, **30**, 1179 (1973).
140. J. E. Houston, *J. Vac. Sci. Tech.* **12**, 255 (1975).
141. L. Yin, I. Adler, T. Tsang, M. H. Chen, and B. Craseman, *Phys. Letters*, **46A**, 113 (1973).
142. C. C. Chang, *Surface Science*, **23**, 283 (1971), **25**, 53 (1971).
143. J. S. Johannessen, W. E. Spicer, and Y. E. Strausser, *Appl. Phys. Letters*, **27**, 452 (1975).
144. T. W. Haas, J. T. Grant, and G. J. Dooley III, *J. Appl. Phys.* **43**, 1853 (1972).
145. L. Sanche and G. L. Schulz, *Phys. Rev.* **A5**, 1672 (1972), **A6**, 69 (1972); *J. Chem. Phys.* **58**, 479 (1973).

146. O. A. Mosher, W. M. Flicker, and A. Kupperman, *J. Chem. Phys.* **62**, 2600 (1975).
147. W. Tam and C. E. Brion, *J. Electr. Spectr.* **3**, 263, 467 (1974), **4**, 139, 149 (1974).
148. R. H. Heubner, R. J. Celotta, S. R. Meilczcuek, and C. E. Kuyatt, *J. Chem. Phys.* **59**, 5434 (1973).
149. G. R. Wight and C. E. Brion, *J. Electr. Spectr.* **1**, 457 (1972), **3**, 191 (1974), **4**, 25 (1974).
150. M. J. Wiel and C. E. Brion, *J. Electr. Spectr.* **1**, 309, 443 (1972).
151. G. R. Branton and C. E. Brion, *J. Electr. Spectr.* **3**, 129 (1974).
152. V. H. Dibeler, J. A. Walker, and H. M. Rosenstock, *J. Res. Natl. Bur. Stds.* **70A**, 459 (1966).
153. I. Nenner and G. J. Schulz, *J. Chem. Phys.* **62**, 1747 (1975).
154. N. Swanson, R. J. Celotta, C. E. Kuyatt, and J. W. Cooper, *J. Chem. Phys.* **62**, 4880 (1975).
155. J. P. Ziesel, I. Nenner, and G. J. Schulz, *J. Chem. Phys.* **63**, 1943 (1975).
156. Courtesy of McPherson Corporation Ltd.
157. W. M. Mularie and T. W. Rusch, *Surface Science*, **19**, 469 (1970).
158. N. V. Smith and W. E. Spicer, *Phys. Rev. Letters*, **23**, 769 (1969).
159. B. M. Hartley and J. B. Swan, *Aust. J. Phys.* **23**, 655 (1970).
160. G. W. Simmons, *J. Colloid and Interface Science*, **34**, 343 (1970).
161. B. D. Powell and D. P. Woodruff, *Surface Science*, **33**, 437 (1972).
162. J. E. Rowe and H. Ibach, *Phys. Rev. Letters*, **31**, 102 (1973).
163. J. E. Rowe, H. Ibach, and H. Froitzheim, *Surface Science*, **48**, 44 (1975).
164. J. E. Rowe, M. M. Traum, and N. V. Smith, *Phys. Rev. Letters*, **33**, 1333 (1974).
165. H. Ibach, *J. Vac. Sci. Tech.* **9**, 7.3 (1971).
166. F. M. Propst and T. C. Piper, *J. Vac. Sci. Tech.* **4**, 53 (1967).
167. H. Froitzheim, H. Ibach, and S. Lehwald, *Phys. Rev. Letters*, **36**, 1549 (1976).
168. G. H. Kramer, "The Norbornyl Cation", in: *Adv. Phys. Org. Chem.* **11**, 177 (1975).

2

Many-electron Theory
of Photoemission

R. L. MARTIN

and

D. A. SHIRLEY

Materials and Molecular Research Division,
Lawrence Berkeley Laboratory, and Department of Chemistry,
University of California, Berkeley, California 94720

I. INTRODUCTION

One of the most important methods of studying the internal structure of atoms and molecules involves the absorption of a quantum of light. For example, microwave and infrared spectroscopy yield information about the rotational and vibrational degrees of freedom in a molecule, while optical and ultraviolet absorption spectroscopy probe vibrational structure as well as the electronic degrees of freedom. These techniques are very similar in that they all involve the resonant absorption of light quanta. This takes the system from an initial state, characterized by a wave-function Ψ_i, to a final state specified by the wavefunction Ψ_f. The difference in energy between the two states $E_f - E_i$ is equal to $h\nu$, the energy imparted by the photon. Absorption takes place only at the resonant frequencies. Thus the experimental techniques used in these types of spectroscopy generally involve exposing the sample to a known photon flux and examining the resultant flux from the sample after the interaction has taken place.

Photoelectron spectroscopy (PES) is in principle very similar to the techniques mentioned above. The major operational difference arises because the final states observed in PES lie in the ionization continuum. Absorption of a photon thus results in the ejection of at least one photoelectron from the system. These electrons are subjected to kinetic-energy analysis, and it is the kinetic-energy spectrum which contains information about the absorption process. As in the other absorption techniques, the energy conservation equation

$$hv = E_f - E_i \tag{1}$$

must be satisfied.

It is helpful to think of a general final state reached by absorption of the photon as a superposition of many *degenerate* states ψ_j.

$$\Psi_f = \sum_j c_{fj} \psi_j \tag{2}$$

Each of these states partitions the total energy, E_f, into two components:

$$E_f = E_j + T_j \tag{3}$$

where E_j is the energy of an ionic state and T_j is the kinetic energy of the ionized electron. The measurement of the kinetic energy of the electron focuses our attention on a particular state ψ_j, and we have

$$hv = (E_j + T_j) - E_i \equiv E_B{}^j + T_j \tag{4}$$

where the quantity $E_B{}^j$ is defined as the binding energy of the photoelectron.

The most commonly used experimental procedure is therefore to fix the photon frequency and scan the photoelectron kinetic energy spectrum for peaks in intensity. Observation of a peak at an energy T_j implies the existence of an excited ionic state separated from the initial state by an energy $E_B{}^j$. This yields information about the ionic states of the sample, and, to some degree, about the properties of the initial state. The probability of observing an electron of energy T_j, given by $|c_{fj}|^2$, is related to the *cross-section* for photoionization. This provides further information about both states involved in the transition. It is important to note that if one simply observed the attenuation of the photon flux, the information obtained would pertain to a combination of absorption processes involving all the ionic states that are energetically accessible to the radiation. The advantage of PES is that it allows the study of *specific* ionic states.

This chapter will deal with the nature of these excited states, their energies, and the transition probabilities for reaching them via photon absorption. In Section II the basic theoretical formalism for the interaction of the radiation field with an N-electron system will be reviewed. The nature of the wavefunctions used to describe electronic states and the means of computing them is presented in Section III. In Section IV, the physical concepts which emerge from a study of the wavefunctions will be used to characterize the nature of the ionic states observed in PES. Section V will then analyze the photoionization cross-section in terms of the logical hierarchy of approximations commonly employed in cross-section calculations. The sum rule which relates the cross-section to the relaxation energy will also be discussed. In Section VI the origin and magnitude of the relaxation energy in a variety of systems will be examined and related to the physical and chemical properties of the species. The chapter ends with a brief discussion of the most commonly used approaches for estimating binding energies.

II. INTERACTION WITH THE RADIATION FIELD AND PHOTOIONIZATION

We begin by briefly reviewing the semiclassical treatment of the interaction of radiation with matter. As Schiff[1] points out, the term "semiclassical" refers to the assumption that the radiation field may be treated classically (within the framework of Maxwell's equations), whereas the system of particles is treated quantum-mechanically. This approximation has the advantage of simplicity and, for the absorption of radiation, gives the same results as quantum field theory.

A. Time-dependent Perturbation Theory and Fermi's Golden Rule

Consider a system of particles in a stationary state of a time-independent electrostatic Hamiltonian H_0. At some time t_1 a time-dependent term is introduced which represents the electromagnetic field. The field is assumed to be weak enough to be considered a small perturbation, but this disturbance may induce transitions to other stationary states of the particle Hamiltonian. The methods of time-dependent perturbation theory can be used to learn the probability that the system will be found in one of these states at some later time t_2.

The stationary states, Ψ_n, of H_0 satisfy the Schrödinger equation

$$H_0 \Psi_n = E_n \Psi_n \tag{5}$$

and have a simple oscillatory evolution in time

$$\Psi_n(t) = \Psi_n \exp\left[-(i/\hbar) E_n t\right] \tag{6}$$

A general solution of the equations of motion

$$i\hbar \frac{d\Psi(t)}{dt} = H_0 \Psi(t) \tag{7}$$

for some arbitrary state $\Psi(t)$ can be written

$$\Psi(t) = \sum_n c_n \Psi_n \exp\left[-(i/\hbar)E_n t\right] \tag{8}$$

The square modulus of the coefficient, $|c_n|^2$, is independent of time. It gives the probability of observing the superposition state, $\Psi(t)$, in some eigenstate Ψ_n.

If a time dependence is present in the Hamiltonian; i.e. if

$$H = H_0 + V(t) \tag{9}$$

then Eq. (8) is no longer a general solution of the wave equation. In fact there are no longer actually any stationary states. However, the form of the Hamiltonian we have chosen [Eq. (9)] implies that it still may be useful to expand the general solution in terms of the complete set of stationary states associated with H_0. Thus the solution is still given by Eq. (8), but we must now consider the expansion coefficients to be time-dependent.

Substitution of (8) into the Schrödinger equation

$$i\hbar \frac{d\Psi(t)}{dt} = H\Psi(t) \tag{10}$$

yields equations of motion governing the expansion coefficients:[2]

$$i\hbar \frac{dc_k(t)}{dt} = \sum_n c_n(t) V_{kn} \exp(i\omega_{kn} t) \tag{11}$$

where V_{kn} is the matrix element of the perturbation between the unperturbed states,

$$V_{kn} = \langle \Psi_k | V | \Psi_n \rangle \tag{12}$$

and

$$\hbar \omega_{kn} \equiv E_k - E_n \tag{13}$$

From this point we proceed as usual in perturbation theory. The coefficients for which we wish to solve are expressed as a power series in the perturbation, usually taken only to first order. Integration of this equation yields a probability amplitude for observing the arbitrary state Ψ'_k.

In anticipation of the nature of the specific perturbation to be considered later, we note that if the system is originally in some eigenstate, Ψ'_i, of H_0, and if the perturbation depends harmonically on the time,

$$V(t) = V_0 \exp(i\omega t) \tag{14}$$

then the probability of finding the system in some other eigenstate Ψ'_k, i.e. $|c_k(t)|^2$, is directly proportional to the time that the perturbation has been active. This implies we should convert our attention to a transition probability per unit time, which finally leads to Fermi's "Golden Rule":

$$P_{k \leftarrow i} = \frac{2\pi}{\hbar} \rho(E_k) |\langle \Psi'_k | V | \Psi'_i \rangle|^2 \tag{15}$$

Here $P_{k \leftarrow i}$ is the transition probability per unit time for the process $\Psi'_i \rightarrow \Psi'_k$. The term $\rho(E_k)$ is the energy density of final states in the neighborhood about Ψ'_k.

B. The Classical Radiation Field and the Photoemission Cross-section

In order to use Eq. (15) to calculate the transition probabilities induced by the electromagnetic field, we must decide upon the form of the perturbation V. It is possible to show by correspondence arguments that the Hamiltonian describing a system of spinless particles of charge $-e$ and mass m in an electromagnetic field is given by

$$H = H_0 + \left(\frac{-i\hbar e}{2mc} \nabla \cdot \mathbf{A} - \frac{i\hbar e}{mc} \mathbf{A} \cdot \nabla + \frac{e^2}{2mc^2} |A|^2 \right) - e\phi \tag{16}$$

Although we are not specifically interested in spinless particles, the interaction between the spin of the electron and the incident light wave is negligible. The operator H_0 represents the Hamiltonian describing the particles in the absence of the field. The vector $-i\hbar\nabla$ is a sum of momentum operators for the individual particles

$$\nabla = \sum_i (\nabla_i) \tag{17}$$

The radiation field itself is described by the vector potential $\mathbf{A}$ and a scalar potential ϕ. These are related to the electric and magnetic field strengths, $\mathbf{E}$ and $\mathbf{H}$, by

$$\mathbf{E} = -\frac{1}{c} \frac{\partial}{\partial t} \mathbf{A} - \nabla \phi \tag{18}$$

$$\mathbf{H} = \nabla \times \mathbf{A}$$

There is some flexibility in choosing the potentials which define the field, and, for fields such as those associated with a light wave, it is common to work in the Coulomb gauge. In this case we have

$$\nabla \cdot \mathbf{A} = 0 \tag{19}$$

$$\phi = 0$$

Since we have assumed that the field is weak, we furthermore neglect the term in $|A|^2$, and finally obtain:

$$H = H_0 - \frac{i\hbar e}{mc} \mathbf{A} \cdot \nabla \tag{20}$$

or

$$V(t) = \frac{-i\hbar e}{mc} \mathbf{A} \cdot \nabla \tag{21}$$

Now the vector potential for radiation propagating in the form of a plane wave of wave vector $\mathbf{q}$ and frequency ω can be written

$$\mathbf{A} = \mathbf{u} A_0 [\exp(-i\mathbf{q} \cdot \mathbf{r}) \exp(i\omega t) + \exp(i\mathbf{q} \cdot \mathbf{r}) \exp(-i\omega t)] \tag{22}$$

where $\mathbf{u}$ is a unit vector specifying the direction of the electric field vector (the polarization), and A_0 is the amplitude of the potential. The intensity associated with this plane wave is

$$I = \frac{\omega^2}{2\pi c} A_0^2 \tag{23}$$

Since the perturbation is harmonic and we are considering a final state Ψ'_k which lies in the continuum, we can substitute Eqs. (21, 22, 23) into Eq. (15) and find that

$$P_{k \leftarrow i} = \frac{4\pi^2 \hbar e^2 I}{m^2 c \omega^2} \rho(E_k) |\mathbf{u} \cdot \langle \Psi'_k | \exp(i\mathbf{q} \cdot \mathbf{r}) \nabla | \Psi'_i \rangle|^2 \tag{24}$$

This is an expression for the transition probability per unit time from state Ψ'_i to state Ψ'_k with $E_k > E_i$. Only the second component of the vector potential [Eq. (22)] has contributed to this probability.

This probability is generally expressed in a somewhat different form. The cross-section, σ, is defined as the total transition probability per unit time divided by the incident photon flux. This flux is simply the intensity of the electromagnetic field divided by the photon energy. A more convenient quantity, however, is the differential cross-section for ejection of an electron in a small solid angle, $d\Omega$, with respect to some axis, e.g. that of the electric field vector. This is given by

$$\frac{d\sigma_{k \leftarrow i}}{d\Omega} = \frac{\pi \hbar^2 e}{m^2 c \omega} \rho(E_k) |\mathbf{u} \cdot \langle \Psi'_k | \exp(i\mathbf{q} \cdot \mathbf{r}) \nabla | \Psi'_i \rangle|^2 \tag{25}$$

where $\rho(E_k)$ is the density of final states corresponding to the given solid angle.

This completes the development of the cross-section for photoemission in a purely formal way. The major assumption which has been made thus far is that the interaction between the electrons and the electromagnetic field is small enough that it can be treated in first order. The final assumption about the field which we have not discussed thus far, but is generally made, involves the exponential factor in the matrix element [Eq. (25)]. It can be expanded in the series

$$\exp{(i\mathbf{q}.\mathbf{r})} = 1 + i\mathbf{q}.\mathbf{r} + \tfrac{1}{2}(i\mathbf{q}.\mathbf{r})^2 + \dots \tag{26}$$

If only the first term in this sum is retained, the resulting simplification is known as the "dipole approximation". Since the momentum of the photon is directly proportional to q, it is sometimes referred to as the neglect of photon momentum; this omission will obviously become less acceptable as the photon energy increases. For the purposes of PES, the dipole approximation should be rather good as long as $q \ll k$, where k is the wavevector of the photoelectron.[3]

III. THE WAVEFUNCTIONS

Let us now consider the wavefunctions Ψ'_k and Ψ'_i, which describe eigenstates of an electrostatic Hamiltonian in the absence of perturbation. In systems containing two or more electrons, exact solutions for these wavefunctions do not exist, and we are forced to seek appropriate approximations. As the structure of the final ionic states and the mechanisms from which they derive oscillator strength are usually interpreted in the language of these approximations, it is helpful to examine in some detail what they imply about the electronic structure of the system and the nature of the ionization process.

The Hamiltonian for which we seek solutions of the Schrödinger equation will be of the non-relativistic electrostatic form for an N-electron system in the field of a nucleus of charge Z,

$$H_0 = \sum_{i=1}^{N}\left[-\tfrac{1}{2}\nabla_i^2 - \frac{Z}{r_i}\right] + \sum_{i}^{N}\sum_{j>i}^{N}\frac{1}{r_{ij}} \tag{27}$$

The first term in brackets represents the kinetic energy and nuclear attraction operators for the ith electron and the last term is the Coulombic interaction between electrons i and j.[4]

Nearly all work on this problem involves the use of the Variation Principle. This approach employs an approximate form for the N-electron wavefunction which contains adjustable parameters that are varied to minimize the expectation value of the Hamiltonian. Because the energy found in this way must be an upper bound to the actual energy, the optimized parameters define the best approximation to the true wavefunction available using a particular model.[5]

A. One-electron Models

The one-electron approximation is nearly always employed to calculate electronic structure. It is assumed that the N-electron wavefunction can be expressed in a form which involves N one-electron functions. The simplest wavefunction of this type is the Hartree product,[6] in which the motion of any one electron is assumed to be completely independent of the others, i.e.

$$\Psi_0(1, 2, ..., N) = \phi_1(1)\,\phi_2(2) ... \phi_N(N) \tag{28}$$

The spin orbital $\phi_1(1)$ is a function of the coordinates of electron 1, and is the product of a spatial function, $\chi_1(r_1, \theta_1, \phi_1)$, and a one-electron spin function, $\alpha(1)$ or $\beta(1)$, where α corresponds to $m_s = +\frac{1}{2}$ and β to $m_s = -\frac{1}{2}$.

If we assume the motion of each electron is governed by a central field, the one-electron functions will be hydrogen-like. The $\{\chi\}$ are thus products of a radial function and a spherical harmonic,

$$\chi_1(1) = \mathbf{R}_{nl}(r_1)\,Y_{lm}(\theta_1, \phi_1) \tag{29}$$

The quantum numbers n, l, and m are the same as those in the hydrogen problem and we therefore speak of the orbitals as being s, p, or d-like, etc. The radial function $\mathbf{R}_{nl}(r)$ is regarded as adjustable and application of the variational technique (subject to the constraint that the radial function should remain normalized to unity) leads to a set of N integro-differential equations which determine the optimum set of orbitals $\{\phi\}$. Each such orbital must satisfy a pseudo-Schrödinger equation for an effective Hamiltonian in which the potential is provided by the nuclear attraction and the spherically averaged Coulombic interaction with all the other electrons. These equations are solved iteratively, since the potential in which a specific electron moves depends on the other, as yet undetermined, orbitals. One originally guesses a set of radial functions. These orbitals are used to generate a potential, which leads to an improved set of functions. These new functions generate a new field, etc. This is continued until all the orbitals change by less than some acceptable threshold from one iteration to the next, and this final potential is known as the self-consistent field.

The Hartree product [Eq. (28)] suffers from the serious drawback that it does not satisfy the requirement of antisymmetry the exact wavefunction must obey; interchange of the coordinates of two electrons does not result in a change in the sign of the wavefunction. The simplest wavefunction for a closed shell atom which preserves the product form of Hartree but satisfies the antisymmetry requirement is given by

$$\Psi'_0(1, 2, \ldots, N) = \mathscr{A}(N)\{\phi_1(1)\, \phi_2(2) \ldots \phi_N(N)\} \tag{30}$$

$\mathscr{A}(N)$ is called the N-electron antisymmetrizer and permutes the coordinates of the electrons in the direct product. Its effect is more explicitly seen in the equivalent form of the Slater determinant:

$$\Psi'_0(N) = \frac{1}{(N!)^{\frac{1}{2}}} \begin{vmatrix} \phi_1(1) & \phi_2(1) & \ldots & \phi_N(1) \\ \phi_1(2) & \phi_2(2) & \ldots & \phi_N(2) \\ \vdots & & & \\ \phi_1(N) & \phi_2(N) & \ldots & \phi_N(N) \end{vmatrix} \tag{31}$$

When the determinantal function above is subjected to the variational technique (constraining the $\{\phi\}$ to remain normalized and orthogonal), the familiar Hartree–Fock equations result:

$$\left[-\tfrac{1}{2}\nabla_1^2 - \frac{Z}{r_1} \right] \phi_i(1) + \sum_{j \neq i} \left[\int \phi_j{}^*(2) \frac{1}{r_{12}} \phi_j(2)\, d\tau_2 \right] \phi_i(1) - \sum_{j \neq i} \delta(m_{si}, m_{sj})$$

$$\times \left[\int \phi_j{}^*(2) \frac{1}{r_{12}} \phi_i(2)\, d\tau_2 \right] \phi_j(1) = \sum_j \varepsilon_{ij} \phi_j(1) \tag{32}$$

The first two terms in brackets arise in Hartree's formulation, but the last one is strictly a result of the antisymmetric form of the wavefunction. This exchange term is too well known to warrant discussion.

The reason we have written out the Fock equations explicitly is to point out the presence of the Lagrangian multipliers, ε_{ij}. It can be shown that if the one-electron spin orbitals are subjected to a unitary transformation, the total wavefunction is unchanged, and the form of the Hartree–Fock equations is also invariant. Therefore the spin-orbitals are *not uniquely determined*, and caution should be employed in placing too great an importance on the "physical nature" of these one-electron functions. The fact that many of the final ionic states important in PES can be described in terms of the ionization of an electron from a specific orbital rests on the success of Koopmans' Theorem[7] as a fairly accurate first approximation to the ionic state. Koopmans, however, realized that there is an optimum set of spin-orbitals for describing ionization; the canonical set which result from that particular unitary transformation of the ϕ's which diagonalizes the Lagrangian multiplier matrix ε_{ij}. It is fortunate that Koopmans' assumption works as well as it does; however, situations arise for which one-electron descriptions are no longer adequate (as is true for the case of satellite structure in PES, to be discussed later). Rigorously, we can only say that photoionization takes a system described by one many-body wavefunction to a final state characterized by another many-body wavefunction. The canonical Hartree–Fock orbitals are "special" for describing this process because they happen to lump most of the "many-body" effects into one orbital.

To illustrate this point, consider the transition from the ground state of the molecule carbon monoxide ($^1\Sigma^+$) to its first ionic state possessing $^2\Pi$ character. This transition can be described fairly accurately in terms of the ionization of an electron from the canonical orbital[8] 1π. The "molecular orbital" extends over both atomic centers;[9] however, the delocalized molecular orbitals of CO can be transformed into a completely equivalent set which are largely localized and correspond to the classical concepts of bonding pairs of electrons.[10] A description of the same transition in terms of these localized orbitals would necessitate talking of ionization of "part of an electron" from a carbon–oxygen "bonding" orbital, another fraction of an electron from the carbon "lone pair", etc. In this representation the transition must be referred to as a many-body process whereas it is adequately described as a one-electron process in the canonical representation. The same arguments apply to Bloch versus Wannier functions when discussing a solid. Although it reduces to a question of semantics, the point has largely been unappreciated by photoelectron spectroscopists, and the question of what constitutes "many-body" effects in ionization is meaningful only within the context of a specific representation.

B. Correlation and Configuration Interaction

We now turn to the final refinement in the form of the wavefunction which allows one, in principle, to approach the exact wavefunction to any degree of accuracy desired.[11] The particular method we shall describe is not the only one available for correcting the shortcomings of the Hartree–Fock function, but it is the one in most common use by quantum chemists. This model is termed Configuration Interaction (CI), so called because in the early days of quantum mechanics it was felt that the Hartree–Fock wavefunction was not exact because of its interaction with low-lying excited states. It has since been recognized that this is not the case. The assumption of the central field and the spherically averaged potential, while accounting for the long-range portion of the Coulombic interaction, does not allow for the description of the instantaneous repulsion between electrons. The CI concepts introduced below will be used in the discussion of the cross-section in Section IV. After the formalism is developed, the types of configurations important for correlating various systems will be discussed.

There are an infinite number of solutions to the Hartree–Fock equations (32) in addition to those which are occupied in the Hartree–Fock determinant. These unoccupied solutions are termed the *virtual* orbitals. Obviously, an infinite number of Slater determinants can be formed by "exciting" electrons from one or more of the Hartree–Fock orbitals into virtual orbitals, and the exact wavefunction can therefore be expanded in this series of Slater determinants. Thus the exact wavefunction can be written

$$\Psi_0(N) = \sum_k C_k \Phi_k \tag{33}$$

where the C_k are the expansion coefficients (again generally determined variationally) and Φ_k represents a specific Slater determinant. This added flexibility usually results in a decrease in the energy of the wavefunction of less than 1%, but even this is often large compared to electron affinities, reaction energetics, and other properties of interest to the chemist. Furthermore, the changes in the charge density brought about by CI are often very important in the computation of dipole moments, the electrostatic field at the nucleus, etc. A recent review of the effects of correlation on many properties of interest has been given by Schaefer.[12]

The exact form and convergence properties of the CI wavefunction (33) are, of course, dependent on the orbital basis employed. For closed-shell systems such as the neon atom, the Hartree–Fock determinant dominates all others. The remaining corrections have been termed "dynamical correlations" by Sinanoğlu[13] and can be shown to primarily reflect short-range correlations in the motion of two electrons. The inclusion of such effects thus keeps the electrons farther apart and reduces the energy. In the ^1S ground state of the neon atom, e.g. this correlation energy has been estimated[14] to be $10 \cdot 37$ eV compared to the Hartree–Fock energy of $3497 \cdot 73$ eV; a difference of approximately $0 \cdot 3\%$.

In open-shell atoms and molecules, fundamentally different types of CI occur. In many cases, it is not even possible (within the usual assumptions of doubly occupied spatial orbitals) to write a single determinant which possesses the correct symmetry for the state in question.[15] Even at this level, the concept of one electron in a particular orbital must be abandoned. The asymmetry of the Coulomb field means that it is no longer even roughly accurate to speak of individual electrons possessing specific angular momenta as was the case for the closed-shell central field. In addition, relatively large CI effects appear which are characterized by excitations from the Hartree–Fock orbitals into virtual orbitals that are "nearly degenerate" with them. For example, the $2s^2 \rightarrow 2p^2$ excitation is very important for correlating the ground state of Be.

The two factors mentioned above fall into the general category of "internal" and "semi-internal" correlation as discussed by Silverstone and Sinanoğlu.[16] In addition to them, the "all-external" or dynamical correlation present in closed-shell systems is also important. Because of these problems it is sometimes hazardous, even in the Hartree–Fock approximation, to speak of the "ionization of the $1s$ electron" in an open-shell atom or molecule. The phenomenon of multiplet splitting in PES is a dramatic example of this.[17,18]

IV. THE FINAL STATE IN PHOTOEMISSION

In Section II we emphasized the fact that photoionization is a transition between two states characterized by N-electron wavefunctions. In order to obtain some physical insight into the processes leading to the final states observed in PES, we must at least begin by discussing the transition in terms of a one-electron orbital model. The particular failures of the one-electron picture will become apparent later.

A. The Primary State

The most intense peaks, or "primary" states, observed in photoelectron spectroscopy involve, to first approximation, the ionization of an electron from a specific canonical spin-orbital in the atom or molecule. These primary states are the ones roughly describable by Koopmans' assumption, in the sense that the electron density in the ionic state resembles the original system with a "hole" in the region of space which characterized the orbital. The most probable continuum function is the one which results when the photoelectron accepts the unit of angular momentum transferred in the absorption process. Thus ionization of the $1s$ electron in neon is well described by a final ionic state of ^2S symmetry, coupled to a continuum function of p symmetry, which gives ^1P symmetry for the entire system. The most important channels in the ionization of the $2p$ electron involve d- and s-symmetry continuum func tions, etc.[19]

Although these one-electron descriptions are often qualitatively satisfying, the "passive" electrons in the final state have actually relaxed; they are not optimally described by the same spin-orbitals as in the ground state.[20] This relaxation, even without explicitly involving CI, constitutes a many-body effect in the sense that the motion of those electrons not directly involved in the ionization are coupled to the influence of the departing photoelectron. This relaxation phenomenon has important consequences for both the energy and intensity of the primary states and will be discussed later.

B. Correlation States

Toward higher binding energy from each of these primary peaks there are generally satellites which have come to be known as reflecting the presence of "shake-up" states. There are, in general, an infinite number of such states associated with each primary state, but only a few of them have observable intensities. They can, in favourable circumstances, be 20–80% as intense as the primary peak. The first satellites observed and assigned in x-ray photo-electron spectroscopy could be described as one-electron excitations accompanying ionization. These excitations followed "one-electron" monopole selection rules, e.g. ionization of the neon $1s$ electron accompanied by the excitation of a $2p$ electron into a $3p$ orbital. This monopole mechanism results in an ionic state of the same angular momentum (^2S) as the primary hole state and a continuum function of p character, yielding the overall ^1P symmetry required by the dipole selection rules.

Although the orbital picture described above is commonly used, compared to the primary states these "shake-up" states are much less favorably described in terms of one-electron transitions. First of all, there are usually two or more final states of the proper symmetry which can be derived from a given one-electron transition. This follows because each one-electron excitation may result in two (or more) unpaired valence electrons which can couple to the unpaired core electron to give two (or more) final states having the same symmetry as the primary state. Each one-electron excitation thus splits, a result analogous to the multiplet splitting phenomenon in the primary states of paramagnetic species. Furthermore, the assumptions of one electron, one orbital often have to be discarded. This is due to the possibility of configuration mixing in the final state, which can lead to many one-electron processes being involved in reaching a given final state. As an example, the $2p \rightarrow np$ and $2p \rightarrow n'p$ processes may both become important in reaching a particular shake-up state in neon. Bagus and Gelius[21] have shown that the energies of the Ne $1s$ satellites are fairly well described in terms of an (optimized orbital) multi-determinantal wavefunction corresponding to a specific ($2p \rightarrow np$) orbital excitation. This would seem to imply that in this case mixing among the various excited configurations is small. The intensities computed in the sudden approximation[22] from these wavefunctions, however, are in poor agreement with experiment.[23] Recent calculations on the F $1s$ satellites in hydrogen fluoride have shown that the most intense satellite state in the spectrum involves strong mixing of both $3\sigma \rightarrow 5\sigma$ (roughly F $2p_\sigma \rightarrow$ F $3s$) and $1\pi \rightarrow 2\pi$ (F $2p_\pi \rightarrow$ F $3p_\pi$) excitations.[24] Any attempt to describe this state as being reached by a *single* one-electron excitation would require, at the least, removal of the restrictions of specific angular momenta for every orbital. We would be forced to speak of the excitation as involving orbitals which have both σ and π character.

More recently, however, it has become increasingly apparent that the criterion for observation of this type of satellite is not that it follows one-electron monopole selection rules, but rather that its dominant configuration has the possibility to mix with, and thereby gain intensity from, the primary hole state. For example, in the XPS spectrum of the argon atom there is a broad feature ~ 10 eV from the $3s$ primary hole state. This peak has an intensity of $\sim 20\%$ of the $3s$ peak. Spears *et al.*[25] have suggested that most of the intensity in this region is due to $Ar^+(3s^2 3p^4 3d; {}^2S)$. Although this configuration differs by two orbitals from $Ar^+(3s^1 3p^6; {}^2S)$, it mixes strongly and "steals intensity" from the latter. Spears *et al.* termed this a "CI" satellite, but noted that in a more general sense the conventional "monopole" satellites also fell into this classification.

The third type of state observed arises from what is called the "conjugate shake-up" mechanism. The transition moments to these states are generally much smaller than the previous two types mentioned. As an example, a conjugate shake-up process accompanying $1s$ ionization in neon might lead to the ^2P final ionic state of $Ne^+(1s^1 2s^2 2p^5 3s)$. The continuum function would then have either s or d symmetry, resulting in the overall ^1P character necessary from the dipole selection rules. This path is termed "conjugate shake-up" since it appears that the one-electron excitation is $2p \rightarrow 3s$, which does not follow the monopole rules proposed for the normal shake-up process. The distinction in this case is complicated again when the many-electron nature of the wavefunction is considered. The same ionic state could be imagined to be reached through ionization of the $2p$ electron accompanied by the monopole excitation $1s \rightarrow 3s$. The important factor is that the final ionic state has ^2P symmetry and cannot mix with the primary ionic state. This state has been observed by Gelius[26] and has an intensity of 0.06% relative to the $1s$ hole state.

The more common conjugate shake-up situation occurs when ionization and excitation occur in the same shell. For example, the final state $1s^2 2s^2 2p^4 3s$ (εp), reached in the one-electron model through the transitions $2p \rightarrow \varepsilon p$, $2p \rightarrow 3s$, cannot be reached via the usual monopole selection rules. The ionic state also has the wrong parity to mix with $2p$ hole state. Wuillemier and Krause[27] have estimated that an upper limit for the intensity of this process relative to the normal case (final electron configuration $1s^2 2s^2 2p^4 3p$) is of the order of 25%. States of this type have also been identified in the He(I) and He(II) spectra of gaseous cadmium, mercury, and lead.[28] The ground state of Hg, for example, is described by the Hartree–Fock determinant

$$[\text{core}] 6s^2(^1S)$$

The $6s$ level primary ionic state

$$[\text{core}] 6s^1(^2S)$$

is observed as well as the conjugate state

$$[\text{core}] 6p^1(^2P)$$

The latter is roughly 1% as intense as the primary peak (at the He(I) photon energy). The conjugate excitation $6s \rightarrow 6p$ is invoked in the one-electron model to explain this final state. Berkowitz et al.[28] have shown that a great deal of the transition moment to this state is caused by admixture of the "nearly degenerate" configuration

$$[\text{core}] 6p^2(^1S)$$

into the *ground-state* wavefunction. Thus the inclusion of correlation into the ground state of Hg is a very important mechanism for contributing to the observed satellite structure.

To summarize this qualitative overview of satellites we again point out that the ionic states observed fall into two classes: those which have the proper symmetry to mix with a nearby primary ionic state and those which do not. It will be shown in the next section that, in the absence of many-electron effects, there would be no satellites at all *observed* in photoelectron spectroscopy. For this reason the satellites are also referred to as "correlation peaks".

V. MANY-ELECTRON EFFECTS ON THE CROSS-SECTION

In this section we examine the specific mathematical form of many-body effects on the photoionization transition moment. The terms which arise in a single-determinantal description of both initial and final state will be dealt with first, followed by the effects of configuration interaction.

A. Relaxation in the Primary State

Let us begin with the single Slater determinants

$$\Psi_i(N) = \frac{1}{(N!)^{\frac{1}{2}}} |\phi_1(1)\ \phi_2(2)\ ...\ \phi_N(N)| \tag{34a}$$

and

$$\Psi_f(N) = \frac{1}{(N!)^{\frac{1}{2}}} |\chi(1)\ \phi_2'(2)\ ...\ \phi_N'(N)| \tag{34b}$$

The orbitals of the final state have been primed to emphasize that they are not necessarily identical to the initial state functions. We have also associated the continuum function, $\chi(1)$, with the orbital ϕ_1, i.e. if the set $\{\phi_2', \phi_3', ...\}$ closely resembles $\{\phi_2, \phi_3, ...\}$ except for the effects of relaxation, then Ψ_f corresponds to the primary state associated with the orbital ϕ_1.

When these wavefunctions are substituted into the transition moment, the result is given by

$$
\begin{aligned}
T_{f \leftarrow i} &= \left\langle \Psi_f(N) \left| \sum_{k=1}^{N} \nabla_k \right| \Psi_i(N) \right\rangle \\
&= \langle \chi | \nabla_1 | \phi_1 \rangle \langle \Psi_f(N-1, \chi, 1) | \Psi_i(N-1, \phi_1, 1) \rangle \\
&\quad + \sum_{j=2}^{N} (-1)^{1+j} \langle \chi | \nabla_1 | \phi_j \rangle \langle \Psi_f(N-1, \chi, 1) | \Psi_i(N-1, \phi_j, 1) \rangle \\
&\quad + \sum_{j=1}^{N} (-1)^{1+j} \langle \chi | \phi_j \rangle \left\langle \Psi_f(N-1, \chi, 1) \left| \sum_{k=2}^{N} \nabla_k \right| \Psi_i(N-1, \phi_j, 1) \right\rangle \tag{35}
\end{aligned}
$$

The notation $\Psi'_j(N-1,\chi,1)$ refers to an $N-1$ electron Slater determinant which is formed from the N-electron determinant by deleting the column containing the orbital χ and the row denoting electron 1; i.e.

$$\Psi'_j(N-1,\chi,1) = \frac{1}{[(N-1)!]^{\frac{1}{2}}} \begin{vmatrix} \phi_2'(2) & \phi_3'(2) & \dots & \phi_N'(2) \\ \phi_2'(3) & \phi_3'(3) & \dots & \phi_N'(3) \\ \vdots & & & \\ \phi_2'(N) & \phi_3'(N) & \dots & \phi_N'(N) \end{vmatrix} \qquad (36)$$

The same notation applies to the wavefunction $\Psi_i(N-1,\phi_j,1)$. It is formed by striking the column containing ϕ_j and the row containing electron 1 from $\Psi_i(N)$. The sums over the index j are over all spin-orbitals. Since χ has either α or β spin (depending on the nature of ϕ_1), certain terms in the sums over j in Eq. (35) vanish by spin orthogonality. For the present, however, we will retain the full expression, but simplify its appearance with the following definitions:

$$S^{1j} \equiv \langle \Psi'_j(N-1,\chi,1) | \Psi_i(N-1,\phi_j,1) \rangle$$

$$p^{1j} \equiv \left\langle \Psi'_j(N-1,\chi,1) \left| \sum_{k=2}^{N} \nabla_k \right| \Psi_i(N-1,\phi_j,1) \right\rangle$$

Equation (35) is then given by

$$T_{f\leftarrow i} = \langle \chi | \nabla_1 | \phi_1 \rangle S^{11} + \sum_{j=2}^{N} (-1)^{1+j} \langle \chi | \nabla_1 | \phi_j \rangle S^{1j}$$

$$+ \sum_{j=1}^{N} (-1)^{1+j} \langle \chi | \phi_j \rangle p^{1j} \qquad (37)$$

The first term of (37) is related to the usual one-electron interpretation. An electron in orbital ϕ_1 makes a dipole transition to the continuum. If the orbital angular momentum of ϕ_1 is given by λ, then $\langle \chi | \nabla_1 | \phi_1 \rangle$ can be non-zero only if χ has $\lambda+1$ or $\lambda-1$ symmetry.

The factor S^{11} multiplying this one-electron moment is the overlap of the "passive orbitals", i.e. those not directly involved in the ionization. This overlap factor is generally between 0·9 and 1·0 for primary states, but much smaller for satellite states. Its effect is to introduce the many-body aspects of relaxation into the cross-section. In fact, it is easy to show that if we had made Koopmans' assumption—i.e. $\phi_2 = \phi_2'$, $\phi_3 = \phi_3'$, etc.—all the sums in Eq. (37) would vanish, S^{11} would be unity, and we would be left with the active electron approximation

$$T_{f\leftarrow i} = \langle \chi | \nabla_1 | \phi_1 \rangle \qquad (38)$$

In addition, if we consider an excited state associated with the primary peak (ϕ_1), the overlap integral S^{11} vanishes and the satellites are therefore forbidden. Relaxation thus introduces a multiplicative factor which reduces the contribution of the one-electron moment in the primary peak and distributes it among the various excited states. This has been discussed by Fadley,[29] who pointed out that for this reason cross-sections computed in the frozen orbital model are not directly comparable to experimental cross-sections found by photoelectron spectroscopy, but are more appropriate for experimental situations where no discrimination on the basis of ionic state energy is made.

The first sum over j in Eq. (37) arises from the antisymmetry requirements on the initial state wavefunction, and brings components into the total transition moment which arise from dipole transitions involving the other orbitals of the initial state. It will be shown in a later example to be interpretable as an ionization accompanied by monopole excitation.

The second line in Eq. (37) arises from the action of the remaining momentum operators, ∇_2 through ∇_N. Here an electron appears to make a monopole transition ($\phi_j \to \chi$) and the passive orbitals have rearranged themselves through a dipole excitation. The form of this term is very similar to that of the conjugate shake-up mechanism proposed by Berkowitz et al.[28]

Each of these three types of processes contribute to the transition moment even in a primary state. For example, consider the neon ($1s$) primary hole state reached by absorption of soft x-ray radiation. The ionic state has ^2S character and the continuum function is p-like. The first term in Eq. (37)

$$\langle \chi_p | \nabla | 1s \rangle S^{1,1s}$$

will dominate. The normal shake-up mechanism is involved in the nonvanishing term

$$\langle \chi_p | \nabla | 2s \rangle S^{1,2s}$$

An electron appears to be ionized from the $2s$ orbital accompanied by the monopole transition $1s \to 2s$. Finally, a nonvanishing contribution

$$\langle \chi_p | 2p \rangle p^{1,2p}$$

involves ionization of the $2p$ electron, accompanied (roughly speaking) by the excitation $1s \to 2p$. All three mechanisms reach the same final state and reflect the many-body nature of photoionization.

In the particular example used here, the second term should be negligible with respect to the first. This can be seen from examination of the ratio

$$R_2 = \frac{\langle \chi_p | \nabla | 2s \rangle S^{1,2s}}{\langle \chi_p | \nabla | 1s \rangle S^{1,1s}} \tag{39}$$

If R_2 is substantial compared to unity, retention of the second term is warranted. Now $(S^{1,2s}/S^{1,1s}) \ll 1$; in fact a rough estimate for this term based on Bagus' results[20] is 10^{-3}. Furthermore $\langle \chi_p | \nabla | 2s \rangle / \langle \chi_p | \nabla | 1s \rangle$ is of the order of magnitude of 10^{-1} for x-rays of approximately 1 keV energy. Thus the second term makes a contribution approximately 10^{-4} that of the first. As a general rule, the ratio of the overlap factors will always be small for any primary state, thereby decreasing the importance of this term. Certain situations might arise, however, when this small factor would be counterbalanced by a large ratio in the one-electron moments and this mechanism could then conceivably make a sizable contribution to the total cross-section.

It is much more difficult to estimate the importance of the third term. Its effect is governed by the ratio

$$R_3 = \frac{\langle \chi_p | 2p \rangle p^{1,2p}}{\langle \chi_p | \nabla | 1s \rangle S^{1,1s}} \tag{40}$$

To estimate the factor $\langle \chi_p | 2p \rangle / \langle \chi_p | \nabla | 1s \rangle$ we note that if we choose a plane wave for χ_p, i.e. $\chi_p \propto \exp(i\mathbf{k} \cdot \mathbf{r})$, then $\langle \chi_p | \nabla | 2p \rangle = ik \langle \chi_p | 2p \rangle$, and

$$\frac{|\langle \chi_p | 2p \rangle|^2}{|\langle \chi_p | \nabla | 1s \rangle|^2} = \frac{1}{k^2} \frac{|\langle \chi_p | \nabla | 2p \rangle|^2}{|\langle \chi_p | \nabla | 1s \rangle|^2}$$

Qualitatively, one would thus expect this term to be very dependent on the photon energy due to the presence of both the $1/k^2$ factor and the ratio of the transition moments. The $p^{1,2p}/S^{1,1s}$ ratio, however, is energy independent. $p^{1,2p}$ is the complex conjugate of the *X-ray emission* transition moment—in the approximation in which relaxed orbitals are used for the initial ($1s$ hole) state and the neutral atom ground-state functions are used to describe the final ($2p$-hole) state. $p^{1,2p}$ in neon is of the order of 10^{-1} bohr^{-1}, while $S^{1,1s}$ is nearly unity. In the general case, the emission transition moment will be dependent on the specifics of the atomic or molecular structure. A ratio of this type has been examined for the F($1s$) hole state of HF and been found to be negligible at XPS energies.[24] Recently Williams[30] has analyzed the various terms for the specific case of Ne $1s$ ionization. In this work a Hartree–Fock continuum function was generated in the field of a fully relaxed core. It was found that the ratio R_2 was less than 10% from threshold to 2000 eV photon energy. R_3, however, was as large as 0·65 a few eV above threshold. It quickly becomes small at higher photon energies and 100 eV above threshold is only $\sim 10\%$. An interesting result of this work is that although the third term is large in the region about threshold, the first term is smaller than the analogous result using Koopmans' Theorem; the net effect is that the relaxed core and frozen orbital results are very similar near threshold and begin to deviate from one another only when the third term has become small.

To summarize, the major many-body effect brought about by relaxation is a reduction of the active electron transition moment by the multiplicative factor S^{11}. Neglect of relaxation would therefore result in a predicted cross-section which is higher than the experimental result by a factor of $(S^{11})^2$. In fact, this tendency toward overestimation has been noted by Wuillemier and Krause[27] in a recent comparison of experimental data for neon with theoretical predictions which disregard relaxation. They have found that the discrepancy is greatest in those cases where relaxation effects should be more important, e.g. for nearly all incident photon energies the calculated $2s$ orbital cross-section is $\sim 20\%$ greater than experiment, whereas the $2p$ orbital cross-section is in much better agreement. More theoretical work is needed to determine if this discrepancy is due to the relaxation effect, or is primarily a result of the need for a more sophisticated wavefunction which explicitly includes configuration interaction.[31]

B. Configuration Interaction

Thus far we have examined the consequences of relaxation on the photo-ionization cross-section. This was done within the assumption of single-determinantal wavefunctions for both the initial and final states. Such wavefunctions suffer from the inadequacies pointed out in Section III, and inclusion of configuration interaction can have significant effects on the cross-section. CI in the initial state appears to be the prime contributor to the observed intensity of the conjugate shake-up peaks in Hg and Cd.[28] The importance of correlation in determining the intensity of normal shake-up peaks in which ionization and excitation occur in the same valence shell has been noted by Åberg,[32] Carlson et al.,[33] and Byron and Joachain.[34] The importance of both initial and final state correlation in core-level satellite spectra has recently been studied by Martin et al.,[24] Manson,[35] and Basch.[36] In the following discussion we will concentrate on the qualitative aspects of CI as it affects core-level satellite structure.

Suppose that the initial state is described by a multiconfiguration wavefunction $\Psi'_0(N)$

$$\Psi'_0(N) = \sum_m D_{0m} \Phi_m(N) \tag{41}$$

where D_{0m} is the coefficient of the configuration Φ_m in the wavefunction Ψ'_0. The configurations may be single Slater determinants expanded in the occupied and virtual orbital set $\{\phi\}$ or, if necessary, sums of determinants chosen to possess the symmetry properties of the ground state. As discussed previously, the coefficient of the Hartree–Fock configuration, D_{00}, will be the leading term in the expansion. For closed-shell atoms or molecules it will usually have a value between 0·9 and 1·0, the rest of the coefficients being 0·2 or less.

Each final state is expanded similarly,

$$\Psi'_f(N) = \sum_n D_{f'n} \Phi_n'(N) \tag{42}$$

where the primes on the configuration denote that they have been formed from a set of orbitals appropriate for the final state. Note that the configurations are functions of all N electrons and we should therefore in principle allow for the possibility of mixing among the various channels in the continuum,[35a] a phenomenon which can have marked effects on absolute cross-sections. In addition, $\{\Phi'(N)\}$ should also include any "bound states" which may be imbedded in the continuum. Mixing of this sort leads to autoionization and greatly enhances the photoionization cross-section. Autoionization also has a very dramatic effect on the satellite intensities in Ba excited by He I radiation.[37] In the case of XPS satellites, however, these effects have thus far not been crucial to an understanding of the structure of the spectrum and we therefore specialize to the case where each configuration contains the same one electron function $\chi_{f'}$, and consider CI in the ion alone. Thus

$$\Psi_{f'}(N) = \chi_{f'}(1) \sum_n C_{f'n} \Phi_n'(N-1) \qquad (43)$$

Again, the primary hole state, $f' = 0$, is characterized by a large C_{00} where $\Phi_0'(N-1)$ is the hole state Hartree–Fock configuration. For the satellite states f', there may be several configurations which mix strongly. This will be dependent to some extent upon the virtual orbitals used to define the excited configurations, but in most cases there will be a small number (~ 1–3) of configurations with coefficients greater than 0·5.

Insertion of Eqs. (41) and (43) into the expression for the transition moment (35) yields

$$T_{f'0} = \sum_{m,n} C_{f'n}^* D_{0m} \langle \chi_{f'} | \nabla | \phi_1 \rangle S_{nm}^{11} + \cdots \qquad (44)$$

We have again assumed that the final state predominantly involves ionization from orbital ϕ_1. S_{nm} is the $(N-1)$ electron overlap integral between configurations n and m and the dots represent the other terms obtained. In view of the previous discussion they should be small for core-level ionization and will be neglected.[38]

The ratio of the transition moments to the primary state and a satellite is thus given by

$$\frac{T_{00}}{T_{f'0}} \approx \frac{\langle \chi_0 | \nabla | \phi_1 \rangle}{\langle \chi_{f'} | \nabla | \phi_1 \rangle} \left(\sum_{m,n} C_{0n}^* D_{0m} S_{nm}^{11} \Big/ \sum_{m,n} C_{f'n} D_{0m} S_{nm}^{11} \right) \qquad (45)$$

If the final states are close in energy, then the one-electron matrix elements should be very nearly the same. The density of final states which enters into the cross-section [Eq. (25)] should also be similar for the two states. These two assumptions lead to the relative intensities of the two states in the overlap approximation,[39]

$$\frac{I_{00}}{I_{f'0}} \approx \left| \sum_{m,n} C_{0n}^* D_{0m} S_{nm}^{11} \right|^2 \Big/ \left| \sum_{m,n} C_{f'n} D_{0m} S_{nm}^{11} \right|^2 \qquad (46)$$

To illustrate these CI effects, we have drawn a state diagram for neon in Fig. 1. On the left is the Hartree–Fock level for the ground state and one of its excited configurations; above these are the primary ionic state and a first approximation to the "shake-up" state. For simplicity, we have suppressed the exchange interaction in the ionic states; i.e. there are actually two ^2S states which result from the configuration $1s\,2s^2\,2p^5\,3p$. In the middle of the diagram

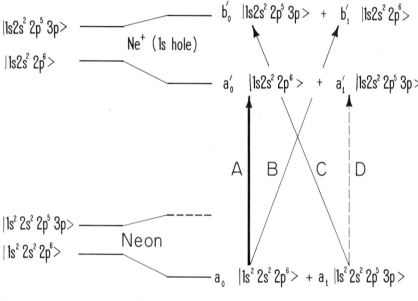

Basic configurations   Eigenstates

Fig. 1. Simple model to illustrate the effects of configuration interaction on correlation-peak intensities in Ne $1s$ photoemission (not to scale). With $1s$ exchange suppressed, the Ne$^+$($1s$ hole) configuration manifold would closely resemble the ground-state manifold (left). Introducing configuration-interaction, this 1 : 1 correspondence would also obtain for the eigenstates (right), and $a_0 \sim a_0'$, $a_1 \sim a_1'$, etc. The main peak arises primarily from path A. Paths B and C arise because the two configurations "look for themselves" in the correlation state. They are roughly equal strength, but the dashed path (D) is weak.

we have allowed the ionic configurations to interact, forming the observable states of the ion. The ground-state function has also been allowed to mix with its excited configurations. On the far right we have assigned coefficients to the configurations in each eigenstate. These have magnitudes

$$a_0,\ a_0',\ b_0' \sim 1$$

$$a_1,\ a_1',\ b_1' \sim 0{\cdot}1$$

The effective intensity of the primary hole state is given in our example by the four contributions to the overlap integral denoted by A, B, C, and D. The total overlap integral for the primary hole state is dominated by the contribution from A because it is a product of two large coefficients and a large determinantal overlap. Analogues to B and C are smaller because they involve a small product of coefficients together with a small overlap integral. This integral is not zero, since the orbitals of the hole state have relaxed somewhat. Finally, the contribution from D is small because, although the determinantal overlap is large, the product of the coefficients is very small.

In the case of the satellites, however, the total overlap is a fraction of that for the primary state and configuration interaction contributions are much more important. A main contributor may be the analog to path A, since the coefficients are both large. Within this overlap picture, the small intensity of the satellites is due to the small determinantal overlap between the shake-up configurations and the ground state. Path B might also contribute an amount of the same order of magnitude since, although the product of coefficients is small, the determinantal overlap is large. This is also presumably the predominant contribution to the argon $(3s^2 3p^4 3d; {}^2S)$ satellite mentioned in Section IV-B. The excited state will have an unusually large coefficient for the $3s^1 3p^6$ configuration because it is "nearly degenerate" with $3s^2 3p^4 3d$. This leads to a large relative intensity via the analog to path A.

The two contributions mentioned thus far could be termed a CI effect in the final state. For similar reasons, the analog of path C is also important for the satellites and it arises through an initial state CI mechanism. The contribution from path D is obviously smaller than the others. An example of the importance of initial-state configuration interaction is found in the case of the Ne $1s$ satellites. We have compared the intensities of the satellites using both the Hartree–Fock and a correlated initial state.[23] In these calculations, the final state wavefunctions were found from a CI calculation on the ion. Table I presents the results of this work. It can be seen that the introduction of path D nearly doubles the satellite intensities and brings them into very good agreement with the experimental results.

To summarize this section, many-body effects on the cross-section arise from two somewhat artificially separate phenomena. The cross-section to a primary hole state is affected predominantly by relaxation in the passive orbitals. This results in an apparent reduction in the cross-section from that computed assuming no relaxation. Additional relaxation effects and the inclusion of CI is expected to be of lesser importance for most primary states, although there may arise situations where it becomes significant (multiplet splitting, closely spaced primary states, etc.). The intensities of satellite peaks, on the other hand, depend entirely upon relaxation and configuration interaction contributions. In a strictly formal vein, of course, there exist only eigenstates of the electrostatic Hamiltonian. The concepts of relaxation and CI arise only when we attempt to form better approximations to those

TABLE I

Neon 1s Correlation-state Energies and Intensities[a]

State	ΔE (theo)[b]	ΔE^c (exp)	$I_{HF}{}^d$ (theo)	$I_{ISCI}{}^e$ (theo)	I^f (exp)
(1s hole state)	0·0 ($-96\cdot694$ au)	0·0	100 (0·824)	100 (0·774)	100
$2p \to 3p$	36·4	37·35 (2)	1·26	2·47	3·15 (10)
$2p \to 3p$	39·9	40·76 (3)	1·68	2·60	3·13 (10)
$2p \to 4p$	41·9	42·34 (2)	0·85	1·48	2·02 (10)
$2p \to 5p$	43·0	44·08 (5)	0·24	0·43	0·42 (06)
$2p \to 6p$	45·2	45·10 (7)	0·05	0·09	~0·15
$2p \to 4p$	46·0	46·44 (5)	0·46	0·70	0·96 (11)
$2p \to 5p$	47·4	48·47 (7)	0·07	0·11	0·17 (05)
$2p \to 6p$	49·5	—	0·04	0·06	—

[a] From refs. 23 and 26.

[b] Absolute energy of the 1s hole state in Hartree atomic units given parenthetically; all others in eV relative to it.

[c] Relative energies (from ref. 26), in eV.

[d] Hartree–Fock initial state wavefunction; the parenthetical number is the square of the actual overlap in the 1s hole state. The relative peak intensities are given as percentages of this value.

[e] Correlated initial state wavefunction; this included double excitations into the Rydberg orbitals for the ground state, i.e. configurations of the form $1s^2\, 2s^2\, 2p^4\, np^2$, etc.

[f] Ref. 26; we have estimated a value for the $(2p \to 6p)$ state from Gelius' figure.

eigenstates than are available within the confines of an independent electron model.

Thus far, we have examined relaxation as it affects peak intensities. It is well known that it also has important consequences for the energies of the final states. The latter part of this chapter will be concerned with the qualitative aspects of the final state stabilization which comes about through the re-arrangement of the passive electrons. This relaxation energy can be related to the intensity of the satellite peaks through an approximate sum rule derived by Manne and Åberg.[40] Although these authors obtained the result from an application of the sudden approximation, it can also be derived in the dipole approximation. We shall not show this, but simply note that it follows from the neglect of an energy dependence in the ratio of the cross-section for the primary state versus its satellites.

In our notation, the sum rule is given by

$$E_R = \sum_{f=1} \left(\frac{I_f}{I_0}\right) \Delta_f \bigg/ \sum_{f=0} \left(\frac{I_f}{I_0}\right) \tag{47}$$

where E_R is the relaxation energy, (I_f/I_0) is the intensity of the satellite peak relative to the main peak and Δ_f is the separation in energy between the satellite and the primary state. The denominator simply reflects a normalization condition so that the intensity units are arbitrary. The summations are taken over discrete satellites; they convert to an integration over any continua.

From Eq. (47) we see that there exists a "lever arm" relationship between the satellite intensities and the relaxation energy. If E_R were zero, no satellites would be observed. In the case that E_R is large, the relaxation manifests itself either as an intense set of satellites "near the main peak", or weak satellites "far from the main peak" or, of course, something in between. The sum rule provides a great deal of qualitative information about the relaxation process. For example, it is a common misconception that there are no satellites in the core-level photoelectron spectrum of metals. It is known, however, that there is a large relaxation energy involved in core ionization in these species, so there must be a fairly large probability for multiple excitation processes. In metals the shake-up (as well as the multiple ionization or shake-off) spectrum is essentially continuous because the excitations are into the conduction band. Thus while no discrete peaks are observed, the relaxation energy is manifested as a broad background on the high-binding energy side of the main peak.

VI. RELAXATION EFFECTS ON BINDING ENERGY

The foregoing discussion related the photoemission spectrum to the photoelectric process *per se*. Two features that were emphasized—the many-electron nature of the process and the multiplicity of final states—should make it clear that "relaxation energy" is a concept without a unique meaning. In a strictly formalistic, many-electron description of the photoemission problem this concept need never arise. However, in most discussions that focus on the properties of real systems, a one-electron description is adopted at some point. "Relaxation energy" (or "polarization energy") then becomes a useful term for describing the energy reduction of the passive electrons in the final state. The relaxation energy $E_R(j)$ accompanying photoemission from one-electron level j is usually defined by

$$E_B(j) = E_f^0(N-1) - E_i^0(N)$$
$$= -\varepsilon_j(N) - E_R(j) + \Delta E_{corr} + \Delta E_{mult} + \Delta E_{rel} \qquad (48)$$

Here $E^0(N-1)$ and $E^0(N)$ are respectively the total energies of the primary final state and the initial state. The orbital energy $\varepsilon_j(N)$ is the energy assigned to the jth orbital in the initial state; by Koopmans' Theorem[7], $-\varepsilon_j(N)$ is the binding energy that orbital j would have if the passive orbitals were unchanged during photoemission (i.e. no relaxation). In referring to $-\varepsilon$ we usually automatically neglect multiplet structure and correlation energy. The former is important only for open-shell systems, and can be corrected in a straightforward way, through the term

$$\Delta E_{\mathrm{mult}} = (E_f - \bar{E}_f) - (E_i - \bar{E}_i)$$

Here $E_f - \bar{E}_f$ is the multiplet energy separation of the final state from the average energy of that configuration, within a single configuration description, and $E_i - \bar{E}_i$ is defined similarly. For most simple atomic configurations these quantities have been tabulated in terms of Slater integrals.[41] The correlation-energy correction ΔE_{corr} accounts for the difference in the energy stabilization of the final and initial states through configuration interaction. Clearly ΔE_{mult} and ΔE_{corr} must be considered together for open-shell systems, in which ΔE_{mult} is non-zero. The last term ΔE_{rel} is an artifact, necessary only if a non-relativistic theory has been used. We shall ignore ΔE_{rel} in the following discussion, as it it is only an avoidable complication. The terms ΔE_{mult} and ΔE_{corr} are sometimes important. No unexceptional general statement can be made about these terms, but they may have either sign and are usually small in magnitude (≈ 1 eV). By contrast the relaxation energy term always lowers the binding energy ($E_R > 0$), and it is large ($\gg 1$ eV) for core levels. This section treats relaxation energy in atoms, molecules, solids, adsorbates, and solutes. In each case the physical origin of E_R will be discussed, its magnitude considered (with examples where available), and relevant applications mentioned.

A. Atoms

Removal of an electron from an atomic orbital creates a positive hole toward which the passive electrons' orbitals relax to minimize the systems' total energy. Within the constraint of a one-determinant wavefunction this relaxation takes place adiabatically; i.e. the electrons' quantum numbers do not change. Hedin and Johansson[42] showed that the relaxation energy, $E_R(j)$, accompanying ionization from orbital j can be treated conveniently as the sum of inner-shell, intra-shell, and outer-shell contributions,

$$E_R(j, n) = E_R(n' < n) + E_R(n' = n) + E_R(n' > n) \tag{49}$$

Here n is the principal quantum number of orbital j and n' is that of the passive electrons.

The inner-shell term $E_R(n' < n)$ is negligible. Hedin and Johansson obtained this result by direct calculation for Na, K, and their ions. To obtain some physical insight into why this is true, we note that the potential inside a hollow charged sphere is constant. Thus the presence of an electron in shell n has little influence on a wavefunction in shell $n' < n$.

Intra-shell relaxation is intermediate in magnitude, usually $\leqslant 5$ eV. It arises through a reduction, during removal of an electron from orbital j, in the average electrostatic repulsion among the passive electrons in shell n. The leading term in $E_R(n' = n)$ involves a decrease in the Slater integral $F^0(nn)$, *not* a change in the coefficient of this integral (which would appear in the orbital energy). The physical picture in this case is that the electrons in shell n are all constrained to lie at essentially the same radius but may distribute themselves on a sphere of that radius to minimize their repulsive interaction. A simple classical model shows that the loss of an electron from an eight-electron s, p shell will lead to a reduction of $\sim 3\%$ in the average pair repulsion between the remaining electrons. If F^0 is reduced by 3%, the value of $E_R(n' = n)$ would be 3·3 eV for the $n = 2$ shell in sodium and 1·9 eV for the $n = 3$ shell in potassium, in rough agreement with the values 2·9 eV and 1·2 eV, respectively, calculated by Hedin and Johansson.

Outer-shell relaxation is easy to understand, and $E_R(n' > n)$ may be very large. An electron in the n shell shields orbitals in the $n' > n$ shells almost completely. Removal of an n-shell electron therefore increases the effective nuclear charge experienced by the n' shell by practically one unit. This essentially quantitative shielding has led to simplified estimates of $E_R(n' > n)$ based on "equivalent core" models, which have proved to be quite accurate.[43] For core-electron binding energies, outer-shell relaxation is by far the largest contributor to $E_R(j, n)$ provided that several electrons occupy a shell with $n' > n$. Thus for atomic potassium, Hedin and Johansson found values of E_R (and percentage arising from $E_R(n' > n)$) of 32·8 eV (96%), 10·8 eV (82%), and 2·2 eV (40%), respectively, for the $1s$, $2s$, and $3s$ orbitals.

A number of estimates of $E_R(j, n)$ for light atoms are available. Bagus did early hole-state calculations.[20] Rosén and Lindgren[44] developed an optimized relativistic Hartree–Fock–Slater method which has been applied by Gelius and Siegbahn[45] to the hole states of elements through Cu ($Z = 29$). Outer-shell relaxation energies can also be calculated by a method[43] that combines the polarization potential approach of Hedin and Johansson with the equivalent-cores model. Table I gives a summary of the total calculated relaxation energies of the orbitals in selected light atoms. Most of the values were taken from ref. 45 for consistency, but many of them are also available in other sources and the agreement between different calculations is very good.

Table II shows that $E_R(j, n)$ decreases monotonically with increasing n, as expected because of the dominance of outer-shell relaxation. For the same reason, $E_R(j, n)$ increases monotonically with Z for a given orbital. In valence orbitals only intra-shell relaxation is important, and it is fairly small in most cases. For s, p shells the value of E_R increases as the shell is filled within a given period but decreases from one period to the next higher one. This is a consequence of the $2p$ shell, for example, being smaller than the $3p$ shell, and consequently having a greater average electron–electron repulsion energy. E_R is slightly larger for ns electrons than for np electrons. These observations will all be valuable below in discussing relaxation in molecules. Even in atoms these results are useful, because they indicate that relaxation energies in valence shells should decrease somewhat in going to heavy atoms.

TABLE II

Calculated Atomic Relaxation Energies for the Orbitals of Light Atoms (eV)[a]

Atom	$1s$	$2s$	$2p$	$3s$	$3p$	$3d$	$4s$
He	1·5						
Li	3·8	0·0					
Be	7·0	0·7					
B	10·6	1·6	0·7				
C	13·7	2·4	1·6				
N	16·6	3·0	2·4				
O	19·3	3·6	3·2				
F	22·1	4·1	3·9				
Ne	24·8	4·8	4·7				
Na	24·0	4·1	4·4				
Mg	24·6	5·2	6·0	0·7			
Al	26·1	6·1	7·1	1·0	0·2		
Si	27·1	7·0	8·0	1·2	0·4		
P	28·3	7·8	8·8	1·3	0·6		
S	29·5	8·5	9·6	1·4	0·9		
Cl	30·7	9·3	10·4	1·6	1·1		
Ar	31·8	9·9	11·1	1·8	1·4		
K	32·8	10·8	12·2	2·2	2·0		
Ti	35·4	13·0	14·4	3·6	3·4	2·0	0·3
Mn	40·1	17·2	18·8	5·1	4·9	3·6	0·4
Cu	48·2	23·7	25·7	7·7	7·2	5·3	0·3

[a] Values are mostly from ref. 45. Some are interpolated.

B. Molecules

The relaxation energy accompanying photoemission from core levels in molecules is nearly always larger than in atoms, because additional electronic charge can flow toward the positive hole. It is convenient, though arbitrary, to consider the total relaxation energy as the sum of atomic plus "extra-atomic" contributions,

$$E_R(j) = E_R^a(j) + E_R^{ea}(j) \tag{50}$$

Naturally the exact partitioning of $E_R(j)$ in this way can be neither un-ambiguous nor unique, but it can be meaningful within any particular molecular orbital model. To gain physical insight into E_R^{ea} we can envision it as arising through polarization of electrons toward the hole. Alternatively, we may think of one unit of positive charge having been added to the molecule.

It would naturally expand repulsively to the outside of the molecule to minimize repulsive interaction. A homonuclear diatomic molecule would be expected to acquire a net charge of approximately $+e/2$ on each atom, because the core hole from atom A would be screened by transfer of $\sim -e/2$ of electronic charge through polarization of the valence orbitals. Similarly, in methane, ionization of the C $1s$ orbital would be accompanied by transfer of charge $\sim -e/4$ from each hydrogen. That these expectations are approximately borne out is shown in Table III, which gives final-state atomic chaiges calculated in the "RPM" approach.[46] This is a model that uses CNDO/2 molecular orbitals[47] and accounts for relaxation.

TABLE III

Charge transfer accompanying core-level ionization
in nitrogen and methane (in eV)[a]

Atom	Orbital	q (initial)	q (final)	Δq (bond)[b]
C in CH_4	C $1s$	-0.05	-0.09	0.25
H in CH_4	C $1s$	$+0.01$	$+0.26$	
N_2 (active N)	N $1s$	0	$+0.38$	0.62
N_2 (passive N)	N $1s$	0	$+0.62$	

[a] From ref. 46.
[b] Charge transfer along each bond.

From this discussion, E_R for a given core level would be expected to increase substantially from the free atom to the diatomic molecule. Since the single additional atom will not provide more than $\sim e/2$ of screening charge, however, this increase in E_R is limited. Experimental results on second-row elements have shown that the $1s$ binding energies are (2–3 eV) lower than the value calculated for free atoms.[43] This is probably the approximate size of the extra-atomic relaxation energy in these molecules. Additional ligands allow further enhancement of E_R because more electrons are available for screening, and the charge build-up on any ligand is small. Even in the second-row hydrides of C, N, and O the $1s$ binding energies are about 6 eV lower than in the free atoms: again most of this difference can be attributed to the E_R^{ea} term. Nearly all the total possible extra-atomic relaxation energy is already realized for these small molecules. Increase of the molecular size even to infinity (a solid) adds only ~ 2–3 eV additional relaxation energy. This is to be expected on the argument that screening leaves a positive charge of $+e$ distributed on the outside of a molecule of radius R where it exerts a repulsive potential of e/R. The largest change in R^{-1} with increasing molecular size has already been realized for the hydrides.

In molecular orbitals relaxation energies must be considered in two classes. *Delocalized* orbitals, in which the electronic charge is distributed more or less uniformly around the molecule, can be treated in the same terms as were valence shells in atoms. Consider, for example, a diatomic molecule in the second row. Ionization from a molecular orbital made up of atomic $2p$ functions will entail essentially the same relaxation among the passive $n = 2$ orbitals that was obtained in the free atom. This implies that E_R terms in molecular orbitals can be estimated by summing over atomic orbital population P_{ij} times appropriate intra-shell relaxation energies for those orbitals:

$$E_R^i(MO) \cong \sum_j P_{ij} E_R^j(AO) \tag{51}$$

where the $E_R^j(AO)$ values could be taken from Table I. This approach was used by Banna *et al.*[48] for molecular orbitals in fluorinated methanes with considerable success. It is conceptually preferable to the common practice of estimating binding energies in molecules simply by reducing the calculated orbital energies by a constant factor. The use of Eq. (51) should yield rather good estimates of molecular-orbital binding energies, although errors due to correlation-energy differences will still be present.

In *localized* molecular orbitals, i.e. lone-pairs or highly polarized orbitals, additional contributions to E_R can arise through extra-atomic relaxation. Atomic (intra-shell) relaxation would still be present, and a relation like Eq. (50) would be applicable. The E_R^{ea} term is not readily calculable, but it is both important and useful, as it is closely related to conventional chemical properties, as discussed below. We note that the presence of additional relaxation in localized molecular orbitals should be of some value in identifying these orbitals, although no applications of this feature have been tested as yet.

Variations in core-level and lone-pair binding energies are closely related to variations in chemical reactivity. In fact, the lone-pair binding energy is essentially the Lewis basicity within a constant additive factor. By extending the concept of Lewis basicity to include *core* lone pairs, core-level binding energies can be included in this statement. The physical significance of this connection can be appreciated by considering the following two reactions of an alcohol molecule in the gas phase:

$$ROH + H^+ \rightarrow ROH_2^+: \quad E = PA \tag{52a}$$

$$ROH \rightarrow RO^*H + e^-: \quad E = E_B(O\ 1s) \tag{52b}$$

In Eq. (52a) a proton approaches the OH group and becomes bound. An O $1s$ electron is lost in Eq. (52b), leaving a positive hole on the oxygen atom. In both cases the system must respond to the addition of a positive point charge, located either within or very near the oxygen atom. The absolute magnitudes of the energies of these two reactions are very different but their variation as the R group is altered is nearly identical,[49]

$$\Delta(PA) \cong -\Delta E_{\mathrm{B}} \qquad (53)$$

This relationship was first observed for alcohols and amines;[49] it has been confirmed and extended to a large number of other molecules,[50, 51] thereby establishing the relationship between Brönsted basicity and core-level binding energy for a given functional group. While not all of the variation in E_{B} arises from variation of $E_{\mathrm{R}}^{\mathrm{ea}}$, calculations[49] show that this is the main contributor.

The Lewis basicity is a more general concept. It does not refer to reactions with any particular acid, but rather to the system's tendency to give up an electron from a lone pair in the valence shell. Now core electron binding energies and those of lone-pair valence electrons on the same atom have been shown to vary together.[52] This is expected, because both variations result from a combination of the initial-state potential and the polarization of this potential during photoemission, measured at the orbital in question. The concept of Lewis basicity can therefore be extended (at least to within a proportionality factor close to unity) to include the variation in binding energy of *core* electron "lone pairs". A stringent test of the transferability of ligand function is then afforded by comparing the variation of E_{B} for core levels of two functional groups when the ligands are changed together. In comparing the series RI and ROH for alkyl iodides and alcohols, the I $3d_{\frac{5}{2}}$ and O $1s$ binding energies were found to vary linearly.[49] Even HI and water lay on the straight line.

The chemical "message" of these results is simple. Reactivity, like binding energy, depends not only on the properties of the initial state but on those of the final state as well. In working with the formalism of extra-atomic relaxation, we say that the binding energy depends on both the effective potential of a given core level in the initial state and on the change in this potential on photoionization. The chemical-reactivity description would use the terms *inductive* and *polarization* effects. It is a mistake to discuss chemical properties in terms of ground-state properties (such as dipole moments) alone: the same is true of binding energies. Since both basicity and binding-energy variation depend on the *same* combinations of inductive and polarization phenomena, it is fortunate that this additional complication is present. Binding-energy shifts appear to possess considerable diagnostic value for the determination of chemically interesting properties, and extensive future application can be expected.

One immediate application of these ideas can be made in the interpretation of the variation of the "ionization potential" of the alkyl alcohols. This is the binding energy of the oxygen lone pair. In sign and approximately in magnitude it is equal (see Fig. 2) to the extra-atomic relaxation energy as predicted by the RPM model.[49] Thus extra-atomic relaxation appears to account for a phenomenon that has received a variety of explanations in the last 30 years.

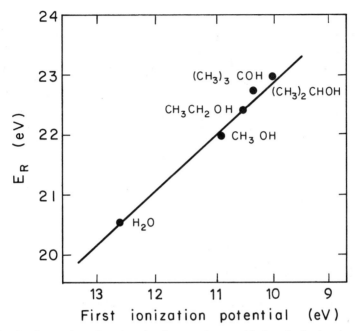

Fig. 2. Comparison between the (oxygen lone pair) first ionization potential in aliphatic alcohols and the relaxation energy due to screening of a positive charge on the oxygen atom. The line has a slope of 0·9. This agreement suggests that the variation in ionization potential, which has been variously attributed to "hyperconjugation" and other effects, is in fact largely a consequence of extra-atomic relaxation.

C. Solids

Relaxation energies in solids are best discussed separately for insulators and conductors. We shall treat insulators first.

When a molecule or a multi-atomic ion is present in a molecular or ionic solid, respectively, the binding energy of an orbital in that moiety is naturally considered as the sum of its local and lattice contributors.[53] The relaxation energy accompanying loss of an electron from orbital i will have a (molecular or ionic) local contribution plus a contribution due to lattice polarization,

$$E_R(i) = E_R(i, \text{loc}) + E_R(i, \text{latt}) \qquad (54)$$

The $E_R(i, \text{loc})$ term has been dealt with in Sections A and B. For large molecules or ions $E_R(i)$ consists mainly of this term and $E_R(i, \text{latt})$ can be neglected. Little evidence is available on the $E_R(i, \text{latt})$ term in molecular crystals. We may safely assume that it is small (< 1 eV) on theoretical grounds.

In the ionic crystal case the experimental situation is reasonably clear. Fadley et al.[53] first discussed the polarization energy term for an ionic lattice. They used a model described by Mott and Gurney[54] to conclude that this term is of the order 1 eV or less for a series of potassium salts. The largest $E_R(i, \text{latt})$ terms would, of course, be expected in monatomic ions and, on balance, the alkali halides appear to present the most suitable salts for study of $E_R(i, \text{latt})$ terms. Citrin and Thomas[55] carried out a careful study of core-level binding energies in eleven alkali halides. By comparing observed binding energies with predictions of a simple theory, they were able to obtain evidence for the existence of an $E_R(i, \text{latt})$ term, plus an estimate of its size. They avoided the troublesome problem of the reference level for an insulator by comparing cation and anion core-levels with free-ion binding energies. Uncertainties in the reference energy shift all the levels in the crystal *together*. This analysis followed that of Fadley et al.[53] including both a Madelung and an *electronic*[56] relaxation energy, but, adding a repulsive term $E(i, \text{rep})$, they gave the equation (in our notation)

$$E_B(i) = E_B(i, \text{FI}) + \Phi(e^2/R) - E(i, \text{rep}) - E_R(i, \text{latt}) \tag{55}$$

Here $E_B(i)$ and $E_B(i, \text{FI})$ are respectively the binding energy of orbital i in the alkali halide lattice and the free ion, while $\Phi e^2/R$ is the Madelung potential energy. Since this latter term is of the order of 10 eV in magnitude, it might appear that errors of ~ 1 eV are incurred in this model by neglecting the $\sim 10\%$ covalency of alkali halide lattices.[57] In fact this is not a serious problem, because the Madelung term tends to cancel the change in core-level binding energy on forming the alkali or halide ion.[58]

Citrin and Thomas calculated $E_R(i, \text{latt})$ using a method given by Mott and Littleton[59] for estimating the polarization energy around a lattice vacancy. This probably represents an upper limit for $E_R(i, \text{latt})$, because the nearest neighbors' polarizabilities must be smaller in a lattice with no vacancy. At any rate, the differences between calculated values of $E_R(i, \text{latt})$ for cations and anions in the same lattice ranged up to 1·2 eV. Inclusion of this term in Eq. (55) improved the differences between calculated and measured cation and anion binding energies. This gives somewhat indirect evidence that their calculated values of $E_R(i, \text{latt})$—which range between 1·45 eV and 2·69 eV—are at least approximately correct. Thus extra-ionic relaxation energies in alkali halide lattices may be taken to vary around ~ 2 eV.

Another rather indirect measure of extra-atomic relaxation energy in alkali halides is given by the relative differences between core-level peak energies and Auger energies in photoemission spectra in free atoms and crystal lattices. The additional electron hole in the Auger final state polarizes neighboring ions more strongly than the photoemission final states. Hence the Auger transition entails additional extra-atomic relaxation. In going from atomic sodium to sodium salts, Kowalczyk et al.[60] found that the Na(KLL) Auger transition energy increased by 4·3 eV in NaI and 3·7 eV in Na$_2$O. A simple lattice-polarization model indicates that the *additional* extra-atomic relaxation in an Auger transition should be about twice that accompanying photoemission.[60] Thus half these observed differences, or ~ 2 eV, can be attributed to $E_R(i, \text{latt})$ in NaI, for example, in good agreement with the results of Citrin and Thomas. We may conclude that the lattice contribution to extra-atomic relaxation energies accompanying core-level ionization in alkali halide crystals is of the order of 2 eV, as the model of Mott et al.[54, 59] would predict.

Uncertainty about reference energies has thus far precluded a definite discussion of the $E_R(i, \text{latt})$ term in semiconductors. In a semi-metal—graphite—it has been possible to calculate the $E_R(i, \text{latt})$ term quantitatively by using empirical binding energies in hydrocarbon molecules as calibration points. Davis and Shirley[46] used a relaxation model with CNDO wavefunctions to calculate C 1s relaxation energies for trigonally bonded carbons in benzene and larger planar hydrocarbons, extrapolated to an infinite lattice, and obtained a C 1s binding energy (284·4 eV) in excellent agreement with experiment (284·7 eV). This theory did not start from first principles but used the experimental C 1s binding energy in benzene as an anchor point. It appears that graphite is one lattice for which an accurate core-level binding energy can be predicted.

Extra-atomic relaxation energy terms in metals are often large, and they can be treated in a straightforward way because there is no reference energy problem. On photoemission of a core electron from an atom in a metallic lattice, the itinerant valence electrons are attracted toward the positive hole thus created. In contrast to the insulator case, screening charge can actually be transferred to the atom from which the photoelectron is ejected. This phenomenon is conveniently discussed in terms of positive phase shifts, δ_l, in the partial l waves that describe the itinerant electrons. Friedel's allow theory[61] is useful here: the photo-excited atom can be treated as an impurity of one unit higher atomic number than the lattice. The Friedel sum rule in the form

$$\sum_l (2l+1)\,\delta_l = \tfrac{1}{2}\pi \tag{56}$$

would apply. This relation states that the excess charge of $+e$ present on the impurity will be screened through phase shifts that lead to a net charge of $-e$ being attracted toward the hole. The l-character of this charge depends on the character of the conduction bands immediately above the Fermi energy. Thus in the 3d transition metals, most of the shielding is by the d wave, while in copper the s and p waves are important. Figure 3 shows a dramatic decrease in the quantity

$$\Delta E_B(3p) = E_B(3p, \text{free atom}) - E_B^V(3p, \text{metal}) \tag{57}$$

for the $3d$ metals at the end of the d shell. Ley et al.[62] explained this behavior in terms of a potential model similar to that of Hedin and Johansson.[42] For the $3d$ metals through Ni, d-wave state density lies just above the Fermi energy, and the resultant d-wave screening is similar to that expected for a $3d$ electron. The potential model gives

$$\Delta E_B(3p) \simeq \tfrac{1}{2}F^0(3p, 3d) \tag{58}$$

where F^0 is a Slater integral. The $3p, 3d$ repulsive interaction F^0 is large because the $3d$ orbital is relatively small. In Cu and Zn the $3d$ shell is full, there

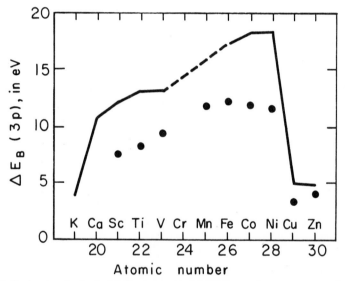

Fig. 3. Reduction in the $3p$ binding energy of $3d$ elements from gas to metal (points) and estimates based on the simple extra-atomic relaxation model (line). The dramatic decrease between nickel and copper is a consequence of the filling of the $3d$ shell.

is little d-wave density above E_F, and the corresponding $F^0(3p, 4s)$ and $F^0(3p, 4p)$ integrals that approximately describe $\Delta E_B(3b)$ for s and p wave screening are much smaller, contributing in part to the sudden drop in ΔE_B between Ni and Cu. Figure 3 shows experimental values of ΔE_B as well as theoretical estimates based on Eq. (57). This relation overestimates ΔE_B because the shielding charge is, of course, not completely localized. Another treatment of extra-atomic relaxation in metals, based on polarizability of the lattice, has been given by Citrin and Hamann.[63] Recently Watson et al.[63a] have pointed out that the large ΔE_B values in the $3d$ series, and their variation with Z, must arise in part from the fact that the valence-shell configurations change from $d^n s^2$ in free atoms to $d^{n+1} s$ in metals. They attributed about half the ΔE_B term, and most of the break between Ni and Cu, to screening effects other than extra-atomic relaxation. At this writing the relative importance of the various contributions to ΔE_B is unresolved.

Valence electrons in metals also experience a substantial E_R^{ea} term. This fact can be obscured by making an oversimplified interpretation of the delocalized nature of these electrons. In fact, the mean binding energies of electrons in the valence bands, $\bar{E}_B(VB)$, is lower than binding energies of the corresponding orbitals in the free atoms, E_A, mainly because of the E_R term. Wigner and Bardeen,[64] in their classic paper on the work function in alkali metals, derived an expression that can be rearranged[65] to

$$\bar{E}_B(VB) = E_A + E_C - \left[\frac{3e^2}{5r_s} - \frac{0 \cdot 458 e^2}{r_s}\right] \tag{59}$$

Here E_C is the cohesive energy. The quantity in brackets can be interpreted as the Coulomb and exchange energy differences imposed by creation of a hole in the valence bands (r_s is the radius of the Wigner–Seitz sphere). A common misinterpretation of the Wigner–Bardeen theory is based on the idea that valence orbitals should show little relaxation because Koopmans' Theorem can be used. It is, of course, true that little intra-shell relaxation would be expected, as in atoms or molecules (see Table I). However, the presence of itinerant valence electrons assures that the analogue of extra-atomic relaxation is also present, in the form of polarization of the electron gas toward the "Coulomb hole" (for Coulomb energy) or "Fermi hole" (for exchange). Unfortunately, the magnitude of E_R^{ea} is quite insensitive to the degree of localization of the final-state hole.[65] It is clear, however, that substantial relaxation does take place on photoemission of a valence electron from a metal, in contrast to an atom or molecule.

D. Adsorbates on Metals

The power of electron spectroscopy for solving problems in surface science and catalysis has led to many applications of photoemission to adsorbed species on metallic substrates, usually with the intention of studying adsorbate–catalyst interactions. Relaxation energy shifts play a rather crucial role in these studies.

In physical adsorption, the adsorbate photoemission peaks have the same structure observed in the free atom or molecule. The binding energy of each peak is lowered relative to the gas-phase value by an additional relaxation energy that arises through polarization of the substrate valence electrons to screen the adsorbate hole state, i.e.

$$E_B{}^F(\text{ads}) = E_B(\text{gas}) - \phi - E_R(\text{ads}) \tag{60}$$

Here the work function ϕ is retained because while it is the vacuum-referenced binding energy, $E_B{}^v(\text{ads}) = E_B{}^F(\text{ads}) + \phi$, that should be compared to $E_B(\text{gas})$, the work function may be altered by the adsorbate's presence. Yates and Erickson[66] have studied xenon adsorbed on a clean tungsten surface, finding an E_R value of 2·6 eV for the $3d_{\frac{5}{2}}$ levels, similar to the value observed for xenon embedded in a metallic lattice.

Core-level binding energies in adsorbed molecules are generally substantially smaller than the gas-phase values. The presence of a valence-electron reservoir in the metal allows the molecular equivalent of outer-shell relaxation, as electron charge is transferred into the molecule's valence orbitals during photoemission. Thus Barber et al.[67] reported an O $1s$ binding energy of 532 eV for oxygen adsorbed on graphite. This corresponds to 537 eV relative to the vacuum level after a work-function correction. This is still some 9 eV lower than the value $E_B \sim 546$ eV expected for free atomic oxygen, indicating a substantial E_R. For oxygen in adsorbed CO the $1s$ binding energies show some relaxation energy relative to gaseous CO,[68, 69] but the experimental situation is generally unclear as yet.

Adsorbate molecules show molecular orbital peaks in photoemission spectra that yield detailed adsorbate–substrate bonding information. Thus Demuth and Eastman[70] found that most of the molecular orbitals shifted to lower binding energy in adsorption of C_2H_2, C_2H_4, and C_6H_6 on nickel. The average shifts of all the σ-orbitals were 3·2, 2·1, and 1·7 eV, respectively. It should be noted that these shifts are smaller than core-level shifts on adsorption, because they arise from the molecular analogue of *intrashell* relaxation. Two other observations should be made. These additional values of E_R from extra-molecular relaxation decrease with increasing molecular size because the hole charge tends to be screened already within the molecule. Also, close examination of the photoemission spectra of adsorbates in this work and elsewhere shows that the strongly bound σ-orbitals show larger relaxation shifts than more weakly-bound π-orbitals, as is the case for molecular orbitals in free molecules, discussed above.

Orbitals that form chemisorption bonds to the substrate tend to oppose the trend and shift to higher binding energy. This is an initial state effect. It has been observed for the least-bound π-orbitals in C_2H_2, C_2H_4, and C_6H_6 by Demuth and Eastman.[70] In CO, a similar shift has been observed for the 5σ-orbital, which merges with the 1π-orbital, yielding an intense combined peak.[71] The 5σ-peak is the carbon "lone pair" orbital. It *should* be perturbed, as CO is believed to stand up as an adsorbate, with the carbon atom bonded to the substrate. Additional evidence for this orientation is provided by the relative oxygen and carbon core-level shifts.[72] The various changes expected in molecular photoemission spectra on adsorption are illustrated in Fig. 4.

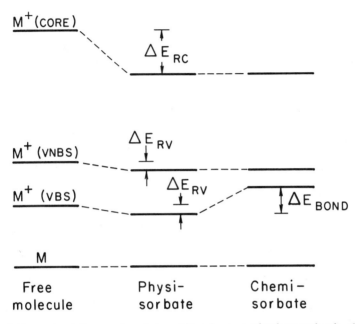

Fig. 4. Energy level diagram for photoemission from an adsorbate molecule, showing ground state M, core-hole state $M^+(c)$, and hole states $M^+(VBS)$ and $M^+(VNBS)$ of valence orbitals that do and do not bond to the substrate, respectively. Vacuum reference levels are used. Hole-state energies (and binding energies) are lowered on physisorption because of screening by the substrate's valence electrons. The core-hole relaxation energy ΔE_{RC} is greater than that of the valence-level hole states, ΔE_{RV}. On chemisorption, the adsorbate-to-substrate bonding orbital is identified by an *increase* in its hole-state (and binding) energy, ΔE_{BOND}.

In an earlier section it was emphasized that binding energies, and particularly relaxation energies, are good indices to chemical reactivity. Chemical properties cannot be understood in terms of initial-state properties alone: the final state (following reaction) or an intermediate state must also be considered. The same is true for binding energies. In adsorbates the chemical reactivity versus binding energy parallelism is not so easily made. It would appear, however, that the same polarization effects that lower the binding energy in an adsorbate molecule should serve to stabilize any activated complex formed by this molecule, thus speeding up the reaction. It would be as naive to attempt to explain catalytic reactions in terms of properties of the reactants alone as it would for any other reactions. Relaxation effects appear to hold one of the keys to an understanding of the complex problem of heterogeneous catalysis.[72]

E. Solutes

Solvation has the effect of providing a reservoir of electrons for screening holes created by photoemission from solute molecules. For this reason the binding energy of a given orbital should be lower if the molecule is in solution than it would be in the gas phase. Liquid–vapor shifts in this direction have been observed.[73] The few data available as yet appear to support the expectation that dielectric liquids will show relaxation-energy shifts about equal to those found in dielectric solids. Enhanced relaxation energies in solutes are well worth study. They are related, for example, to variations in the order of reactivity between gaseous and dissolved molecules. Thus the difference in basicity between methanol and t-butanol should be reduced or even reversed in solution, because the solvent can more easily assist in lowering the energy of the protonated methanol ion. Such a reversal is already known for the acidity of alcohols.[74] Further connections of this type between E_B and reactivity provide motivation for photoemission studies on solutes.

VII. APPROXIMATE METHODS FOR COMPUTING BINDING ENERGIES

The last section dealt with the theory of electron binding energies. For completeness, some approximate methods are mentioned below. Only fairly rigorous methods fall within the scope of this discussion. Other methods are discussed in appropriate chapters.

Any attempt to calculate binding energies quantitatively must include relaxation effects. The potential energy of the active orbital due to the entire charge distribution of the molecule must also be considered, not just the local "atomic charge". Beyond these constraints the model employed for a given study will depend on the parameters of that study. The most straightforward approach is an accurate SCF-CI calculation on both initial and final states. Unfortunately, this is seldom feasible for large molecules or large numbers of molecules. More approximate models are therefore nearly always employed. In comparing these models a few simple points can be made:

1. There is little to be gained in practice by using "polarization potential" theories; since *ab initio* hole-state calculations are still required, one may well compare total energies of initial and final states.

2. For large molecules or for problems involving many types of molecules, even single-determinantal *ab initio* calculations are likely to become impractical. Less exact molecular-orbital models such as CNDO, INDO,[47] etc. must then be employed.

3. In applying "intermediate-level" molecular orbital models, it is inadvisable to compare total energies of hole states and ground states. The errors entailed in these models preclude reasonable accuracy via this approach.

Binding energies of molecular orbitals can probably be estimated to sufficient accuracy by using orbital energies from *ab initio* calculations on molecular ground states, with relaxation corrections as discussed in Section VI-B. Orbital energies from intermediate-level models are less satisfactory in predicting the order of orbital binding energies.[48]

Core-level binding energies cannot, of course, be calculated using CNDO, etc. models because they do not explicitly consider core orbitals. However, Basch[75] and Schwartz[76] showed that core-level orbital energy shifts (i.e. shifts in the orbital energy of a given orbital, say C $1s$, from one molecule to another) are nearly equal to (minus) shifts in the potential at the nucleus,

$$\Delta \varepsilon \cong -\Delta V_n$$

Since ΔV_n is easily (and even rather accurately) calculable from CNDO wavefunctions, it is possible to calculate orbital-energy shifts at the CNDO level, a great simplification compared to using *ab initio* calculations. This approach was first used in a predictive fashion by Davis *et al.*[77] who found quite good agreement with experimental binding energies in small molecules. It is termed a "potential model" because the electrostatic potential, rather than the total energy, is used to estimate shifts in binding energy. This specific approach was later termed GPM, because only the ground-state potential was calculated.[78, 46]

Relaxation effects can be taken into account by extending this model, applying the "equivalent cores" approximation of Jolly[79] and modifying Hedin and Johansson's polarization potential theory to apply to the potential at the nucleus. Davis and Shirley[80] found that this "relaxation potential model" (RPM) gave somewhat better agreement with experiment than the GPM approach. The essential feature of the RPM is that binding energy shifts are given by

$$\Delta E_{\mathrm{B}} = -\Delta V_n - \Delta V_{\mathrm{R}}$$

The relaxation term is given by

$$\Delta V_{\mathrm{R}} = \tfrac{1}{2}[V_n(Z+1) - V_n]$$

that is, the change in potential at the nucleus on ionization of a core electron due to outer electron relaxation is approximated by the change that would occur if the nuclear charge were increased by one unit. This RPM approach is both accurate and easy to use. It gives results in good agreement with experiment[78] and yields ΔV_{R} values in fairly good agreement with *ab initio* estimates.[46] It has been observed[81] that RPM gives systematic differences in both ΔE_B and ΔV_R compared to *ab initio* values. The very good overall agreement of RPM predictions with experiment offers encouragement that these differences could be reduced by adjustments in the CNDO parameters.

We close this chapter by noting that further discussions of various models for binding energy shifts are given in the appropriate chapter. In particular, Jolly discusses the equivalent-cores approximation in Chapter 3. The essential equivalence of this approximation and the potential model has been shown.[82] Finally, two additional techniques for estimating binding energies are being evaluated as this is written: the $X_{\alpha\beta}$ method[83] and a propagator approach.[84] Both are promising but it is too early to evaluate them.

FOOTNOTES AND REFERENCES

1. Nearly every introductory quantum mechanics text presents a discussion of the semiclassical treatment. Many also include sections on quantum field theory. See, for example, L. I. Schiff, "Quantum Mechanics" (McGraw-Hill, New York); E. Merzbacher, "Quantum Mechanics" (John Wiley & Sons, New York).
2. Here, and elsewhere, the summation sign indicates a summation over discrete states and, when appropriate, integration over continuous variables.
3. The two most common expressions for the transition moment in the dipole approximation are the velocity and length forms. For *exact* wavefunctions, the two are equivalent and related by

$$\langle \Psi_k | \nabla | \Psi_i \rangle = \frac{-m(E_k - E_i)}{\hbar^2} \langle \Psi_k | \mathbf{r} | \Psi_i \rangle$$

4. We have used the atomic units for which $e = \hbar = m = 1$. The unit of energy is given by the Hartree (27·2097 eV) and distance is measured in bohrs (0·5292 Å).
5. This is true as long as certain conditions are met. See J. K. McDonald, *Phys. Rev.* **43**, 830 (1933).
6. For an excellent discussion of these topics, the reader is referred to the two volumes by Slater, "Quantum Theory of Atomic Structure" (McGraw-Hill Book Co., New York, N.Y., 1960).
7. T. Koopmans, *Physica* **1**, 104 (1934).
8. Although we have been discussing the Fock equations for an atomic system, upon the assumption of the Born–Oppenheimer approximation all principles carry over into the molecular case. The only modification is that numerical integration of the Fock equations becomes very impractical and we are generally forced to resort to still another expansion.

$$\phi_i = \sum_k C_{ik} \, \eta_k$$

The functions $\{\eta\}$ make up what is known as the basis set. It is generally regarded as fixed, and the variation is performed on the expansion coefficients $\{C\}$. This leads to the Hartree–Fock–Roothaan equations.[9] In the limit of a complete basis set, the orbitals found by this method approach the Hartree–Fock orbitals. In actual practice, however, the basis set must be of a very limited size and thus selection of a basis which is flexible enough to describe all the $\{\phi\}$ accurately is a very important step in the calculation.

9. C. C. J. Roothaan, *Rev. Mod. Phys.* **23**, 69 (1951).
10. W. England, L. S. Salmon, and K. Ruedenberg, *Fortschr. Chem. Forsch.* **23**, 31 (1971).
11. P. O. Löwdin, *Rev. Mod. Phys.* **32**, 328 (1960).
12. H. F. Schaefer, "The Electronic Structure of Atoms and Molecules" (Addison-Wesley, Reading, Mass., 1972).
13. O. Sinanoğlu, *J. Chem. Phys.* **36**, 706, 3198 (1962).
14. A. Veillard and E. Clementi, *J. Chem. Phys.* **49**, 2415 (1968).
15. Many methods exist for bypassing this problem without specifically resorting to a CI calculation (spin unrestricted Hartree–Fock, spin-extended Hartree–Fock, etc.). They all involve removing certain restrictions from the standard closed-shell formalism and are discussed in ref. 12.
16. H. J. Silverstone and O. Sinanoğlu, *J. Chem. Phys.* **44**, 1899 (1966).
17. K. Siegbahn, C. Nordling, G. Johansson, J. Hedman, P. F. Hedén, K. Hamrin, U. Gelius, T. Bergmark, L. O. Werme, R. Manne, and Y. Baer, "ESCA Applied to Free Molecules" (North-Holland, Amsterdam, 1969).
18. C. S. Fadley and D. A. Shirley, *Phys. Rev.* **A2**, 1109 (1970). *See also* D. W. Davis, R. L. Martin, M. S. Banna, and D. A. Shirley, *J. Chem. Phys.* **59**, 4235 (1973), and references therein.
19. R. J. W. Henry and L. Lipsky, *Phys. Rev.* **153**, 51 (1967).
20. P. S. Bagus, *Phys. Rev.* **139**, A619 (1965).
21. P. S. Bagus and U. Gelius, data referred to in ref. 17.
22. T. Åberg, *Phys. Rev.* **156**, 35 (1967).
23. R. L. Martin and D. A. Shirley, *Phys. Rev.* **A13**, 1475 (1976).
24. R. L. Martin and D. A. Shirley, *J. Chem. Phys.* **64**, 3685 (1976); R. L. Martin, B. E. Mills and D. A. Shirley, *J. Chem. Phys.* **64**, 3690 (1976).
25. D. P. Spears, J. H. Fischbeck, and T. A. Carlson, *Phys. Rev.* **A9**, 1603 (1974); for additional examples of this sort, see G. K. Wertheim and A. Rosencwaig, *Phys. Rev. Lett.* **26**, 1179 (1971); S. P. Kowalczyk, L. Ley, R. A. Pollack, F. R. McFeely, and D. A. Shirley, *Phys. Rev.* **B7**, 4009 (1973); P. S. Bagus, A. J. Freeman, and F. Sasaki, *Phys. Rev. Lett.* **30**, 850 (1973).
26. U. Gelius, *J. Electr. Spectr. and Related Phenomena*, **5**, 985 (1974).
27. F. Wuillemier and M. O. Krause, *Phys. Rev.* **A 10**, 242 (1974).
28. S. Süzer and D. A. Shirley, *J. Chem. Phys.* **61**, 2481 (1974) (Cd); J. Berkowitz, J. L. Dehmer, Y. K. Kim, and J. P. Desclaux, *J. Chem. Phys.* **61**, 2556 (1974) (Hg); S. Süzer, M. S. Banna, and D. A. Shirley, *J. Chem. Phys.* **63**, 3473 (1974) (Pb).
29. C. S. Fadley, *Chem. Phys. Letters*, **25**, 225 (1974); C. S. Fadley, *J. Electr. Spectr. and Related Phenomena*, **5**, 895 (1974).
30. R. S. Williams, private communications.
31. Amus'ya *et al.* have computed the neon orbital cross-sections within a many-body formalism and the results are in much better agreement with experiment. Their particular method, however, does not distinguish between relaxation and CI effects as we have defined them and it is difficult to know the relative importances of the two. See M. Ya. Amus'ya, N. A. Cherepkov, and L. V. Chernysheva, *Soviet Physics JETP*, **33**, 90 (1971).
32. T. Åberg, *Phys. Rev.* **A2**, 1726 (1970).
33. T. A. Carlson, M. O. Krause, and W. G. Moddeman, *J. Phys.* (*Paris*), **32**, C4–76 (1971).
34. F. W. Byron, Jr. and C. J. Joachain, *Phys. Rev.* **164**, 1 (1967).

35. S. T. Manson, *J. Electr. Spectr. and Related Phenomena*, **9**, 21 (1976).

35a. U. Fano, *Phys. Rev.* **124**, 1866 (1961).

36. H. Basch, private communication.

37. B. Brehm and K. Höfler, *Int. J. Mass. Spectrom. Ion Phys.* **17**, 371 (1975); H. Hotop and D. Mahr, *J. Phys.* **B8**, L301 (1975); U. Fano, *Comments on Atom. and Mol. Phys.* **D4**, 119 (1973).

38. There may exist configurations in the ground state CI expansion in which the orbital ϕ_1 is not occupied. In that case, the sum over m will not always contain a matrix element involving the orbital and we should therefore regard expression (44) as a sum over only those determinants which contain the function ϕ_1.

39. All variations in the intensity due to the angular distribution in the differential cross-section have been neglected.

40. R. Manne and T. Åberg, *Chem. Phys. Lett.* **7**, 282 (1970).

41. J. C. Slater, "Quantum Theory of Atomic Structure" (McGraw-Hill, New York, 1960), Vol. II, p. 286.

42. L. Hedin and A. Johansson, *J. Phys.* **B2**, 1336 (1969).

43. D. A. Shirley, *Chem. Phys. Lett.* **16**, 220 (1972).

44. A. Rosén and I. Lindgren, *Phys. Rev.* **176**, 114 (1968).

45. U. Gelius, *Physica Scripta*, **9**, 133 (1974).

46. D. W. Davis and D. A. Shirley, *J. Electr. Spectr. and Related Phenomena*, **3**, 137 (1974).

47. J. A. Pople and D. L. Beveridge, "Approximate Molecular Orbital Theories" (McGraw-Hill, New York, 1970).

48. M. S. Banna, B. E. Mills, D. W. Davis, and D. A. Shirley, *J. Chem. Phys.* **61**, 4780 (1974).

49. R. L. Martin and D. A. Shirley, *J. Amer. Chem. Soc.* **96**, 5299 (1974).

50. T. X. Carroll, S. R. Smith, and T. D. Thomas, *J. Amer. Chem. Soc.* **97**, 659 (1975).

51. B. E. Mills, R. L. Martin, and A. D. Shirley, *J. Amer. Chem. Soc.* **98**, 2380 (1976).

52. J. E. Hashmall, B. E. Mills, D. A. Shirley, and A. Streitwieser, Jr., *J. Amer. Chem. Soc.* **94**, 4445 (1972).

53. C. S. Fadley, S. B. M. Hagstrom, M. P. Klein, and D. A. Shirley, *J. Chem. Phys.* **48**, 3779 (1968).

54. N. F. Mott and R. W. Gurney, "Electronic Processes in Ionic Crystals" (Clarendon Press, Oxford, 1948), 2nd ed.

55. P. H. Citrin and T. D. Thomas, *J. Chem. Phys.* **57**, 4446 (1972).

56. Nuclear relaxation is slower and need not be considered. *See* ref. 53.

57. S. P. Kowalczyk, F. R. McFeely, L. Ley, R. A. Pollak, and D. A. Shirley, *Phys. Rev.* **B9**, 3573 (1974), and references therein.

58. Ref. 53, p. 3788. *See also* ref. 55.

59. N. F. Mott and N. J. Littleton, *Trans. Faraday Soc.* **34**, 485 (1938).

60. S. P. Kowalczyk, L. Ley, F. R. McFeely, R. A. Pollak, and D. A. Shirley, *Phys. Rev.* **B9**, 381 (1974).

61. J. Friedel, *Adv. Phys.* **3**, 446 (1954).

62. L. Ley, S. P. Kowalczyk, F. R. McFeely, R. A. Pollak, and D. A. Shirley, *Phys. Rev.* **B8**, 2392 (1973).

63. P. H. Citrin and D. R. Hamann, *Chem. Phys. Letts.* **22**, 301 (1973).

63a. R. E. Watson, M. L. Perlman, and Jan Herbst, *Phys. Rev.* **B13**, 2358 (1976).

64. E. Wigner and J. Bardeen, *Phys. Rev.* **48**, 84 (1935).

65. L. Ley, F. R. McFeely, S. P. Kowalczyk, J. G. Jenkin, and D. A. Shirley, *Phys. Rev. B*, to be published.
66. J. T. Yates, Jr. and N. E. Erickson, *Surface Science*, **44**, 489 (1974).
67. M. Barber, E. L. Evans, and J. M. Thomas, *Chem. Phys. Lett.* **18**, 423 (1973).
68. T. E. Madey, J. T. Yates, Jr., and N. E. Erickson, *Chem. Phys. Lett.* **19**, 487 (1973).
69. S. J. Atkinson, C. R. Brundle, and M. W. Roberts, *Chem. Phys. Lett.* **24**, 175 (1974).
70. J. E. Demuth and D. E. Eastman, *Phys. Rev. Lett.* **32**, 1123 (1974).
71. D. Menzel observed this spectrum (private communication) as did D. E. Eastman. The interpretation is not universally accepted.
72. A more detailed discussion is available in "X-ray photoemission and surface structure", by D. A. Shirley, *J. Vac. Sci. Technol.* **12**, 175 (1975).
73. H. Siegbahn and K. Siegbahn, *J. Electr. Spectr. and Related Phenomena*, **2**, 319 (1973).
74. J. I. Brauman and L. K. Blair, *J. Amer. Chem. Soc.* **92**, 5986 (1970).
75. H. Basch, *Chem. Phys. Lett.* **5**, 337 (1970).
76. M. E. Schwartz, *Chem. Phys. Lett.* **6**, 631 (1970).
77. D. W. Davis, D. A. Shirley, and T. D. Thomas, *J. Chem. Phys.* **56**, 671 (1972).
78. D. W. Davis, M. S. Banna, and D. A. Shirley, *J. Chem. Phys.* **60**, 237 (1974).
79. W. L. Jolly and D. N. Hendrickson, *J. Amer. Chem. Soc.* **92**, 1863 (1970).
80. D. W. Davis and D. A. Shirley, *Chem. Phys. Lett.* **15**, 185 (1972).
81. M. E. Schwartz, private communication.
82. D. A. Shirley, *Chem. Phys. Lett.* **15**, 385 (1972).
83. J. C. Slater and J. H. Wood, *Int. J. Quant. Chem. Symp.* **4**, 3 (1971); J. W. D. Connolly, H. Siegbahn, U. Gelius, and C. Nordling, *J. Chem. Phys.* **58**, 4265 (1973).
84. O. Goscinski, B. T. Pickup, and G. Purvis, *Chem. Phys. Lett.* **22**, 167 (1973); B. Kellerer, L. S. Cederbaum, and G. Hohlneicher, *J. Electr. Spectroscopy and Related Phenomena*, **3**, 107 (1974).

3

The Application of X-Ray Photoelectron Spectroscopy in Inorganic Chemistry

WILLIAM L. JOLLY

*Chemistry Department, University of California, and
Inorganic Materials Research Division, Lawrence Berkeley Laboratory,
Berkeley, California 94720*

I. INTRODUCTION

Excellent comprehensive reviews of the inorganic aspects of x-ray photoelectron spectroscopy have appeared in recent volumes of *Electronic Structure and Magnetism of Inorganic Compounds*[1] (from the Chemical Society series of Specialist Periodical Reports), and a review emphasizing applications of coordination chemistry was recently published.[2] Therefore in this chapter it was decided not to attempt a complete review of x-ray photoelectron spectroscopic studies of inorganic compounds, which would somewhat duplicate existing reviews, but rather to discuss certain important methods of systemization and to point out promising areas of research.

II. THE EQUIVALENT CORES APPROXIMATION

A. Introduction and History

It is generally recognized that atomic core electrons are closely bound to the nucleus and that they are relatively inaccessible and inert compared to valence electrons. Therefore it is reasonable to assume that valence electrons are affected by the ionization of a core electron essentially the same as they would be by the addition of a proton to the nucleus. This approximation, in which an atomic core which lacks one electron is considered chemically equivalent to the complete core of the next element in the periodic table, is called the *equivalent cores* approximation.

The approximation was used as long ago as 1932 by Skinner,[3] who calculated the $1s$ binding energy of lithium as the sum of three terms:[4]

$$
\begin{array}{lll}
\text{Li} \longrightarrow \text{Li}^+ + e^- & & 5 \cdot 39 \text{ eV} \\
\text{Li}^+ \longrightarrow \text{Li}^{2+} + e^- & & 75 \cdot 62 \text{ eV} \\
\text{Li}^{2+} + e^- \longrightarrow \text{Li}^+(1s \text{ hole}) & & -18 \cdot 21 \text{ eV (estimated)} \\
\hline
\text{Li} \longrightarrow \text{Li}^+(1s \text{ hole}) + e^- & & 62 \cdot 80 \text{ eV}
\end{array}
$$

The third term, the negative of the $2s$ binding energy of a lithium ion with the configuration $(1s, 2s)$, was assumed to be equal to the negative of the ionization potential of Be^+. The experimental value[5] for the lithium $1s$ binding energy is $64 \cdot 85$ eV, in moderately good agreement with the estimated value, $62 \cdot 8$ eV. This type of calculation can give relatively poor estimates of $1s$ binding energies for atoms of higher atomic number.[6] Thus in the case of neon, we calculate

$$
\begin{array}{lll}
\text{Ne} \longrightarrow \text{Ne}^{9+} + 9e^- & & 2149 \cdot 4 \text{ eV} \\
8e^- + \text{Ne}^{9+} \longrightarrow \text{Ne}^+(1s \text{ hole}) & & -1300 \cdot 9 \text{ eV (estimated)} \\
\hline
\text{Ne} \longrightarrow \text{Ne}^+(1s \text{ hole}) + e^- & & 848 \cdot 5 \text{ eV}
\end{array}
$$

The energy of the second reaction is assumed to be the negative of the energy required to strip 8 electrons from Na^+. The estimated $1s$ binding energy is considerably lower than the experimental value, $870 \cdot 2$ eV.[7] Better results are obtained when calculations are made for a core which is well shielded by other electrons. For example, the $2s$ binding energy of sodium may be estimated as follows:[8]

$$
\begin{array}{lll}
\text{Na} \longrightarrow \text{Na}^{2+} + 2e^- & & 52 \cdot 43 \text{ eV} \\
\text{Na}^{2+} \longrightarrow \text{Na}^{2+}(2s2p^6) & & 32 \cdot 79 \text{ eV} \\
\text{Na}^{2+}(2s2p^6) + e^- \longrightarrow \text{Na}^+(2s2p^63s) & & -15 \cdot 03 \text{ eV (estimated)} \\
\hline
\text{Na} \longrightarrow \text{Na}^+(2s2p^63s) + e^- & & 70 \cdot 19 \text{ eV}
\end{array}
$$

The energy of the third reaction is assumed to be the same as the negative of the ionization potential of Mg^+. In this case the estimated binding energy is within 1 eV of the experimental value,[9] 71·1 eV.

More recently, Best[10] has pointed out that it should be possible to correlate x-ray absorption edge data with thermodynamic data using the equivalent cores approximation. However, a lack of the required thermodynamic data has prevented such correlations. Use of the equivalent cores approximation in the prediction of chemical *shifts* of core binding energies was first reported[11-13] in 1970, and since then the approximation has been widely applied in the systematization of core ionization data.

B. Prediction of Chemical Shifts

1. *Prediction from Chemical Reaction Energies.* A chemical shift in core binding energy can be considered as the energy of a chemical reaction. For example, the difference between the nitrogen $1s$ energies of gaseous molecular nitrogen and nitrogen dioxide is the energy of the reaction

$$NO_2 + N_2^{+*} \longrightarrow NO_2^{+*} + N_2, \quad \Delta E = \Delta E_B \qquad (1)$$

The asterisks indicate $1s$ holes. To transform this equation into one involving ground-state chemical species, we apply the equivalent cores approximation. We assume that the interchange of the $+6$ nitrogen core (with a $1s$ hole) in NO_2^{+*} and the $+6$ oxygen core in NO^+ involves no energy.

$$NO_2^{+*} + NO^+ \longrightarrow O_3^+ + N_2^{+*}, \quad \Delta E = 0 \qquad (2)$$

Addition of Eqs. (1) and (2) yields Eq. (3).

$$NO_2 + NO^+ \longrightarrow O_3^+ + N_2, \quad \Delta E = \Delta E_B \qquad (3)$$

Thus we find that the binding energy shift can be expressed as the energy (or heat) of an ordinary chemical reaction. The experimental binding energy shift, $E_B(NO_2) - E_B(N_2)$, is 3·0 eV;[14] this may be compared with the energy of reaction (3) as calculated from thermodynamic data, 3·3 eV. Similar comparisons for other nitrogen compounds are shown graphically in Fig. 1, where experimental shifts are plotted against the corresponding thermochemical energies. The straight line has unit slope; the average deviation of the points from the line is ± 0.24 eV.

It should be noted that the type of approximation made in Eq. (2), i.e. that the *interchange* of equally charged cores involves no energy change, is not as severe an approximation as the assumption that the cores are chemically equivalent. Chemical equivalence corresponds to zero energy change for all substitution reactions of the following type.

$$NO_2^{+*} + O^{6+} \longrightarrow O_3^+ + N^{6+*}$$

$$N_2^{+*} + O^{6+} \longrightarrow NO^+ + N^{6+*}$$

However when calculating chemical shifts, it is only necessary to assume that all such reactions involving the replacement of cores of a particular element have the *same* energy change, so that the *difference* in energy for any pair of reactions is zero.

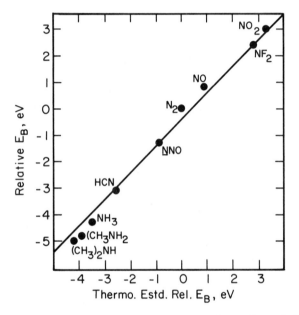

Fig. 1. Plot of nitrogen $1s$ binding energies versus thermodynamically estimated energies. (Reproduced from ref. 16.)

The technique that we have just described—that is, the combination of the equivalent cores approximation with thermodynamic data to predict binding energy shifts—has also been successfully applied to gaseous compounds of boron,[15] carbon,[16] and xenon.[13, 16] Relatively poor correlations were obtained for compounds of oxygen[16] and fluorine,[16] principally because of the inaccuracy of the thermodynamic data employed. Application of the method to solid compounds is more complicated, requires assumptions, and is therefore less accurate than the application to gaseous compounds. However, fairly good correlations have been obtained for solid compounds of boron, carbon, nitrogen, and iodine.[11, 16] In each case the correlation was restricted, because of the nature of the assumptions involved, to molecular compounds or to compounds in which the core-ionized atoms were in anions. It is hoped that further study will yield methods for treating all types of solids by this thermodynamic method.

2. *Prediction from Estimated Chemical Reaction Energies.* When the experimental thermodynamic data required to calculate a binding energy shift by the equivalent cores method are unavailable, the data can sometimes be estimated or calculated with sufficient accuracy for use in the estimation of the binding energy shift. Various non-quantum mechanical methods for obtaining approximate thermochemical data have proved useful for estimating binding energy shifts for compounds of boron,[15] carbon,[12] fluorine,[16] and xenon.[16] It has recently been shown that, for a series of alcohols and a series of amines, the shifts in proton affinity are essentially equal to the corresponding negative shifts in the oxygen $1s$ and nitrogen $1s$ binding energies (see Figs. 2 and 3). These results were taken as evidence that similar relaxation effects occur upon the attachment of a proton to an atom and upon the loss of a core electron from an atom. We shall now show that these results may also be rationalized by a simple extension of the equivalent cores method. We take as an example the oxygen $1s$ shift between two alcohols, ROH and R'OH. This binding energy shift may be taken as the energy of the sum of two reactions.

$$ROH + R'OH_2^+ \longrightarrow ROH_2^+ + R'OH \qquad (PA)_{R'OH} - (PA)_{ROH}$$
$$ROH_2^+ + R'FH^+ \longrightarrow RFH^+ + R'OH_2^+$$
$$\overline{ROH + R'FH^+ \longrightarrow RFH^+ + R'OH \qquad E_B(ROH) - E_B(R'OH)}$$

The energy of the first reaction is simply the proton affinity of R'OH less the proton affinity of ROH. The second reaction, for which we assume $\Delta E = 0$, corresponds to interchanging the isoelectronic groups H_2O and HF between the species ROH_2^+ and $R'FH^+$. The assumption of $\Delta E = 0$ does not require that $\Delta E = 0$ for each of the following processes,

$$ROH_2^+ + HF \longrightarrow H_2O + RFH^+$$
$$R'OH_2^+ + HF \longrightarrow H_2O + R'FH^+$$

but rather that the energies of the two processes are equal. The assumption seems reasonable, particularly if the groups R and R' are similar. The fact that Fig. 2 shows $\Delta(PA) \approx -\Delta E_B$ indicates that the assumption is justified.

It has been shown that even qualitative data can be useful for predicting which of two binding energies is the greater.[19] For example, in the case of chlorine trifluoride, core ionization of the two different types of fluorine

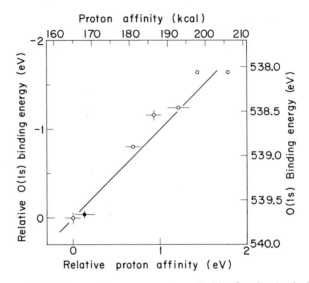

Fig. 2. Oxygen 1s binding energies versus proton affinities for simple alcohols (open circles) and CF_3CH_2OH (filled circle). (Reproduced from ref. 17.)

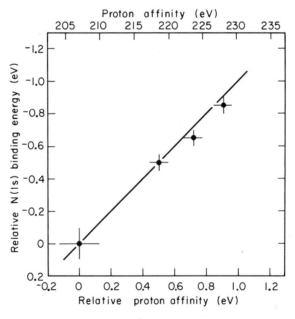

Fig. 3. Nitrogen 1s binding energies versus proton affinities for NH_3 and amines (Reproduced from ref. 17.)

atoms corresponds to the formation of two different neon-containing cations, as shown below.

Equivalent-cores representation of ClF_3 in which the equatorial fluorine has lost a $1s$ electron:

$$F^{-\frac{1}{2}}$$
$$:\overset{\cdot}{\underset{\cdot}{Cl}}{}^{2+} \ Ne$$
$$F^{-\frac{1}{2}}$$

Equivalent-cores representation of ClF_3 in which an axial fluorine has lost a $1s$ electron:

$$F$$
$$:\overset{\cdot}{Cl}{}^{+}-F$$
$$Ne$$

As a rough but reasonable approximation, we may assume that only one orbital of the chlorine atom (a $3p$ orbital) is involved in bonding to the axial ligands and that the $3s$ and two $3p$ orbitals of the chlorine atoms are used for the lone pairs and for bonding to the equatorial ligand. We also assume that, in each case, the bonding between the neon atom and the $ClF_2{}^+$ ion is negligibly weak. Therefore the relative energies of the neon-containing cations are essentially the relative energies of the linear and bent form of $ClF_2{}^+$. Inasmuch as the linear $ClF_2{}^+$ contains only two half-bonds, whereas the bent $ClF_2{}^+$ contains two full bonds, it is clear that the linear ion has the higher energy. Hence one predicts correctly that the binding energy of the equatorial fluorine is higher than that of the axial fluorines. The experimental values[19] are 694·76 eV and 692·22 eV, respectively, corresponding to a 2·54 eV energy difference between the two ClF_2Ne^+ ions.[20]

Quantum mechanical methods for estimating chemical shifts, using the equivalent cores approximation, are attractive for at least three reasons: (1) in general, only closed-shell calculations are required; (2) the difference in energy between two isoelectronic species can be calculated more accurately than the absolute energy of either species, and (3) the calculations can be made for isoelectronic species with identical geometries, corresponding to a photoelectric process in which the nuclei remain fixed. The CNDO/2[16, 21] and MINDO/1[22] methods have given calculated binding energy shifts which are linearly correlated with the experimental shifts but which are in poor quantitative agreement with the experimental shifts. The poor quantitative agreement is probably due to the failure of the semiempirical methods to give accurate energies for the ionic species. On the other hand, *ab initio* calculations have generally given very good results[23-25] In comparisons of experimental data with predictions based on Koopmans' Theorem, hole-state calculations, and equivalent core calculations, the best results are consistently obtained with the equivalent cores calculations. The only poor result by the latter method was obtained for the carbon $1s$ shifts between CO and transition metal carbonyl complexes.[26] This result was probably due to the inadequacy of the computational methods used for the relatively complicated transition metal carbonyl complexes.

C. Other Aspects of the Method

1. *Relaxation Energy.* When a core electron is ejected from an atom in a molecule, the other electrons in the molecule tend to migrate toward the positive hole. In the early days of x-ray photoelectron spectroscopy, it was not known with certainty whether or not this electronic relaxation is complete in the time required for the ejected electron to leave the vicinity of the molecule—that is, it was not known whether or not the measured binding energy corresponds to the difference in energy between the ground state and a relaxed hole state. The equivalent cores thermodynamic method for estimating chemical shifts involves the use of thermodynamic data for ordinary ground-state ("relaxed") chemical species. Therefore we may conclude that, at least to the accuracy with which chemical shifts can be predicted by the equivalent cores method, measured binding energies do correspond to the formation of electronically relaxed hole states.[11]

Three-dimensional perspective plots of electron density, obtained from *ab initio* calculations, permit a qualitative visualization of the electronic relaxation accompanying core electron ejection. In Fig. 4 is shown a plot of the *difference* in electron density between the ground-state carbon monoxide molecule and the same molecule with an oxygen $1s$ hole.[27] The two atoms lie on the grid plane; surfaces above the plane correspond to regions where electron density decreases upon ionization, and surfaces below the plane correspond to regions where electron density increases upon ionization. It can be seen that considerable electron density is lost from a shell immediately around the nucleus and that smaller amounts of electron density are lost from the internuclear region and from the outer periphery of the oxygen atom. Most of the increase in electron density occurs in the valence shell of the oxygen atom. Figure 5 is a plot of the *valence* electron density of CO^{+*} less the *valence* electron density of CF^+. The changes are rather small except in the region of the $1s$ shell. If the O^{7+*} and F^{7+} cores were exactly equivalent, the surface of Fig. 5 would be perfectly flat.

2. *Comparison with the Potential Model.* Binding energy shifts are largely caused by changes in the electrostatic potential experienced by core electrons on going from one molecule to another. On this basis, Basch[28] and Schwartz[29] developed a quantum mechanical potential model for correlating chemical shifts. Shirley[30] has pointed out that, to a certain approximation, the potential

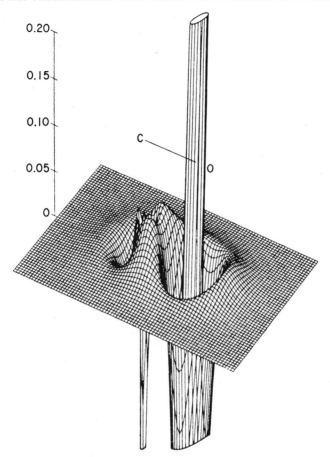

Fig. 4. Electron density difference plot, $CO - CO^{+*}$, derived from the molecule and ion SCF wave functions. The vertical axis is in electrons per cubic atomic units. (Reproduced from ref. 27.)

model and the equivalent cores thermodynamic method are based on the same assumption. However, the potential model derived by Schwartz involves ground-state potentials, whereas the thermodynamic method takes electronic relaxation into account; hence the two methods are not equivalent. As we shall later show, the potential model *can* be modified to include relaxation energy—but by an application of the equivalent cores approximation!

3. *Calculations for Monatomic Ions.* Core binding energies for gaseous ions have not yet been measured. However, quantum mechanical calculations can give binding energies for gaseous atoms and ions which, although somewhat uncertain on an absolute scale, can be used to calculate accurate chemical shifts. Thus Siegbahn *et al.*[31] have calculated the $1s$ binding energy shifts for

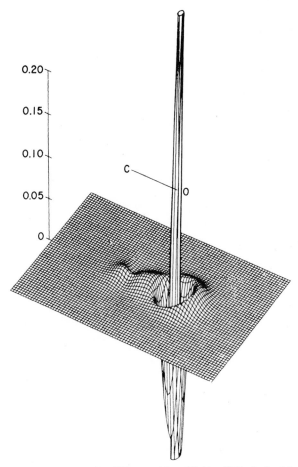

Fig. 5. Valence electron density difference plot, $CO^{+*} - CF^+$, derived from SCF wave-functions. The vertical axis is in electrons per cubic atomic units. (Reproduced from ref. 27.)

several ions of sulfur and chlorine. However, the same chemical shifts can be very easily obtained by the equivalent cores method using optical spectro-scopic data.[4] For example, the chemical shift between Cl^+ and Cl is taken as the difference between the second ionization potential of argon and the first ionization potential of chlorine:

$$Cl^+ + Ar^+ \longrightarrow Cl + Ar^{2+}$$

In Table I the calculated chemical shifts, both with and without inclusion of electronic relaxation energy, may be compared with the corresponding

TABLE I

Calculated $1s$ Shifts for Gaseous Ions

Atom	Q	SCF–HFS calculations[31]		Equivalent-cores calculation
		Without relaxation	With relaxation	
S	-1	-9.2	-11.3	-10.9
	0	0	0	0
	$+1$	12.3	13.8	13.4
	$+2$	27.1	29.8	29.9
	$+3$	44.0	48.1	48.4
	$+4$	62.9	68.5	68.9
	$+6$	109.1	115.0	119.4
Cl	-1	-10.3	-12.4	-12.1
	0	0	0	0
	$+1$	13.4	14.9	14.6
	$+2$	29.2	32.2	31.7
	$+3$	47.2	51.5	51.6
	$+5$	89.2	96.4	96.6
	$+7$	141.0	148.4	153.1

equivalent-cores values. As expected, the values calculated including relaxation energy are in better agreement with the equivalent-cores values. The average deviation between corresponding values (omitting the S^{6+} and Cl^{7+} values, which are in poor agreement) is ± 0.25 eV. This good agreement is surely not fortuitous and attests to the general accuracy of both methods.

4. *Vibrational Excitation of Molecules.* The thermodynamic equivalent-cores method for predicting chemical shifts for molecules involves an approximation not required in the case of monatomic ions. Although the photoelectric process forms hole-state molecule-ions with geometries the same as those of the ground-state molecules, thermodynamic data for the equilibrium geometries of the isoelectronic species are used in the calculations. In general, the equilibrium geometries of the isoelectronic species are different. For example, the equilibrium bond distances in CN^-, CO, and N_2 are 1.14, 1.128, and 1.098 Å, respectively. From the potential energy-bond distance data for CO and N_2 we calculate that, because of the neglect of the strain energy of the excited states, the thermodynamic equivalent-cores method is in error by about 0.04 eV and 0.13 eV in the case of the nitrogen $1s$ and carbon $1s$ binding energies of CN^-, respectively.[11] Because most of the bond distance changes accompanying the core ionization of a given element are in the same

direction, errors of this type tend to cancel when binding energy *shifts* are calculated. However, Shaw and Thomas[32] point out a case in which the error may be fairly serious. The fluorine $1s$ shift between HF and F_2 corresponds to the following reaction.

$$F_2 + HNe^+ \longrightarrow NeF^+ + HF$$

Calculations[33] indicate that the equilibrium bond length of HNe^+ differs from that of HF by about 7%, whereas the bond length of NeF^+ differs from that of F_2 by about 16%. From the calculated potential energy curve for NeF^+, one calculates that NeF^+ with a bond distance equal to that of F_2 would be 0·7 eV above its ground-state energy. It should be noted that the calculated bond distances of NeF^+, as well as the potential energy curve, are quite uncertain. Unfortunately the energy of formation of NeF^+ is not known with enough certainty to permit a meaningful comparison of the equivalent-cores estimate with the experimental chemical shift. However, Shaw and Thomas did find essentially perfect agreement between the equivalent-cores estimate and the experimental shift in the case of the fluorine $1s$ shift between HF and the gaseous F^- ion.

Probably the most important effect of the change in equilibrium geometry accompanying core ionization of a molecule is the broadening of the photo-line. Spectra for molecules frequently show asymmetric lines under high resolution. This asymmetry has been attributed to Franck–Condon transitions between ground and hole states having different geometries.[34, 35] For example, highly accurate calculations[36] for methane have shown that the loss of a carbon $1s$ electron should cause a reduction in the equilibrium C—H bond distance of 0·05 Å. This displacement is sufficient to cause the appearance of two excited vibrational bands in the spectrum and accounts for the asymmetry of the XPS line. Application of the equivalent-cores approximation would have made the calculations much simpler and would probably have given results of comparable accuracy. The equilibrium bond distance of NH_4^+ is 0·057 Å shorter than that of CH_4, and accurate potential energy-bond distance data are available for both species.

The equivalent-cores approximation, in combination with appropriate potential energy data, has been used to account quantitatively for the fact that the carbon $1s$ line of CO is much broader (FWHM 0·82 eV) than the nitrogen $1s$ line of N_2 (FWHM 0·46 eV).[34, 35] In each case, the equivalent-cores stand-in for the hole-state ion is NO^+. The equilibrium bond distance of NO^+ is 0·066 Å shorter than that of CO and 0·036 Å shorter than that of N_2. The fact that the very small difference in bond distance between CO and N_2 causes a large difference in the line width is due to the fact that vibrational broadening is quadratically, not linearly, related to the relative displacement of the potential curves.

In general, a core ionization causes a contraction of the valence electron cloud in the immediate vicinity of the atom, corresponding to a reduction of both the atomic size and the equilibrium bond distances. This contraction of the equilibrium distances is favored when the atom which undergoes core ionization is negatively charged and hence has an easily contracted electron cloud. Thus we can account for the fact that the carbon $1s$ line of CH_4 is broader than that of CF_4, that the sulfur $2p$ line of the terminal sulfur atom in the $S_2O_3^{2-}$ ion is broader than that of the central sulfur atom in that anion, and that the nitrogen $1s$ line of NH_4^+ is broader than that of NO_3^{-} [34] However, it is important to recognize that this correlation of atomic charge with line width is valid only when the formal bond orders of all the bonds in the molecule are unchanged upon ionization of the core. Two examples in which the latter condition is not fulfilled are discussed in the following paragraphs.

When a core electron is lost from the terminal nitrogen atom of N_2O, the equilibrium N—N bond distance probably increases and the equilibrium N—O bond distance probably decreases. This conclusion is reached by consideration of the equilibrium bond distances in N_2O ($r_{N-N} = 1·129$ Å and $r_{N-O} = 1·187$ Å) and NO_2^+ ($r_{N-O} = 1·154$ Å). The same conclusion may be reached by consideration of the valence-bond resonance structures for these isoelectronic species. In N_2O, the N—N bond order is between 2 and 3 and the N—O bond order is between 1 and 2 because of the predominance of the resonance structures

$$N{\equiv}N^+{-}O^- \longleftrightarrow {}^-N{=}N^+{=}O$$

In $O{=}N^+{=}O$, the N—O bond order is exactly 2. Thus we conclude that the ionization of the terminal nitrogen atom of N_2O produces an ion in a strained configuration (in a steep region of the potential energy surface) and that the photoline should be vibrationally broadened. On the other hand, core ionization of the middle nitrogen atom of N_2O produces a species for which we may write the following resonance structures, using the equivalent-cores approximation

$$N{\equiv}O^{2+}{-}O^- \longleftrightarrow {}^-N{=}O^{2+}{=}O$$

The relative weights of these structures are probably similar to those for the analogous structures of N_2O. For this reason, and also because the middle nitrogen atom is positively charged and not subject to much contraction, the core-hole ion is produced in a relatively unstrained state and the photoline for the middle nitrogen atom is not expected to be strongly vibrationally broadened. Indeed, these predictions are in accord with the facts: the FWHM of the terminal nitrogen line is $0·70$ eV; that for the middle nitrogen atom is $0·59$ eV.[34]

The carbon $1s$ spectrum of C_3O_2 consists of a relatively broad line corresponding to the outer carbonyl carbon atoms and a relatively narrow line corresponding to the middle carbon atom.[37] The ground state of the molecule can be represented by the valence bond structure $O{=}C{=}C{=}C{=}O$, with small contributions from the resonance structures

$$^+O{\equiv}C{-}C^-{=}C{=}O \longleftrightarrow O{=}C{=}C^-{-}C{\equiv}O^+$$

The ion formed by core ionization of the middle carbon atom can be represented, using the equivalent-cores approximation, by the structure $O{=}C{=}N^+{=}C{=}O$. However, because of the greater electronegativity of the nitrogen atom, the contribution of the resonance structures

$$^+O{\equiv}C{-}N{=}C{=}O \longleftrightarrow O{=}C{=}N{-}C{\equiv}O^+$$

is somewhat greater than the contribution of the analogous structures in the ground state. This change in the weighting of the resonance structures tends to increase the equilibrium $C{-}C$ bond lengths. However, this tendency is counteracted by the tendency for the middle carbon atom (which is more negatively charged than the outer ones) to contract and shorten the equilibrium $C{-}C$ bond distances. The result is a negligible change in the equilibrium $C{-}C$ distances, a slight shortening of the equilibrium $C{-}O$ distances, and a relatively narrow photoline. The ion formed by ionization of an outer carbon atom can be represented mainly by the structure $O{-}C{=}C{=}N^+{=}O$. However, because of electrostatic interactions, the contribution of the resonance structure $^+O{\equiv}C{-}C{\equiv}N^+{-}O^-$ is probably greater, and that of the resonance structure $O{=}C{=}C^-{-}N^+{\equiv}O^+$ is probably less, than the contributions of the analogous structures in the ground-state C_3O_2 molecule. The outer carbon atoms of C_3O_2 are relatively positively charged and would not tend to contract much on core ionization. Hence the ionization produces the ion in a strained configuration, causing a vibrationally broadened line.

5. *Estimation of Chemical Reaction Energies.* Instead of using thermodynamically derived chemical reaction energies to estimate core binding energy shifts, the procedure can be reversed, and binding energy shifts can be used to estimate thermodynamic data.[11, 16] In this way one can use chemical shift data to calculate the energies of a wide variety of unusual species. For example, the carbon $1s$ chemical shift between CH_4 and CH_3NH_2 corresponds to the energy of the following reaction.[38]

$$CH_3NH_2 + NH_4^+ \longrightarrow NH_2NH_3^+ + CH_4$$

In this reaction, the heats of formation are known for all the species except $NH_2NH_3^+$. By combining the known heats of formation with the experimental binding energy shift, we calculate $\Delta H_f^0 = 169$ kcal/mol for $NH_2NH_3^+$. From this value and the known heats of formation of N_2H_4 and H^+, we calculate the proton affinity of hydrazine to be 221 kcal/mol.

The absorption spectrum of N_2 in the 30 Å region shows lines corresponding to the excitation of $1s$ electrons to outer-shell orbitals of the nitrogen molecule.[39] By using the equivalent-cores approximation, these excited levels have been interpreted as the ground-state and excited levels of the NO molecule. Similar interpretations have been made for energy-loss spectra for 2500 eV electrons in N_2,[40] CO,[40] CH_4,[41] NH_3,[41] and H_2O.[41] From the measured energies for the latter three species, it was possible to estimate the heats of formation of the NH_4, OH_3, and H_2F radicals.[41]

III. CORRELATION WITH ATOMIC CHARGE

A. Atomic Charge and the Potential Model

It was early recognized that a core binding energy of an atom should be related to the effective charge of the atom. From simple electrostatic considerations, one would expect that the energy for removing an electron from an atom would be inversely related to the valence electron density on the atom; that is, that binding energy would increase with increasing positive atomic charge and decrease with increasing negative atomic charge. Consideration of a simple shell model of the atom leads to the linear relation[31]

$$E_B = kQ + l \tag{4}$$

where E_B is the binding energy for a particular core level in an atom, Q is the charge of the atom, and k and l are empirical constants. Many sets of binding energy data can be correlated using this equation. When the compounds involved have closely related structures and similar types of bonding, fairly good correlations are obtained using even extremely crude methods for estimating atomic charges. When the compounds involved differ markedly in structure and bonding, relatively poor correlations are obtained and little improvement in the correlation is obtained by using more sophisticated methods for estimating atomic charges.

In general, to obtain a good correlation of binding energy with atomic charge, one must take account of the electrostatic potential due to the charges on all the other atoms in the compound. In other words, one must consider not only the work to remove the electron from the atom which loses the core electron, but also the work to remove the electron from the field of the surrounding charged atoms. This can be accomplished by using the so-called potential model equation[7]

$$E_B = kQ + V + l. \tag{5}$$

where V is the coulomb potential energy at the hypothetical vacated site of the atom in the midst of the other atoms of the compound. This equation gives good results when (a) a reasonably good method for estimating atomic charges is used and (b) the electronic relaxation energies of the compounds are very similar in magnitude. When one correlates compounds for which the relaxation energies differ significantly, the equation must be modified to include a term that accounts for the difference in relaxation energy:[42]

$$E_B = kQ + V + l + E_R \tag{6}$$

B. Solid Compounds

1. *Approximate Correlations.* The Pauling method[43] for estimating atomic charges involves the assumption that the partial ionic character of a bond is given by the relation

$$I = 1 - e^{-0.25(\Delta_x)^2}$$

where Δ_x is the difference between the electronegativities of the bonded atoms. Many investigators have used this method, or modifications of it, to calculate atomic charges for the correlation of binding energies according to Eq. (4). For example, Matienzo *et al.*[44] found a good correlation between the calculated charge on nickel atoms in some simple salts of nickel(II) and the corresponding nickel $2p_{\frac{3}{2}}$ binding energies. A plot of their data is shown in Fig. 6. The success of this correlation is probably partly due to the facts that the nickel ions in most of the compounds are similarly coordinated (octahedrally) and the ligands are relatively simple. Hughes and Baldwin[45] found a good correlation between molybdenum atom charges in a variety of triphenylphosphine molybdenum complexes and the molybdenum $3d_{\frac{3}{2}}$ binding energies. A plot of their data is given in Fig. 7. In this case the good correlation is probably attributable to the fact that the compounds are all neutral complexes with similar coordination numbers.

Remarkably small chemical shifts are observed for monatomic ions and for atoms in polyatomic ions upon going from one salt to another by changing the counter-ion. For example, in a series of sixteen different potassium salts, the overall spread in the K $2p$ binding energy (between KCl and $K_2[Pt(NO_2)_4-Cl_2]$) is only 1·7 eV.[46] Both the N $1s$ and P $2p$ binding energies of a wide variety of salts containing the bis(triphenylphosphine)iminium cation, $N[P(C_6H_5)_3]_2^+$, differ by only a few tenths of an electron volt.[47] The F $1s$ binding energies in UF_4 and LiF differ by only 1·7 eV.[48] The minor effect of crystal environment on chemical shift, illustrated by these data, is particularly surprising when the changes in the theoretical point-charge lattice potentials are considered. In the case of simple salts such as KCl and LiF, if it is assumed

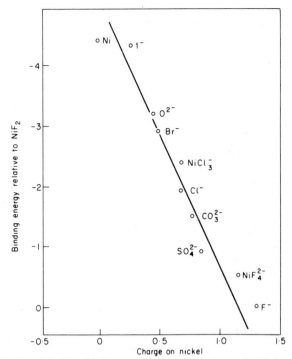

Fig. 6. Relative nickel $2p_{\frac{3}{2}}$ binding energies versus estimated nickel atom charges for some simple nickel(II) compounds. (Reproduced from ref. 44.)

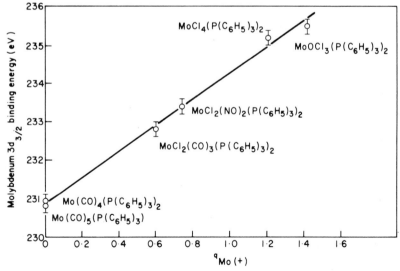

Fig. 7. Molybdenum $3d_{\frac{3}{2}}$ binding energies versus estimated molybdenum atom charges for some molybdenum complexes. (Reproduced from ref. 45.)

that the ions bear unit charges, chemical shifts between two different salts of the same cation or between two different salts of the same anion should be readily calculable from the appropriate sums of Q/r terms for the other ions in the lattice. However, the calculated differences in these Madelung potentials differ greatly from the observed chemical shifts. Citrin et al.[49] have shown that the discrepancies can be markedly reduced by considering the mutual polarization of the ions. That is, the use of unit point charges is not valid. The inclusion of polarization effects is essentially equivalent to the use of fractional atomic charges combined with the calculation of chemical shifts by means of Eq. (5). However, calculations which take these effects into account are quite complicated, particularly for more complicated compounds, and most investigators do not attempt to account for chemical shifts in ionic compounds using Eq. (5).

The fact that small chemical shifts are observed for an atom in an ion on going from one salt to another in spite of large differences in the calculated Madelung potentials illustrates what appears to be a general rule: by a combination of the effects of crystal packing and charge transfer by polarization, a given ion achieves approximately the same potential energy in all its salts. This rule applies even to ions which occupy structurally different sites within the same crystal. For example, Hayes and Edelstein found no apparent broadening of the F $1s$ line of LaF_3 even though the solid contains crystallographically different fluoride ions for which the simple point charge lattice potentials differ by more than 3 eV.[48]

In some complexes, it is possible to distinguish bridging and terminal ligand atoms. Thus Hamer and Walton[50] have observed separate peaks for bridging and terminal chloride ions in cluster complexes of rhenium(III) and molybdenum(II).

2. *Internal Referencing.* The comparison of a core binding energies of atoms in different solid compounds is fraught with difficulty. In the case of nonmetallic samples, one does not know what the effective Fermi levels are, and therefore one cannot put the binding energies on the basis of a common reference level. The usual, probably poor, solution to the problem is to assume that the work functions of all the samples are identical. Nonmetallic samples are also subject to an unknown amount of electrostatic charging, and although various techniques have been devised to minimize errors due to this effect, absolute values of binding energies for solids are always uncertain to some extent because of this charging. Both the work function problem and the electrostatic charging problem can be completely avoided by measuring chemical shifts between core levels of atoms in the same sample. Several studies based on such internal referencing are discussed in the following paragraphs.

The topic of "mixed valence compounds" has become very popular in recent years.[51] X-Ray photoelectric ionization is believed to take place in a time interval of about 10^{-18} s; therefore separate binding energy peaks are possible for atoms of different oxidation states in structurally equivalent sites if the lifetime of a given electronic configuration is greater than about 10^{-18} s. The higher the barrier to electron exchange between the equivalent sites, the longer the lifetime of a given state. R. E. Connick[52] has pointed out to the author that in all mixed valence compounds one would expect electron transfer processes between atoms to be slower than the x-ray photoelectric process. This conclusion is based in part on the fact that the kinetic energy of a valence electron is of the order of 10–20 eV, whereas the kinetic energy of the ejected core electron is usually much higher. If this is true, then one should always observe two core lines for a mixed valence system except in cases of poor resolution.

The antimony $3d$ spectrum of Cs_2SbCl_6 shows two peaks, separated by 1·80 eV, presumably identifiable with the +3 and +5 oxidation states.[53] The fact that the antimony atoms are not all of +4 oxidation state is a particularly interesting result. In this salt, the Sb atoms occupy sites which are almost, but not exactly, identical.[54]

Citrin[55] obtained the ruthenium core spectra shown in Fig. 8 for salts containing the ions $[(NH_3)_5Ru(pyr)Ru(NH_3)_5]^{4+, 5+, 6+}$ and $[Cl(bipy)_2-Ru(pyr)Ru(bipy)_2Cl]^{2+, 3+, 4+}$ (pyr = pyrazine; bipy = bipyridine). In these symmetric complexes the ruthenium oxidation states can be represented as [II, II], [II, III], and [III, III] The chemical shift between the [II, II] and [III, III] peaks is greater in the case of the penta-amine complexes than in the case of the bipyridine complexes, presumably because of greater ruthenium valence electron delocalization in the bipyridine [II, II] complex than in the penta-amine [II, II] complex. Both of the [II, III] complexes show separate peaks for the two different oxidation states. In the case of the penta-amine complexes, the separation between the [II, III] peaks is less than that between the [II, II] and [III, III] peaks. This result is probably caused by the delocalization of the extra electron of the [II, III] complex principally on the pyrazine molecule coordinated to both ruthenium atoms. In the case of the bipyridine [II, III] complex, in which the extra electron density can be delocalized onto the bipyridine ligands attached to the ruthenium(II) atom, the peak separation is essentially the same as that between the corresponding [II, II] and [III, III] peaks.

Su and Faller[56, 57] have shown that the geometric aspects of the bonding of sulfoxides (R_2SO) and nitrosyl (NO) groups to metal atoms can be correlated with differences in the binding energies of different atoms in the ligands. In the case of sulfoxide complexes, which contain either metal—oxygen or metal—sulfur bonds, the absolute S $2p_{\frac{3}{2}}$ and O $1s$ binding energies show very

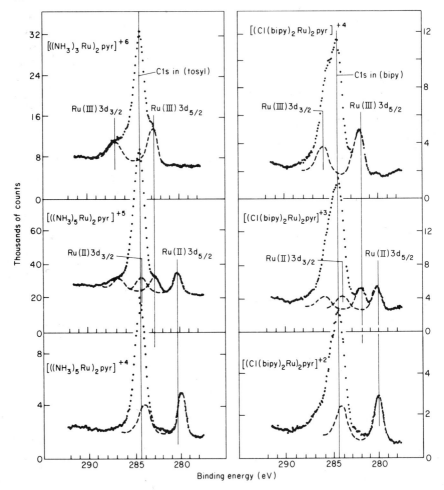

Fig. 8. x-Ray photoelectron spectra of $[(NH_3)_5Ru(pyr)Ru(NH_3)_5]^{4+, 5+, 6+}$ and $[Cl(bipy)_2Ru(pyr)Ru(bipy)_2Cl]^{2+, 3+, 4+}$ salts. The C 1s electron binding energies in both systems have been assigned as 284·4 eV for purposes of comparison. (Reproduced from ref. 55.)

little correlation with the mode of bonding.[56] However, the *difference* between the O 1s and S $2p_{\frac{3}{2}}$ binding energies does show a marked correlation, as shown by the data plotted in Fig. 9. When the metal is bonded to the oxygen atom of the sulfoxide, the O $1s - S\ 2p_{\frac{3}{2}}$ difference is $\sim 365·8$ eV, whereas when the metal is bonded to the sulfur atom of the sulfoxide, the difference is $\sim 365·0$ eV. This type of correlation is reasonable in view of the formal charges on the oxygen and sulfur atoms in the two geometries:

$$\begin{array}{ccc} O^- & & R \\ | & & | \\ R-S^{2+}-M^- & & R-S^+-O-M^- \\ | & & \\ R & & \end{array}$$

In the case of nitrosyl complexes, which can be "linear" (that is with the metal—N—O bond angle near 180°) or "bent" (with the bond angle near 120°), the O 1s binding energy minus the N 1s binding energy is in the range of 132 ± 1 eV for linear nitrosyls and 128 ± 2 eV for bent nitrosyls.[58] It is difficult to rationalize these data on the basis of the formal charges in the two valence bond structures as one would ordinarily write them:

$$^-M-N^+\equiv O^+ \longleftrightarrow M=N^+=O$$

$$^-M-N{\atop\diagdown\atop O}$$

From the formal charges in these structures one might have predicted that the O 1s−N 1s difference would be greater for the bent nitrosyls rather than for the linear nitrosyls. Perhaps the observed behavior is due to an extraordinarily large relaxation energy for the oxygen atom in the bent nitrosyls. By applying the equivalent cores approximation, we may represent the O 1s ionized state of the bent complex as follows:

$$^-M-N{\atop\diagdown\atop F^+} \longleftrightarrow M=N{\atop\diagdown\atop F}$$

The left-hand structure has formal charges which contradict the relative electronegativities of the metal and fluorine, and there would be a strong tendency for electron density to flow from the metal d orbitals to make the right-hand structure dominant. This electron flow corresponds to a large (negative) E_R and a low O 1s binding energy.

C. Gases

1. *Qualitative Interpretation of Chemical Shifts.* It is often possible to interpret core binding energy data in terms of valence electron distribution without direct recourse to a theoretical method for calculating binding energies. Such interpretation is usually at least implicitly based on a relation such as Eq. (6). However, by the appropriate choice of reference binding energies the changes in the potential term, V, and the relaxation energy term, E_R, can be assumed to be very small. Thus shifts in binding energy can be directly, although qualitatively, related to changes in atomic charge.

There has been considerable discussion regarding the nature of the bonding in transition metal carbene complexes.[58] A carbene complex in which the

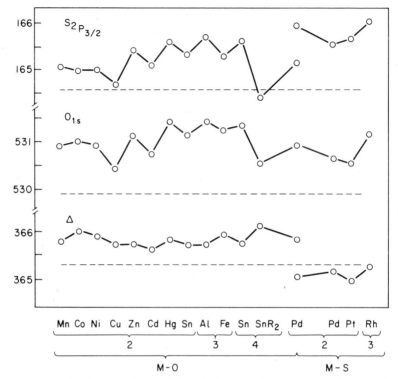

Fig. 9. Core binding energies, and their differences, for some sulfoxide complexes. (Reproduced from ref. 56.)

carbene carbon atom is bonded to a heteroatom with nonbonding electrons, such as $(OC)_5CrC(OCH_3)CH_3$, may be represented as a resonance hybrid of the following structures, I–III:

$$(OC)_5Cr=C\begin{smallmatrix}OCH_3\\CH_3\end{smallmatrix} \longleftrightarrow (OC)_5Cr\overset{-}{=}C\begin{smallmatrix}\overset{+}{O}CH_3\\CH_3\end{smallmatrix} \longleftrightarrow (OC)_5Cr\overset{-}{=}C\begin{smallmatrix}\overset{+}{O}CH_3\\CH_3\end{smallmatrix}$$

I   II   III

Carbon-13 NMR data,[59] the strongly electrophilic reactivity behavior of the carbene carbon atom,[60] and other chemical and physical data[58] have been interpreted as evidence for the importance of structure III, that is, as evidence for an exceptionally positive charge on the carbene carbon atom. However, the C 1s spectrum of the gaseous molecule shows that the binding energies of all three carbon atoms of the $C(OCH_3)CH_3$ group are lower than the binding energy of the carbonyl carbon atoms.[61] Therefore one concludes that the carbene carbon atom is not exceptionally positively charged. This conclusion is based on the reasonable assumption that the potential term and electronic relaxation energy for the carbene carbon atom are essentially the same as those for the carbonyl carbon atoms. The data underscore the facts that NMR chemical shifts of carbon atoms bound to transition metals are very difficult to interpret[62] and that chemical reactivity data do not always correlate well with the properties of ground-state molecules. In this study, the C $1s$ spectrum was obtained for a gaseous sample rather than a solid sample to avoid the spurious C $1s$ line due to hydrocarbon contamination which is generally observed in the spectra of solid samples.

The study of the core binding energies of an isoelectronic series of compounds has the advantage that the nature of the bonding in such compounds changes in a fairly systematic way with changes in the atomic numbers of the atoms. To a good approximation, the electronic relaxation energy for an atom of a particular element in a series of isoelectronic, isostructural compounds can be assumed to be constant. Therefore chemical shifts can be ascribed to changes in atomic charge and potential. Consider the five series of isoelectronic compounds listed in Table II.[63] In the first four series, halogen

TABLE II

Oxygen $1s$, Chlorine $2p_{\frac{3}{2}}$, Carbon $1s$, and Fluorine $1s$ Binding Energies[63]

	E_B (eV)			
Compound	O $1s$	Cl $2p_{\frac{3}{2}}$	C $1s$	F $1s$
TiCl$_4$		205·66 }0·16		
VOCl$_3$	538·73 }0·16	205·93 }0·08		
CrO$_2$Cl$_2$	538·89 }0·54	206·01 }1·27		
MnO$_3$Cl	539·43	207·28		
SiCl$_4$		206·77 }0·39		
POCl$_3$	537·80 }1·54	207·16 }0·16		
SO$_2$Cl$_2$	539·34	207·32		
Si(CH$_3$)$_4$			289·61 }0·96	
P(CH$_3$)$_3$O	535·88 }1·79		290·57 }1·01	
S(CH$_3$)$_2$O$_2$	537·67		291·58	
SiF$_4$			694·56 }0·8	
POF$_3$	538·9 }1·4		695·4 }0·0	
SO$_2$F$_2$	540·3 }0·7		695·4 }−1·4	
ClO$_3$F	541·0		694·0	
SiF$_4$			694·56 }0·8	
POF$_3$			695·4 }−0·4	
SNF$_3$			695·0	

atoms or methyl groups are replaced with oxygen atoms as the atomic number of the central atom is increased stepwise. The last series of compounds is generated by the hypothetical stepwise transfer of protons from one of the fluorine nuclei of SiF_4 to the silicon nucleus of that molecule. It is reasonable to assume that, on progressing down through any one of the five series, the charge of the central atom increases. This increase in charge of the central atom would be expected to cause increased electron withdrawal from a given type of ligand atom and is probably responsible for the increase in the oxygen, chlorine, and carbon binding energies within each series.

Notice, however, that the fluorine $1s$ binding energies do not follow the predicted trend. In both the fourth and fifth series, the fluorine binding energies increase to a maximum value and then decrease. The increase between SiF_4 and POF_3 is probably caused by the dominating effect of the increased positive charge and electronegativity of the central atom. The decrease in fluorine binding energy between SO_2F_2 and ClO_3F and between POF_3 and SNF_3 is apparently due to an increase in the negative charge of the fluorine atoms. This increase in negative charge tends to decrease the binding energy; clearly this effect is greater than and opposed to the effect of the increase in potential caused by the increase in positive charge on the central atom.

The increase in the negative charge of the fluorine atoms can be explained by the dominating effect of a large increase in the extent of hyperconjugation. Hyperconjugation (sometimes called "no-bond resonance") corresponds to the contribution of resonance structures of the following type.

$$\begin{array}{ccc} F & F & O^- \\ | & | & | \\ O{=}P^+{-}F & O{=}S^{2+}{-}O^- & O{=}Cl^{3+}\ F^- \\ F^- & F^- & | \\ & & O^- \end{array}$$

This type of bonding transfers some of the negative formal charge of oxygen atoms to fluorine atoms. On going from POF_3 to ClO_3F, the number of oxygen atoms which can transfer negative charge increases from one to three and the number of fluorine atoms which can accept negative charge decreases from three to one. Consequently, the fluorine atoms acquire an increasing amount of negative formal charge in this series. The effect on the actual charge of the fluorine atoms is most marked between SO_2F_2 and ClO_3F. On going from POF_3 to SNF_3, the hyperconjugation increases markedly because of the increased donor character (decreased electronegativity) of the nitrogen atom compared with that of the oxygen atom.

The effect of hyperconjugation on atomic charges is more apparent in the fluorine compounds than in the chlorine and methyl compounds because of the greater electronegativity of fluorine atoms and the corresponding greater ability of fluorine atoms to accept negative charge. It is possible that hyperconjugation occurs even in methyl compounds such as $P(CH_3)_3O$ and $S(CH_3)_2O_2$, but that the effect is not strong enough to cause a reversal in the carbon binding energies like that observed in the fluorine binding energies. Hyperconjugation is probably not significant in transition metal compounds such as $VOCl_3$ and CrO_2Cl_2. Such compounds, in which metal *inner d* orbitals are importantly involved in the bonding, can be represented by structures such as the following

$$\begin{array}{ccc} & Cl & & Cl \\ & | & & | \\ O=V-Cl & & O=Cr=O \\ & | & & | \\ & Cl & & Cl \end{array}$$

The relatively small increases in oxygen $1s$ binding energies observed for the first series of compounds in Table II are consistent with this interpretation.

2. *The "Transition State" Method.* Explicit inclusion in Eq. (6) of the relaxation energy term, E_R, can be obviated by using modified values of Q and V corresponding to a hypothetical "transition state" molecule which has a valence electron distribution halfway between that of the initial molecule and that of the core-ionized molecule.[64-66]

$$E_B = kQ + k\Delta Q^* + V + \Delta V^* + l$$

Here ΔQ^* and ΔV^* are the changes in Q and V on going from the initial to the transition state. Hence

$$E_R = k\Delta Q^* + \Delta V^* = k(Q_f - Q - 1)/2 + (V_f - V)/2 \qquad (7)$$

where Q_f and V_f are the values of Q and V for the core-ionized molecule. These values can be estimated using the approximation of equivalent cores.

A method similar to the transition-state method described above has been used by Davis and Shirley,[64, 67] in combination with the CNDO method, to calculate relaxation-corrected binding energy shifts for compounds of carbon, nitrogen, and oxygen. The correlation of nine nitrogen compounds was significantly better with the inclusion of the relaxation energy (standard deviation 1·30 eV) than without inclusion of the relaxation energy (standard deviation 2·35 eV). The improvements of the correlations of the carbon and oxygen compounds were not as marked. Shirley[68] has also shown that the transition-state method, combined with calculated or experimental data for ground-state atoms, can be used to estimate relaxation energies for the core ionization of atoms. Howat and Goscinski[69] have shown that calculations for the transition-state method can be simplified by calculating, for each molecule, the charge distribution for the hypothetical transition state rather than for both the ground state and the core-ionized state. They used "pseudo-atom" interpolated CNDO parameters for the atom which undergoes core ionization.

Jolly and Perry[65, 66] have used the transition state method and the CHELEQ electronegativity equalization procedure for calculating atomic charges to correlate the binding energies of a wide variety of gaseous molecules. Equation (6) was used for the correlations; the E_R values were obtained from Eq. (7). The binding energies included 64 carbon $1s$, 20 nitrogen $1s$, 24 oxygen $1s$, 28 fluorine $1s$, 11 silicon $2p$, 13 phosphorus $2p_{\frac{3}{2}}$, 16 sulfur $2p_{\frac{3}{2}}$, 16 chlorine $2p_{\frac{3}{2}}$, 8 germanium $3p_{\frac{3}{2}}$, 10 bromine $3d_{\frac{5}{2}}$, and 5 xenon $3d_{\frac{5}{2}}$ binding energies. The least-squares evaluated k values, the corresponding correlation coefficients, and the standard deviations (with and without inclusion of E_R) are listed in Table III. The correlations, as measured by the standard deviations and the correlation coefficients, are quite good and attest the usefulness of the CHELEQ method. By comparison of the standard deviation with and without E_R it can be seen that, in most cases, inclusion of the relaxation energy markedly improved the correlation.

TABLE III

Parameters of Potential Model Correlations[66]

Element	k	Correlation coefficient	Standard deviation	Standard deviation no E_R
C	30·50	0·972	0·62	0·81
N	30·69	0·987	0·60	1·01
O	25·50	0·896	0·73	0·59
F	27·95	0·934	0·34	0·26
Si	17·29	0·964	0·47	0·51
P	19·28	0·953	0·89	1·33
S	18·63	0·983	0·70	0·74
Cl	18·24	0·988	0·44	0·52
Ge	15·87	0·984	0·34	0·47
Br	13·32	0·993	0·31	0·38
Xe	12·06	0·992	0·39	0·37

The calculation of the atomic charges of a molecule by the CHELEQ method requires that one first write a single valence bond structure for the molecule.[65, 66] In the case of a molecule for which there is more than one satisfactory valence bond structure, a suitable resonance hybrid structure must be written. Inasmuch as the transition state correlations involve the calculation of atomic charges for the core-ionized molecules as well as the ground-state molecules, resonance structures for the core-ionized molecules must be considered when appropriate. For example, although the structure

$$\begin{matrix} & O \\ & \parallel \\ CH_3-&C-OH \end{matrix}$$

is a satisfactory representation of the ground-state acetic acid molecule, the analogous structure for the molecule in which the carbonyl oxygen atom has lost a $1s$ electron is inadequate. By applying the equivalent cores approximation, we see that two resonance structures are important:

$$\begin{matrix} F^+ & & F \\ \parallel & & \vert \\ CH_3-C-OH & \longleftrightarrow & CH_3-C={}^+OH \end{matrix}$$

It is not possible to predict the relative weights of these resonance structures, and therefore the carbonyl oxygen binding energy of acetic acid could not be included in the data used to obtain the oxygen k and l values. However, using Eq. (6) and the oxygen k and l values obtained from compounds with unambiguous valence structures, one can calculate the weighting of the resonance structures which gives perfect agreement between the experimental and calculated values of the carbonyl oxygen binding energy of acetic acid. The resonance hybrid structure for the core-ionized molecule, calculated in this way, is

$$\begin{matrix} & & F^{+0\cdot 68} \\ & & _{1\cdot 68}\vert \\ CH_3 & \underline{} & C \underline{}^{1\cdot 32} O^{+0\cdot 32} \\ & & H \end{matrix}$$

where the numbers next to the bonds are the bond orders. Similar treatment of the oxygen binding energies of other carbonyl compounds yields analogous resonance hybrid structures for the core-ionized molecules.[70] The order of the bond between the core-ionized oxygen atom (or fluorine atom, in the equivalent-cores approximation) and the carbon atom can be taken as a measure of the π-donor character of the groups attached to the carbonyl group. These calculated bond orders for various carbonyl compounds are listed in Table IV. The lower the bond order, the more negative E_R and the greater the combined π-donor characters of the groups bonded to the carbonyl group.

The question of hyperconjugation in molecules such as POF_3 can be attacked by the transition-state method in combination with CHELEQ charge calculations. Let us assume that the ground state of POF_3 can be represented as a resonance hybrid of the structures

$$
\underset{\underset{F}{|}}{\overset{\overset{F}{|}}{F-P^+-O^-}} \longleftrightarrow \underset{\underset{F}{|}}{\overset{\overset{F^-}{}}{F-P^+=O}} \longleftrightarrow \underset{\underset{F}{|}}{\overset{\overset{F}{|}}{F-P^+=O}} \longleftrightarrow \underset{\underset{F^-}{}}{\overset{\overset{F}{|}}{F-P^+=O}}
$$

and that the oxygen $1s$ and fluorine $1s$ core-hole ions of POF_3 can be represented by resonance hybrids with exactly the same weightings of the individual resonance structures. Then one obtains agreement between the experimental oxygen binding energy and the value calculated from Eq. (6) (using the appropriate k and l values) if one assumes the resonance hybrid structure[70]

$$
\underset{\underset{F^{-0\cdot16}}{\big| {\scriptstyle 0\cdot84}}}{\overset{\overset{F^{-0\cdot16}}{\big| {\scriptstyle 0\cdot84}}}{^{-0\cdot16}F\underline{0\cdot84}P^+\underline{1\cdot48}O^{-0\cdot52}}}
$$

The analogous calculation, using the fluorine binding energy, yields the following resonance structure.

$$
\underset{\underset{F^{-0\cdot15}}{\big| {\scriptstyle 0\cdot85}}}{\overset{\overset{F^{-0\cdot15}}{\big| {\scriptstyle 0\cdot85}}}{^{-0\cdot15}F\underline{0\cdot85}P^+\underline{1\cdot45}O^{-0\cdot55}}}
$$

The fact that both binding energies give essentially the same resonance hybrid structure gives credence to the assumptions involved.

TABLE IV

Calculated C—O Bond Orders in C $1s$-ionized Carbonyl Compounds

Compound	C—O bond order
H_2CO	2·0
CH_3CHO	1·79
$(CH_3)_2CO$	1·70
Cl_2CO	1·70
CH_3CO_2H	1·68
F_2CO	1·64
$(C_6H_5)_2CO$	1·58
$(CH_3O)_2CO$	1·40

Acknowledgement

This work was supported by the National Science Foundation (Grant GP–41661x) and the Energy Research and Development Agency.

REFERENCES

1. A. Hamnett and A. F. Orchard, *Electronic Structure and Magnetism of Inorganic Compounds*, **1**, 36 (1972); S. Evans and A. F. Orchard, *ibid.* **2**, (1973); A. Hamnett and A. F. Orchard, *ibid.* **3**, 218 (1974).
2. W. L. Jolly, *Coord. Chem. Rev.* **13**, 47 (1974).
3. H. W. B. Skinner, *Proc. Roy. Soc.* (*London*), **A135**, 84 (1932). *Also see* E. M. Baroody, *J. Opt. Soc. Am.* **62**, 1528 (1972).
4. We have used modern values for the atomic energy levels, taken from C. E. Moore, "Atomic Energy Levels", Vol. I, NSRDS–NBS 35, National Bureau Standards, Washington, D.C., December 1971, and "Ionization Potentials and Ionization Limits Derived from the Analyses of Optical Spectra", NSRDS– NBS 34, National Bureau of Standards, Washington, D.C., September 1970.
5. Obtained from data in ref. 4, giving weights of 1 and 3 to the $1s2s\ ^1S$ and $1s2s\ ^3S$ states of Li^+.
6. D. B. Adams and D. T. Clark, *J. Electron Spectr. Rel. Phen.* **2**, 201 (1973); D. B. Adams, *J. Electron Spectr. Rel. Phen.* **4**, 72 (1974).
7. K. Siegbahn *et al.* "ESCA Applied to Free Molecules" (North-Holland Publ. Co., Amsterdam, 1969).
8. Adams and Clark[6] assumed that an ion of configuration $(1s^22s^1)$ is a core as far as the $2p$ electrons are concerned. Making this very poor assumption, they concluded that the equivalent cores approximation is invalid for calculating $2s$ binding energies for second-row elements.
9. S. P. Kowalczyk, L. Ley, F. R. McFeely, R. A. Pollak, and D. A. Shirley, *Phys. Rev.* **B8**, 3583 (1973).
10. P. E. Best, *J. Chem. Phys.* **47**, 4002 (1967); *ibid.* **49**, 2797 (1968).
11. W. L. Jolly and D. N. Hendrickson, *J. Amer. Chem. Soc.* **92**, 1863 (1970).
12. W. L. Jolly, *J. Amer. Chem. Soc.* **92**, 3260 (1970).
13. J. M. Hollander and W. L. Jolly, *Acc. Chem. Res.* **3**, 193 (1970).
14. P. Finn, R. K. Pearson, J. M. Hollander and W. L. Jolly, *Inorg. Chem.* **10**, 378 (1971).
15. P. Finn and W. L. Jolly, *J. Amer. Chem. Soc.* **94**, 1540 (1972).
16. W. L. Jolly, in "Electron Spectroscopy", D. A. Shirley, ed. (North-Holland Publ. Co., Amsterdam, 1972), pp. 629–645.
17. R. L. Martin and D. A. Shirley, *J. Amer. Chem. Soc.* **96**, 5299 (1974).
18. D. W. Davis and J. W. Rabalais, *J. Amer. Chem. Soc.* **96**, 5305 (1974).
19. R. W. Shaw, T. X. Carroll, and T. D. Thomas, *J. Amer. Chem. Soc.* **95**, 5870 (1973).
20. The energy difference between the ClF_2^+ ions is actually greater than 2·54 eV because the equatorial neon atom is weakly bonded to the chlorine atom and the axial neon atom is probably completely nonbonded. If we restrict the chlorine $3s$ orbital to the lone pairs, then the linear forms of ClF_2^+ lacks an octet whereas the bent form has a complete octet.
21. D. T. Clark and D. B. Adams, *Nature Phys. Sci.* **234**, 95 (1971).

22. D. C. Frost, F. G. Herring, C. A. McDowell and I. S. Woolsey, *Chem. Phys. Letters*, **13**, 391 (1972).
23. D. T. Clark and D. B. Adams, *J. Chem. Soc. Faraday II*, **68**, 1819 (1972); *J. Electron Spectro. Rel. Phen.* **1**, 302 (1972); also see *Discuss. Faraday Soc.* **54**, 43 (1972).
24. D. B. Adams and D. T. Clark, *Theoret. Chim. Acta*, **31**, 171 (1973).
25. L. J. Aarons and I. H. Hillier, *J. Chem. Soc. Faraday II*, **69**, 1510 (1973).
26. J. A. Connor, M. B. Hall, I. H. Hillier, and W. N. E. Meredith, *J. Chem. Soc. Faraday II*, **70**, 1677 (1973).
27. J. Cambray, J. Gasteiger, A. Streitwieser, and P. S. Bagus, *J. Amer. Chem. Soc.* **96**, 5978 (1974).
28. H. Basch, *Chem. Phys. Letters*, **5**, 337 (1970).
29. M. E. Schwartz, *Chem. Phys. Letters*, **6**, 631 (1970).
30. D. A. Shirley, *Chem. Phys. Letters*, **15**, 325 (1972).
31. K. Siegbahn *et al.*, "ESCA; Atomic, Molecular and Solid-state Structure studied by means of Electron Spectroscopy" (Almqvist and Wiksells AB, Uppsala, 1967).
32. R. W. Shaw and T. D. Thomas, *Chem. Phys. Letters*, **22**, 127 (1973).
33. J. F. Liebman and L. C. Allen, *J. Amer. Chem. Soc.* **92**, 3539 (1970).
34. U. Gelius, *J. Electron Spectr. Rel. Phen.*, **5**, 985 (1974).
35. U. Gelius, S. Svensson, H. Siegbahn, E. Basilier, A. Faxälv, and K. Siegbahn, *Chem. Phys. Letters*, **28**, 1 (1974).
36. W. Meyer, *J. Chem. Phys.* **58**, 1017 (1973).
37. U. Gelius, C. J. Allan, D. A. Allison, H. Siegbahn, and K. Siegbahn, *Chem. Phys. Letters*, **11**, 224 (1971).
38. J. S. Jen and T. D. Thomas, *J. Electron Spectr. Rel. Phen.* **4**, 43 (1974).
39. M. Nakamura *et al.*, *Phys. Rev.* **178**, 80 (1969).
40. G. R. Wight, C. E. Brion, and M. J. Van der Wiel, *J. Electron Spectr. Rel. Phen.* **1**, 457 (1972/3).
41. G. R. Wight and C. E. Brion, *Chem. Phys. Letters*, **26**, 607 (1974).
42. U. Gelius, *Physica Scr.* **9**, 133 (1974).
43. L. Pauling, "The Nature of the Chemical Bond", 3rd edn., Cornell University Press, Ithaca, N.Y., 1960. Also see 2nd edn., 1940, pp. 65–66.
44. L. J. Matienzo, L. I. Yin, S. O. Grim, and W. W. Swartz, *Inorg. Chem.* **12**, 2762 (1973).
45. W. B. Hughes and B. A. Baldwin, *Inorg. Chem.* **13**, 1531 (1974).
46. W. E. Moddeman, J. R. Blackburn, G. Kumar, K. A. Morgan, M. M. Jones, and R. G. Albridge, in "Electron Spectroscopy", D. A. Shirley, ed. (North-Holland, Publ. Co., Amsterdam, 1972), pp. 725–32.
47. W. E. Swartz, J. K. Ruff, and D. M. Hercules, *J. Amer. Chem. Soc.* **94**, 5227 (1972).
48. R. G. Hayes and N. Edelstein, in "Electron Spectroscopy", D. A. Shirley, ed. (North-Holland Publ. Co., Amsterdam, 1972), pp. 771–779.
49. P. H. Citrin, R. W. Shaw, A. Packer, and T. D. Thomas, in "Electron Spectroscopy", D. A. Shirley, ed. (North-Holland Publ. Co., Amsterdam, 1972), pp. 691–706; also see *J. Chem. Phys.* **57**, 4446 (1972).
50. A. D. Hamer and R. A. Walton, *Inorg. Chem.* **13**, 1446 (1974).
51. M. B. Robin and P. Day, *Advances Inorg. Chem. Radiochem.* **10**, 247 (1967).
52. R. E. Connick, University of California, Berkeley, private discussion with author.

53. P. Burroughs, A. Hamnett, and A. F. Orchard, *J. Chem. Soc. Dalton*, 565 (1974); also see C. K. Jørgensen, *Chimia*, **25**, 213 (1971) and M. J. Tricker, I. Adams, and J. M. Thomas, *Inorg. Nucl. Chem. Letters*, **8**, 633 (1972).
54. A. T. Jensen and S. E. Rasmussen, *Acta Chem. Scand.* **9**, 708 (1955).
55. P. H. Citrin, *J. Amer. Chem. Soc.* **95**, 6472 (1973).
56. C.-C. Su and J. W. Faller, *Inorg. Chem.* **13**, 1734 (1974).
57. C.-C. Su and J. W. Faller, *J. Organomet. Chem.* **84**, 53 (1975).
58. D. J. Cardin, B. Cetinkaya, M. J. Doyle, and M. F. Lappert, *Chem. Soc. Rev.* **2**, 99 (1973); F. A. Cotton and C. M. Lukehart, *Prog. Inorg. Chem.* **16**, 487 (1972).
59. L. F. Farnell, E. W. Randall, and E. Rosenberg, *Chem. Commun.* 1078 (1971); G. M. Bodner, S. B. Kahl, K. Bork, B. N. Storhoff, J. E. Wuller and L. J. Todd, *Inorg. Chem.* **12**, 1071 (1973).
60. A. Davison and D. L. Reger, *J. Amer. Chem. Soc.* **94**, 9237 (1972).
61. W. B. Perry, T. F. Schaaf, W. L. Jolly, L. J. Todd, and D. L. Cronin, *Inorg. Chem.* **13**, 2038 (1974).
62. J. Evans and J. R. Norton, *Inorg. Chem.* **13**, 3043 (1974).
63. S. C. Avanzino, W. L. Jolly, M. S. Lazarus, W. B. Perry, R. R. Rietz, and T. F. Schaaf, *Inorg. Chem.* **14**, 1595 (1975).
64. L. Hedin and A. Johansson, *J. Phys. B, Ser. 2*, **2**, 1336 (1969); W. L. Jolly, *Discuss. Faraday Soc.* **54**, 13 (1972); D. W. Davis and D. A. Shirley, *Chem. Phys. Lett.* **15**, 185 (1972).
65. W. L. Jolly and W. B. Perry, *J. Amer. Chem. Soc.* **95**, 5442 (1973).
66. W. L. Jolly and W. B. Perry, *Inorg. Chem.* **13**, 2686 (1974).
67. D. W. David and D. A. Shirley, *J. Electron Spectr. Rel. Phen.* **3**, 137 (1974).
68. D. A. Shirley, *Chem. Phys. Lett.* **16**, 220 (1972).
69. G. Howat and O. Goscinski, *Chem. Phys. Lett.* **30**, 87 (1975).
70. T. F. Schaaf and W. L. Jolly, unpublished data and calculations, 1975.

4

Ultraviolet Photoelectron Spectroscopy: Basic Concepts and the Spectra of Small Molecules

W. C. PRICE

King's College, London

I. HISTORICAL DEVELOPMENT

Progress in our understanding of the manner in which electrons exist within the positive nuclear framework of matter has largely come from experiments in which the electrons are excited within or are removed from their nuclear environment. Much of the early investigation on diatomic molecules was carried out by emission spectroscopy and by the late twenties a fairly detailed understanding of their spectra and structure had been achieved. However, emission techniques were not successful in the study of polyatomic molecules because when subjected to electrical discharges large molecules became dissociated into diatomic fragments. The less disruptive techniques of absorption spectroscopy, however, brought some success and when the vacuum ultraviolet absorption spectra of many polyatomic molecules were investigated systems of bands were found which could be fitted into Rydberg series converging towards the first ionization potentials of these molecules. In this way accurate values were obtained for the binding energies of electrons in the π outer orbitals of acetylene, ethylene, and benzene. Sharp Rydberg series were also found for small molecules with non-bonding lone-pair $p\pi$-type electrons such as in H_2O, H_2S, HCl, HBr, HI, CO_2, CS_2, CH_3I etc.[1] These spectra were important since they provided confirmation of the molecular orbital theory of polyatomic molecules which was being developed by Mulliken between 1930 and 1940 and also afforded accurate ionization potentials which were later used as standards. The absorption spectra of the larger molecules were found to have broader and more diffuse bands than those of small molecules and because of their overlapping such bands could not be separated out into series converging to ionization limits. Even in the case of small molecules the bands which occurred at higher energies than the first ionization limit were also found to be so diffuse as to preclude their use for determining the binding energies of *inner* electrons.

Some progress in circumventing this impasse was made by Watanabe and his associates[2] who measured the ion currents produced as the frequency of the photons incident upon a molecule was increased. They used a vacuum ultraviolet monochromator to scan photons of gradually increasing energies over the entrance to an ionization chamber containing the substance being studied. The wavelength at which photoionization set in was found to be relatively sharp even when the ultraviolet absorption bands were broad and diffuse. Thus the method yielded good first ionization potentials for molecules like methane, ethane and large polyatomic molecules such as, for example, substituted aromatics which did not have sufficiently well-defined band systems to show Rydberg series. The Watanabe method was not, however, capable of determining the ionization energies of inner orbitals any better than vacuum ultraviolet absorption spectroscopy, since it was found that as the photon energy was increased beyond the ionization threshold, large increases in ion current were obtained from processes associated with the autoionization of superexcited states, i.e. interactions between energetically equivalent states in which the excitation energy of an inner electron is transferred to ionize an electron from the outer orbital. These processes were of such a magnitude as to obscure any increases in ion current that might have arisen from the onset of ionization from the inner orbital. In order to avoid the effects of autoionization a further change of technique was required. This involved using a photon with an energy greater than that of any superexcited state of the electrons in the range being considered and a measurement of the energy of the ejected photoelectron, this energy being then subtracted from the energy of the incident photon in order to obtain the orbital binding energy. In this way it has been found possible to observe and locate the inner orbitals of molecules without interference from electrons in outer orbitals. Although measurements of the energies of photoelectrons produced in photoionization studies of molecules were first carried out by Vilesov *et al.* in 1961[3] these workers were restricted in their studies to energies less than 11 eV because they employed a fluorite window to separate their photon source from their ion chamber. It was the use by Turner and his co-workers of the 21·22 eV resonance line of helium as a photon source and the avoidance of a window by utilizing differential pumping which formed the basis of the highly successful technique of photoelectron spectroscopy as it is known today. The details of this technique have been described elsewhere in this book and the present chapter will be mainly concerned with the ultraviolet photoelectron spectroscopy of small molecules. However, the brief account of x-ray photoelectron spectroscopy (XPS) which follows is desirable in order to emphasize the relationship between it and ultraviolet photoelectron spectroscopy (UPS) and to discuss common features in subsequent sections.

II. X-RAY PHOTOELECTRON SPECTROSCOPY:
MAIN FEATURES

The use of x-ray sources to eject inner-core electrons was pioneered by Kai Siegbahn and his co-workers at Uppsala in the late fifties. They used their experience in β-ray spectrometry to make the necessary technical development. Extensive details of the work of this school are given in two major publications.[4, 5] The soft $K\alpha$ x-ray lines of Mg or Al (1254 and 1487 eV respectively) are most frequently used because these are narrower than, for example, the lines of Cu $K\alpha$ and they give photoelectrons whose energies are lower and can therefore be measured more accurately. These lines permit examination of the electronic structure down to 1000 eV which covers the $1s$ shells of the light atoms. The XPS lines from the K shells of the atoms in the second period of the periodic table are shown in Fig. 1. Their binding energies increase from lithium (55 eV) to neon (867 eV), the differences between successive elements gradually increasing from 56 eV (Li–Be) to 181 eV (F–Ne). The photoelectron lines are thus well separated from one another and can be used as a method of identifying the elements present in the sample. For the heavier atoms the binding energies of electrons in L, M and higher inner shells can be used for

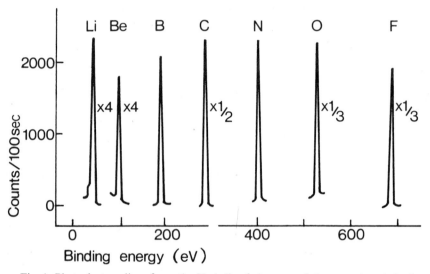

Fig. 1. Photoelectron lines from the K shells of elements of the second period when irradiated with Al Kα plotted against binding energy (i.e. 1487 eV minus the photoelectron energy).

this purpose and thus all the elements of the periodic table can be identified by x-ray photoelectron spectroscopy. It was for this reason that Siegbahn called his method ESCA, the letters standing for "electron spectroscopy for chemical analysis". However, this is a rather restrictive title in view of the wider use now being made of the spectra.

Although the binding energy of an electron in an inner shell is mainly controlled by the charge Ze within the shell, it is also affected to a lesser extent by the cloud of valence electrons lying above it. The XPS line of an element is found to vary over a few electron volts according to the state of chemical combination of the atom and the nature of its immediate neighbors. This is illustrated in the spectra of acetone, $(CH_3)_2CO$, ethyl trifluoracetate, $CF_3CO_2C_2H_5$, and sodium azide, shown in Fig. 2. The C(1s) line of the two methyl carbons in acetone is 2–3 eV lower in binding energy than that of the more positive carbonyl carbon and is, as might be expected from the number of C atoms, twice as strong. Similar remarks apply to the two outer and the

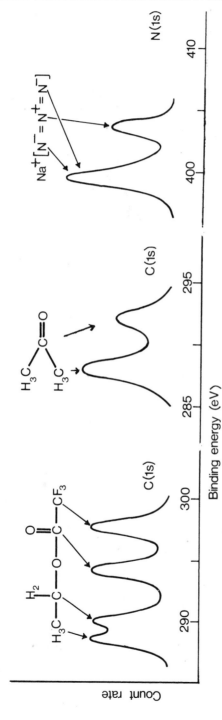

Fig. 2. Spectra of C(1s) core electrons in acetone, ethyl trifluoroacetate, and the N(1s) electrons in sodium azide.

inner N($1s$) electrons of the azide ion. The binding energies of the four different carbons in ethyl trifluoroacetate similarly reflect the different environments of the four carbon atoms in this molecule (note that the formula in the diagram is placed so that the lines lie below the carbon atoms with which they are associated). These "chemical shifts" have been related to the electronegativity difference of the atom from its neighbors and provide a means of estimating the effective charge on the core of each atom. The method has been extended to complicated molecules such as vitamin B_{12} where, using Al(Kα), the photoelectron line of one cobalt atom could be studied in the presence of the lines of 180 other atoms. Similarly both the sulfur and the iron electron line in the enzyme cytochrome c can be recorded. It is clear that the technique is a powerful direct method of studying chemical electronic structure.

In addition to the work on mainly covalent materials, ionic and metallic solids have been studied. For metals the conduction band structure can be found and the photoelectron energy distribution curves compared with calculated densities of states. Because of the small depth of penetration of the radiation and the necessity for the photoelectrons to emerge from the material the method is clearly a surface effect. Depths down to about 2 nm (20 Å) are involved in the process and ultra-high vacuum techniques must be employed to eliminate the effects of surface contamination. Surface properties such as are important in catalytic and adsorption processes have received considerable attention from workers in the field. It should be mentioned that although XPS can produce some valuable information about the electrons in the valence shell, the cross-sections for this radiation of valence shell orbitals are almost an order of magnitude less than those of core orbitals. For example, in hydrocarbons the cross sections of C($2s$) orbitals are about a tenth those of C($1s$) orbitals, and the cross-sections of orbitals built from C($2p$) and H($1s$) are still lower by almost another order of magnitude. This is a limitation not suffered by UPS, where the cross-sections particularly of $2p$ orbitals are generally appreciably higher than those of $2s$ orbitals. This fact is likely to affect seriously the prospects of XPS for detailed outer electron studies.

III. ULTRAVIOLET PHOTOELECTRON SPECTRA OF SMALL MOLECULES

A. Monatomic Gases

For monatomic gases only electronic energy is involved in photoionization and the HeI photoelectron spectrum of atomic hydrogen consists of a single sharp peak corresponding to a group of photoelectrons with energies of 7·62 eV, i.e. equal to the difference in energy of the helium resonance line

(21·22 eV) and the ionization of the H(1s) electron (13·60 eV). For argon the outer electrons are in p-type orbitals and the $(3p)^5$ configuration of the ion has the two states $^2P_{\frac{3}{2}}$ and $^2P_{\frac{1}{2}}$ with statistical weights in the ratio 2 : 1 and energies differing by spin–orbit coupling. The photoelectron spectrum (see Fig. 3)

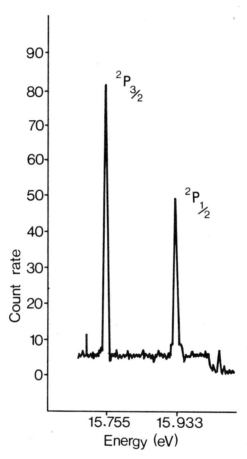

Fig. 3. Photoelectron spectrum of argon ionized by Ne (736 Å).

corresponding to this $(3p)^{-1}$ ionized state is thus a doublet, one component of which is twice as strong as the other. Similar doublets are found for the other inert gases, the spin–orbit coupling increasing with their atomic weight. Narrow line spectra corresponding to ionization of atoms to the lower states of their ions (i.e. ionization from each of the outer orbitals) have also been obtained for N, O, F, Cl, Br, I, Hg, Cd, and Zn.[6, 7, 8, 9]

For the photoejection of an electron from an orbital in a molecule we have the equation

$$I_0 + E_{\mathrm{vib}} + E_{\mathrm{rot}} = h\nu - \tfrac{1}{2}mv^2 \tag{1}$$

where I_0 is the "adiabatic" ionization energy (i.e. the pure electronic energy change) and E_{vib} and E_{rot} are the changes in vibrational and rotational energy which accompany the photoionization. To explain the nature of the information which can be obtained we shall discuss the photoelectron spectra of some simple molecules.

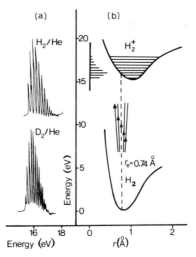

Fig. 4. (a) Photoelectron spectra of H_2 and D_2, (b) potential energy curves of H_2 and H_2^+ showing spectra plotted along ordinate.

Figure 4(b) gives the potential energy curves of H_2 and H_2^+. When an electron is removed from the neutral molecule the nuclei find themselves suddenly in the potential field appropriate to the H_2^+ ion but still separated by the distance characteristic of the neutral molecule. The most probable change is thus a transition on the potential energy diagram from the internuclear separation of the ground state to a point on the potential energy curve of the ion vertically above this. This is the Franck–Condon principle and it determines to which vibrational level of the ion the most probable transition (strongest band) occurs. The energy corresponding to this change is called the vertical ionization potential I_{vert}. Transitions to vibrational levels on either side of I_{vert} are weaker, the one of lowest energy corresponding to the vibrationless state of the ion corresponding to I_{adiab}. The photoelectron spectra of H_2 and D_2 are given in Fig. 4(a) and that of H_2 is also plotted along the ordinate of Fig. 4(b). It is worth noting that because one of the two bonding electrons is removed, the bonding energy at this internuclear distance should be halved. Thus I_{vert} should be equal to

$$I(H) + \tfrac{1}{2}D(H_2) = 13 \cdot 595 + \tfrac{1}{2}(4 \cdot 478) = 15 \cdot 834 \text{ eV}.$$

This agrees as closely as I_{vert} the maximum of the band envelope can be estimated for H_2. However, in other molecules, e.g. HCl, the agreement is not so good due to compensating movements in other shells, i.e. deviations from Koopmans' Theorem,[10] which assumes that the orbital binding energies of electrons are equal to the negative vertical ionization potentials.

It is possible to calculate the change in internuclear distance on ionization from the intensity distribution of the bands in the photoelectron spectrum. Clearly when a bonding electron is removed, the part of the photoelectron spectrum corresponding to this will show wide vibrational structure with a frequency separation that is reduced from that of the ground-state vibration. The removal of relatively nonbonding electrons, on the other hand, will give rise to photoelectron spectra rather similar to those of the monatomic gases and little if any vibrational structure will accompany the main electronic band. The type of vibration associated with the pattern obtained when a bonding electron is removed can usually be identified as either a bending or a stretching mode and this can throw light on the function of the electron in the structure of the molecule, that is, either as angle forming or distance determining. From the band pattern it is frequently possible to calculate values of the changes in angle as well as changes in internuclear distance on ionization, and so the geometry of the ionic states can be found if that of the neutral molecule is known. Changes in bond lengths or angles can be calculated with about 10% accuracy for photoelectron band systems which show structure by using the semi-classical formula

$$(\Delta r)^2 = l^2(\Delta\Theta)^2 = 0 \cdot 543[I_{vert} - {}_{adiab}]\mu^{-1}\omega^{-2} \tag{2}$$

where r and l are in Å, Θ in radians, μ in atomic units, and ω the mean progression spacing in 1000 cm^{-1}.

B. Photoelectron Spectra of Hydrides Isoelectronic with the Inert Gases

As examples of the features mentioned above and also to illustrate how the orbitals of isoelectronic systems are related to one another we shall now discuss the photoelectron spectra of those simple hydrides for which the united atoms are the inert gases Ne, Ar, Kr, and Xe—that is, the atoms obtained by condensing all the nuclei into one central positive charge.

Hydrides such as HF, H_2O, NH_3 and CH_4 can be thought of as being formed from Ne by successively partitioning off protons from its nucleus. The solutions of the wave equation—the wavefunctions or orbitals—arising from the progressive subdivision of the positive charge pass continuously from the atomic to the molecular cases with a gradual splitting of the p^6-orbital degeneracy as the internal molecular fields are set up. Schematic diagrams of the orbital changes are indicated in Fig. 5. The photoelectron spectra shown in Fig. 6(a), (b), (c), and (d) reveal this orbital subdivision in a striking way.

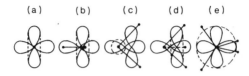

Fig. 5. Schematic diagrams of 2p-orbital structures of (a) Ne, (b) HF, (c) H_2O, (d) NH_3, and (e) CH_4.

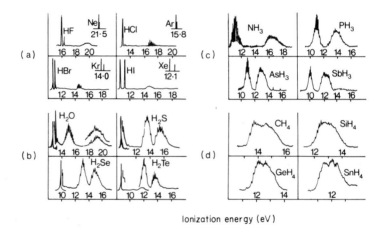

Ionization energy (eV)

Fig. 6. Photoelectron spectra of the hydrides isoelectronic with Ne, Ar, Kr, and Xe obtained with He (21·22) eV) radiation.

Figure 6(a) shows the spectra of the halogen acids which is the first stage in this process. The spectra of the corresponding inert gas (i.e. the united atom) is inserted on the records for comparison. The triply degenerate p^6 shell is split into a doubly degenerate π^4 shell and a singly degenerate σ^2 shell. The π^4 shell is nonbonding and represented by two bands in the photoelectron spectrum which are of equal intensity and show little vibrational structure. The σ shell is formed from the p orbital along which the proton is extracted and gives a negative cloud which binds the proton to the residual positive charge. This $p\sigma$ orbital therefore gives rise in the photoelectron spectrum to a simple progression of bands with a separation corresponding to the vibration frequency of the $^2\Sigma^+$ state of the ion except where the structure is lost by predissociation.

The second stage of partition leads to the molecules H_2O, H_2S, H_2Se, and H_2Te. If two protons were removed co-linearly from the united atom nucleus the linear triatomic molecule so formed would not have maximum stability, since only the two electrons in the p orbital lying along the line will then be effective in shielding the protons from the repulsion of the core. By moving off this line, the protons can acquire additional shelter from the central charge and from themselves through shielding by one lobe of a perpendicular p orbital (see Fig. 5c). The electrons in this orbital then become "angle determining" as distinct from the two previously mentioned p electrons which are mainly effective in determining the bond separations of the hydrogen atoms from the central atom. The remaining p orbital, because of its perpendicular orientation to the plane of the bent H_2X molecule, can affect neither the bonding nor the angle, that is, its electrons are nonbonding. These expectations are borne out by the photoelectron spectra shown in Fig. 6(b). These spectra show how in all H_2X molecules the triple degeneracy of the p^6 shell of the united atom is completely split into three mutually perpendicular orbitals of different ionization energies. The lowest ionization band is sharp with little vibrational structure. The second has a wide vibrational pattern which turns out to be the bending vibration of the molecular ion. The third also has a wide vibrational pattern with band separations greater than those of the second band. These separations can be identified as the symmetrical bond stretching vibrations of the ion. The geometries of these three ionized states can be calculated from the band envelopes and pattern spacings. Only small changes are associated with the band of lowest i.p. Large changes of angle accompany ionization in the second band, which in the case of H_2O causes the equilibrium configuration of this ionized state to be linear. In H_2S, H_2Se, and H_2Te the spectra show that the angles are 129°, 126°, and 124°, respectively. The third ionized state is one in which the internuclear distances are increased but little change occurs in the bond angle. The integrated intensities of all three bands are roughly equal indicating that each originates from the ionization of a single orbital.

The partitioning of three protons from the nucleus of an inert gas molecule leads to a pyramidal XH_3 molecule. One p orbital is directed along the axis of symmetry and provides the shielding which causes the molecule to have a nonplanar geometry, that is, it is angle determining. The other two p orbitals are degenerate and provide an annular cloud of negative charge passing through the three XH bonds and thus mainly determine the bond distances. The structure pattern on the first band of the photoelectron spectrum (Fig. 6c) can be assigned to bending vibrations and indicates that the molecule flies to a symmetrical planar configuration without much change in the bond distances when an electron is ionized from this orbital. The second band in the photoelectron spectrum shows Jahn–Teller splitting which is consistent with its being doubly degenerate. Although it has limited structure, such structure as can be observed corresponds to changes in bond stretching without much change in bond angle.

Finally the photoelectron spectra of the XH_4-type molecules show that the p^6 shell of the united atom has changed to another triply degenerate shell with orbitals of tetrahedral symmetry. The contour shows the presence of Jahn–Teller splitting, and the structure on the low-energy side shows that on ionization the molecule moves by contraction along one side of the enveloping cube toward a square coplanar configuration. The structure on the high-energy side indicates movement in the opposite direction toward two mutually perpendicular configurations in which opposite XH_2 angles are roughly 90°.

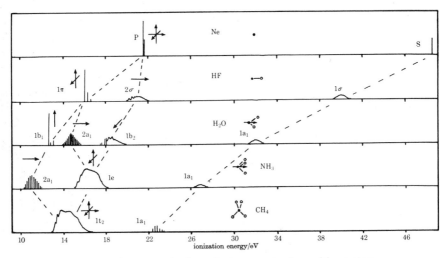

Fig. 7. Diagrammatic photoelectron spectra formed by proton.

In the heavier molecules, for example, SnH_4, the large spin–orbit splitting of the heavy atom influences the structure. The spectra also show that the movement toward coplanarity is progressively less pronounced. Further details on the spectra of these hydride molecules are given by Potts and Price.[11, 12]

It should be mentioned that in the process of partitioning off protons from the neon nucleus to form the second two hydrides, the $2s$ orbitals are distorted though to a lesser extent than the $2p$ orbitals. They are deformed from spherical symmetry in the direction of the extracted proton. They thus have "bumps" in the bond directions and in this way provide an "s" contribution to the bond. This is evident in the vibrational structure of the photoelectron bands associated with these orbitals. Figure 7 shows how the orbitals of the hydrides of F, N, O, and C are formed by proton withdrawal from neon.

C. Photoelectron Spectra of the Halogen Derivatives of Methane

It is convenient to discuss at this point the photoelectron spectra of some "single-bond" molecules in which some of the atoms have additional groups of nonbonding electrons. The bromomethanes, whose photoelectron spectra are illustrated in Fig. 8, are good examples of this molecular type. It can be seen that some bands in their spectra are relatively sharp and therefore can be associated with the nonbonding electrons, while others are broad and clearly

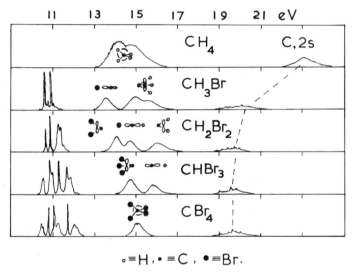

Fig. 8. Photoelectron spectra of the bromomethanes.

arise from electrons in strongly bonding orbitals. The orbital assignment of the latter bands are indicated schematically in the figure. A comparison of these bands with those of the hydrides of the same symmetry, e.g. CH_3Br, and NH_3, CH_2Br_2, and H_2O, etc., shows that the orbitals around the carbon atom are split up in a very similar way as in the hydrides by the departure from tetrahedral symmetry brought about by the halogen substitution. This gives some support to the old concept that around an atom in a stable molecule there should be a closed (inert gas) shell of electrons. It shows further how the degeneracy of the p^6 group of these electrons is split by the fields arising from the different substituents, the splitting of the degeneracy being complete in the case of the methylene halide as it is for water in the hydrides. The sharp bands in the region of 11 eV can be readily associated with orbitals containing $4p$ Br electrons which are split by spin–orbit interaction. In the case of methyl bromide two sharp bands (with very weak accompanying vibrational structure) are split by the magnitude of spin–orbit coupling constant which is

0·32 eV for bromine. Further splitting occurs as the number of bromine atoms increases and eight bands can be seen in CBr_4 corresponding to the eight nonbonding "p" orbitals present in this molecule. A detailed account of the analysis of the photoelectron spectra of the halides of elements in groups III, IV, V, and VI of the periodic table has been given by Potts et al.[13]

D. Photoelectron Spectra of "Multiple" Bonded Diatomic Molecules

To discuss "multiple" bonded systems in which molecular orbitals are formed by p atomic orbitals combining in the "broadside-on" as well as in the "end-on" configuration we shall take as examples the molecules N_2, NO, O_2, and F_2. Their photoelectron spectra are shown in Fig. 10. The orbitals upon which their electronic structures are built are the in-phase and out-of-phase combinations of the appropriate $2s$ and $2p$ orbitals. These are given schematically in the insert. For N_2 the electronic configuration is $(\sigma_g 2s)^2 (\sigma_u 2s)^2 (\pi_u 2p)^4 (\sigma_g 2p)^2$, the orbitals being in order of decreasing ionization energy. With the exception of $(\sigma_u 2s)$, all of these might be expected to provide excess negative charge density between the two nuclei and thus to account for the strength of the N_2 "triple" bond. The additional electron of NO has to go into a π^*2p orbital which is largely outside the nuclei and therefore antibonding, and so in a loose analogy N_2 is the inert gas of diatomic molecules and NO the corresponding alkali metal, since it contains one electron outside a closed shell of bonding electrons. An inspection of the photoelectron spectrum of N_2 in which the ionized states corresponding to removal of the different orbital electrons are marked, shows that by far the most vibration accompanies the removal of the π_u electron. Thus at the internuclear distance of neutral N_2, the nuclei are held mainly by the negative cloud of the $(\pi_u 2p)^4$ electrons. The short-distance bonding character of these π electrons results in the nuclei being pulled through the $(\sigma 2p)^2$ cloud so that this σ orbital is as much outside the nuclei as between them and therefore supplies no bonding at the N—N equilibrium separation. It can be understood readily from the geometry of their overlap that the bonding of p electrons in the broadside on (π) arrangement optimizes at shorter internuclear distances than that in the end on (σ) position. In N_2 the nuclei are in fact on the inside of the minimum of the partial potential energy curve associated with the $(\sigma 2p)^2$ electrons. This orbital is thus in compression, and its electrons have both their bonding and their binding (ionization) energies reduced. In NO, O_2, and F_2 the presence of the additional electrons in antibonding $\pi 2p$ orbitals causes the internuclear distances to be relatively longer than they are in N_2, and it can be seen from Fig. 9 that the $\sigma 2p$ bands of these molecules have progressively more vibrational structure as the internuclear distance approaches more closely that separation for which the bonding of the

$(\sigma\,2p)^2$ orbital is optimized. The associated σ bands move through the $\pi 2p$ systems to higher ionization energies in accord with their increased effective bonding power.

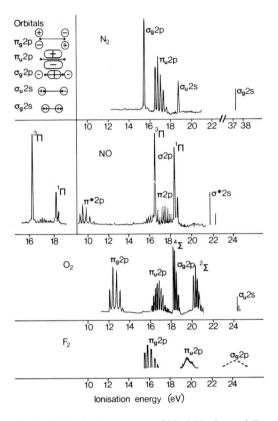

Fig. 9. Photoelectron spectra of N_2, NO, O_2, and F_2.

The features discussed above can be illustrated by considering orbital potential energy curves as illustrated in Fig. 10. In the orbital approximation of molecular electronic structure, the complete dissociation energy curve of a molecule can be split up into the partial orbital bonding curves which give internuclear distances. Different types of orbital have their potential minima at different internuclear distances. These do not coincide with the actual equilibrium internuclear distance r_e of the molecule which is determined by the minimum of the sum of the orbital energies taken at each value of r. At any particular r the electrons in different orbitals are at different relative positions in their orbital binding energy curves. These orbital curves are not of course directly determinable but have been drawn to be as far as possible

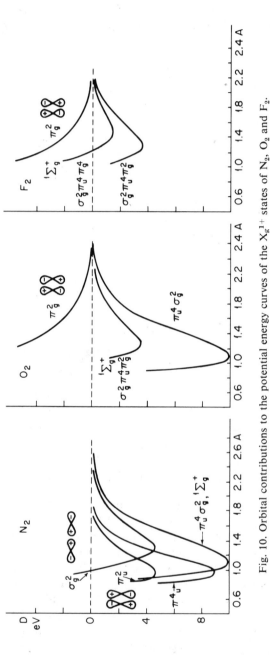

Fig. 10. Orbital contributions to the potential energy curves of the X_g^{1+} states of N_2, O_2 and F_2.

consistent with the experimental facts. For instance $(\sigma_g 2p)^2$ binding curve has been drawn as a near Morse curve to have an r_e of about 1·4 Å, a dissociation energy of 4 eV, and an $\omega_e = 1200$ cm^{-1}. The curve for $(\pi_u 2p)^4$ can then be found for $r > r_e$ by subtracting the σ_g curve from the observed dissociation curve of N_2 on the assumption that at large distances the $(\sigma_g 2s)^2 (\sigma_u 2s)^2$ orbitals do not contribute to total bonding. For the present purpose it has been assumed that this is also true at smaller values of r. The $(\pi_u 2p)^2$ curve can then be obtained by halving the $(\pi_u 2p)^4$ curve obtained by the above subtraction. The $(\pi_g 2p)^2$ repulsive curve can be obtained by plotting the difference between O_2, $X^1\Sigma$ and $N_2 X^1\Sigma$ which should check with the $(\pi_u 2p)^2$ curve of which it is a reflection at large r. The antibonding power of the antibonding orbital is only slightly greater than the bonding power of the bonding orbital (i.e. the former is proportional to $S(1-S)^{-1}$ and the latter to $S(1+S)^{-1}$, where S is the overlap integral). Similarly the $(\pi_g 2p)^2$ curve of fluorine can be obtained by plotting the difference between F_2, $X^1\Sigma_g^+$ and O_2, $X^1\Sigma_g^+$ or alternatively by halving the differences between the N_2, $X^1\Sigma_g$ and the F_2, $X^1\Sigma_g^+$ curves.

It will be noted that the optimum bonding energies of $(\sigma_g 2p)^2$ and $(\pi_u 2p)^2$ are roughly equal. This might be thought to contradict the fact that the single, double, and triple bond energies of ethane, ethylene, and acetylene are not in the ratio of 1 : 2 : 3, but in the ratio 1 : 1·76 : 2·41. However, it is readily appreciated that the figures indicate that the σ bond must lose about 60% of its bonding by compression to smaller internuclear distances when forming part of the triple bond. Because of the steep slope of the repulsive part of its curve, this loss would rapidly increase with further reduction in internuclear distance. Thus in N_2 the $(\sigma_g 2p)$ orbital has lost nearly all its bonding power at the equilibrium internuclear distance and its photoelectron band has the features characteristic of ionization from a nonbonding orbital. The rapid increase in its bonding power with increasing r is evident from the increasing vibrational structure of the $(\sigma 2p)^{-1}$ systems in NO and O_2 (see Fig. 9). On the other hand, the $\pi_u 2p$ electrons in N_2 find themselves at internuclear distances only slightly larger than those of their potential minimum and are thus strongly bonding. As the equilibrium internuclear distance increases in passing from N_2 to NO, O_2, and F_2 as electrons are added in antibonding repulsive orbitals, the $\sigma_g 2p$ orbital acquires bonding ultimately becoming the basic single bond in F_2. The value of the dissociation energy of F_2 (1·6 eV) is less than the optimum $(\sigma_g 2p)^2$ bond energy of about 4 eV, because, as already mentioned, the bonding power of the $(\pi_u 2p)^4$ orbitals is more than offset by the antibonding power of the $(\pi_g 2p)^4$ orbitals which are filled in this molecule.

Other interesting points illustrated by Fig. 9 are that the spacing in the $2p$ antibonding bands in NO and O_2 (first systems) are larger than those of the $2p$

bonding bands (second systems). This is to be expected since the removal of an antibonding electron increases the vibration frequency while that of a bonding electron decreases this frequency. Another interesting feature is that the separation of the first and second bands, which reflects the overlap between the out-of-phase and in-phase ($\pi 2p$) orbital, rapidly decreases with increase in internuclear distance. For NO, O_2, and F_2 these separations are 7·5, 4·5, and 3·0 eV, respectively. This indicates the reduction in "multiple" (π) bonding as the effect of the increasing number of π antibonding electrons reduces and annuls the bonding of π bonding electrons by increasing the interatomic distance and reducing the orbital overlap.

The insert of the 16–18 eV region of NO in Fig. 9 shows the $^3\Pi$ and $^1\Pi$ bands with an intensity ratio of 3 : 1, and thus illustrates how closely the intensities follow the statistical weights of the ionized states, agreeing with the number of channels of escape open to the electrons. Similar remarks apply to the $^4\Sigma$ and $^2\Sigma$ bands of O_2, the integrated intensities of which are in the ratio of 2 : 1. A further interesting feature which these multiplets illustrate is the greater bonding of the states of lower multiplicity. Since the lower multiplicity corresponds to the states of antisymmetric spin function, it is associated with the symmetric (summed) space coordinate wavefunction of the orbitals between which the spin interaction is occurring. The additional overlap to which this gives rise results in greater bonding relative to states of higher multiplicity where the total space coordinate wavefunction is obtained by subtracting those of the interacting states.

The photoelectron spectrum contains enough information to plot a rough potential energy diagram of the states of the molecular ion provided the internuclear distance of the neutral species is known. The shape of curve is obtained from the vibrational spacing. By placing the energy point corresponding to the maximum of the vibrational pattern at the internuclear distance of the neutral molecule its position on the horizontal scale can be fixed. That on the vertical scale is fixed by the ionization energy found for the $v' = 0$ band. Internuclear separations for excited states can then be read off the diagram. This is illustrated for NO^+ in Fig. 11 where the photoelectron spectrum is plotted along the OY axis of the potential energy diagram.

The variation of orbital energies of isoelectronic molecules with increasing mass of the constituent atoms is illustrated by the spectra of O_2, SO, S_2, and Te_2 shown in Fig. 12. The smaller orbital overlap arising from the greater spread of the orbitals of the heavier atoms reduces both their respective bonding and the antibonding characters. This is evident from the reduced widths of the corresponding bands.

The "building up" of the electronic structures of molecules by adding electrons to the unfilled orbitals as an atom is changed to one with greater Z follows very closely the "aufbau" process in atoms. Although we have considered this only for diatomic molecules, it can be applied to triatomic and polyatomic systems and extends the shell concept to even the valence electrons of molecules. The difference between atoms and molecules is simply that instead of the positive charge being concentrated at one point it is split into

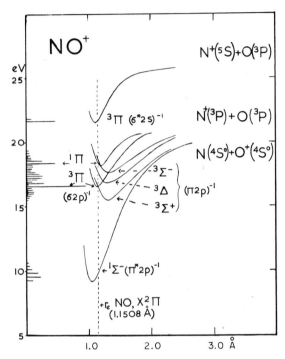

Fig. 11. Potential energy curves of NO⁺ showing relation to the photoelectron spectrum plotted along OY.

several different centers. It might be asked whether we should now replace the bond structures of chemists by the molecular orbital description. No spectroscopist, for example, has even seen an electron in a hybridized sp^3 orbital—a concept the chemist uses constantly. No real conflict exists, since the bond between two atoms is measured by the amount of negative charge lying between the two positive centers. From the orbital point of view this is made up by contributions from two orbitals, for example, in CH_4 from the $(2pt_2)^6$ orbitals and from the $(2sa_1)^2$ orbital. As pointed out in the previous discussion of this molecule, the latter acquires bonding character by being distorted (by charge extraction from Ne) in the bond directions. The sp^3 hybridized notation is a shorthand way of saying this. When, however, changes in energy are involved, even, for example, in a mechanical model with certain normal vibrations being excited by a periodic force, this force must be resolved with respect to the axis of symmetry of the system and will only feed energy into a normal vibration of its own frequency. In exactly the same way, the interaction of the electric vector of the light wave with the electrons in a molecule must be

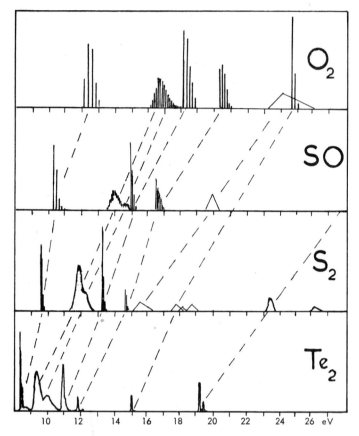

Fig. 12. Photoelectron spectra of the isoelectronic diatomic molecules O_2, SO, S_2, and Te_2.

with the orbitals which are the irreducible representations of the states involved in the transition moment and the transitions can only occur between states represented by these orbitals. Because of the indistinguishability of electrons a particular electronic state of an atom or a molecule can only be described in terms of a probability distribution of negative charge surrounding the positive framework. The orbital structure is manufactured as a necessary consequence of sampling this with dipole radiation.

E. Spectra and Structure of Some "Multiple" Bonded Polyatomic Molecules

In the same way that the spectra and orbital structure of the simple hydrides can be understood by partitioning off protons from the central charge of the inert gas atoms, the orbital structure of some simple polyatomic molecules can be followed by partitioning off protons from the nuclei of the diatomic molecules with which they are isoelectronic. This is illustrated in Fig. 13. Starting with the well-understood spectrum of N_2, we pass to HCN by

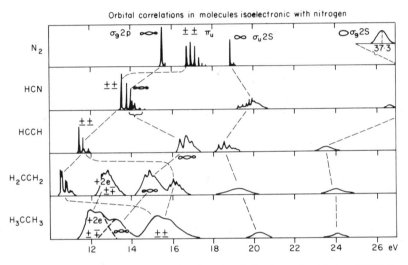

Fig. 13. Photoelectron spectrum showing the derivation of orbitals of HCN and C_2H_2 from the isoelectronic diatomic modecule N_2 also related orbitals in C_2H_4 and C_2H_6.

displacing a unit positive charge from one of the nitrogen nuclei to a point further along the bond axis. The movement of the orbitals is indicated by the broken lines. It can be seen that the $\sigma 2p$, the $\pi 2p$, and the $\sigma_g 2s$ orbitals have their ionization energies reduced by this lessening in the positive charge in the region in which they are mainly located. Another important effect which is apparent from comparison of the spectra is that the $\pi_u 2p$ electrons lose a considerable amount of their bonding as the nuclei move further away from the position of optimum $p\pi$ orbital overlap. This can be seen from the change in the intensity distribution of the vibration band pattern (i.e. I_{max} moves toward I_{adiab}). It is also apparent from the relatively large decrease in ionization energy of this bond orbital. The change from N_2 to HCN affects the $\sigma_u 2s$ orbital, which now has one of its lobes in the region between the H and the C positive centers. It consequently changes from an antibonding NN to a strongly bonding CH orbital. Its ionization energy is therefore increased and the vibrational pattern of the band changes to one which contains many vibrations of the CH bond reduced in frequency to 1690 cm^{-1} from its value in the neutral molecular ground state (3311 cm^{-1}).

A similar partitioning off of charge from the second nitrogen atom leads to the molecule acetylene, in which there is a further increase in central bond distance. The π_u 2p-orbital energy is again lowered by the positive charge reduction and also by loss of bonding due to further removal from its optimum bonding overlap position. The latter effect can be seen in the progressive lowering of the relative intensities of the vibrational bands to the first (zero) band of the system. The σ_g 2s and σ_u 2s are reduced in ionization energy by the charge removal, and the latter shows by its vibrational pattern that it is strongly CH bonding, while the former is probably CC bonding though because of its diffuseness evidence from vibrational pattern in support of this is not forthcoming. The greatest change in orbital character occurs in the σ_g 2p orbital which becomes strongly CC bonding as the internuclear separation increases toward the optimum value for $p\sigma$-type bonding. This is indicated by the characteristic bonding envelope of the band pattern in which the CC frequency, reduced to about 1400 cm^{-1} from its ground-state value of 1983 cm^{-1}, is strongly excited. It also now exerts some CH bonding, and both combine to raise the ionization energy by nearly 3 eV.

By bringing up two hydrogen atoms on to acetylene and suitably distorting the molecular framework one can form ethylene. The spectra show that one of the π_u CC orbitals now becomes the strongly bonding in-phase $b_{2u}(CH_2)$ orbital, increasing its binding energy by about 4 eV, and the two additional electrons go into the corresponding out-of-phase $b_{1g}(CH_2)$ orbital. Bringing up two more hydrogen atoms causes the remaining π-electron pair to go into the degenerate in-phase $e_{1u}(CH_3)$ orbital and the two additional electrons to enter the corresponding out-of-phase orbital. In order to cover the complete energy range, these spectra have been taken from recordings with 30·4 nm (40·1 eV) radiation. It can be seen that the separations of the bands corresponding to in-phase and out-of-phase 2s C combinations (i.e. σ_g 2s and σ_u 2s, respectively) diminishes with increasing CC distances in the molecules C_2H_2, C_2H_4, and C_2H_6. This is a direct result of the reduction in 2s C orbital overlap as their separation increases.

A similar diagram (Fig. 14) shows how the orbitals of formaldehyde and ethylene are derivable from those of oxygen. Of course in this case it is necessary to start with oxygen in its $^1\Sigma_g^+$ state, and although the relevant photoelectron spectrum has not been obtained, it can be constructed with confidence from the known spectroscopic data of the state and the photoelectron spectra of $^3\Sigma_g^-$ and $^1\Delta_g$. The spectra show how the antibonding π2p orbital of oxygen becomes the carbonyl "lone pair", the π_u^4 orbital splits into the carbonyl π and the $b_2(CH_2)$, and the σ2p becomes the $a_1(CO)$ bond orbital. Their consistent correlation with the orbitals of ethylene gives confidence to the general assignment of the orbitals.

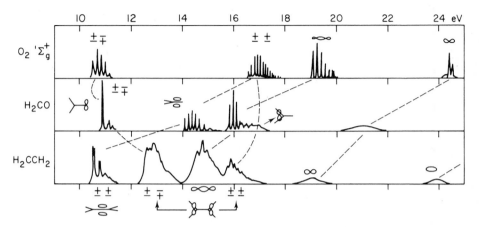

Fig. 14. Photoelectron spectra showing relation of orbitals of H_2CO and C_2H_4 to those of O_2, $b\,^1\Sigma_g{}^+$.

F. Spectra and Structure of Triatomic Molecules

Whether a triatomic molecule will be linear or bent in its ground state depends on which configuration will afford best coverage of the positive charge by the electrons in the orbitals associated with a particular geometry. The various ways in which the energies of different orbitals are affected by changing the angle from 180° to 90° were discussed by Walsh[14] before the advent of photoelectron spectroscopy and are qualitatively described in Walsh diagrams. It is found that molecules with 16 electrons in the valence shell are linear in their ground states (e.g. CO_2, N_2O, etc.). When the number is greater than 16 (e.g. NO_2, O_3, etc.) the molecule has a bent ground state, and this situation persists until the valence shell is completely filled as in OF_2 (20 valence electrons). The structure becomes linear again for XeF_2 (22 outer electrons). For molecules formed from first-row elements the orbitals can be regarded as being derived from the in-phase and out-of-phase combinations of the $2p$ orbitals of the outer atoms and the $2p$ and $2s$ orbitals of the central atom. The molecular orbitals so constructed for a bent molecule are shown diagrammatically in Fig. 15. Those for a linear molecule are very similar and can be visualized by changing the angle to 180°.

In carbon dioxide the orbital of lowest ionization energy corresponds to the out-of-phase combination of $2p$ (0) orbitals which can be written as $(\pi_0 - \pi_0,$ $\pi_g{}^4)$. This splits on bending into $3b_2$ (in-plane) and $1a_2$ (out-of-plane) orbitals. As can be seen from the photoelectron spectra shown diagrammatically in

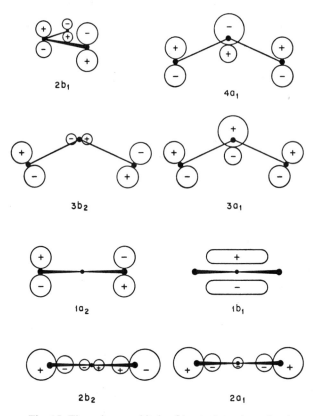

Fig. 15. The valence orbitals of bent triatomic molecules.

Fig. 16 this orbital is nonbonding in the linear triatomic molecules. The next filled orbital is the in-phase $(\pi_O + \pi_C + \pi_O, \pi_u^4)$ combination which, as expected and indicated by the structure of its band system, is strongly bonding. It splits into $3a_1$ (in-plane) and $1b_1$ (out-of-plane) orbitals in the bent molecule. The next inner orbital is that corresponding to the out-of-phase $(\sigma_O + p\sigma_C - \sigma_O, \sigma_u^2)$ orbital. It is nonbonding and followed by the in-phase $(\sigma_O + s\sigma_C + \sigma_O, \sigma_g^2)$ bonding orbital. The similar orbital structure in the isoelectronic molecule N_2O is reflected in its similar photoelectron spectrum. The π_g and σ_u orbitals of this molecule are lower than in CO_2 because they are located near N and O rather than O and O atoms. Similarly the π_u and σ_g orbitals are higher in N_2O than in CO_2 because of the higher energy contribution of the orbitals from the central atoms—N and C, respectively.

In NO_2 the additional electron has to go into the hitherto unfilled $4a_1$ orbital and the molecule becomes bent (ONO = 134°), finding better charge coverage in this geometry. The $(4a_1)^{-1}$ band appearing on the low-energy side of the

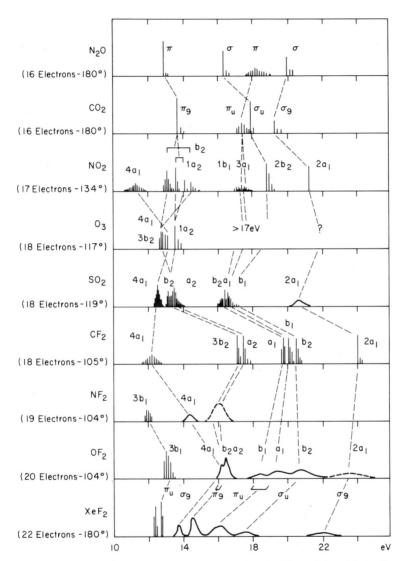

Fig. 16. Diagrammatic photoelectron spectra showing the filling of the orbitals of some triatomic molecules.

photoelectron spectrum shows a wide vibrational pattern involving the bending motion, since, as might be expected, on ionization from this orbital the molecule moves to the linear structure associated with the 16 electrons then remaining in the valence shell. The orbitals which were doubly degenerate in CO_2 are split into in-plane and out-of-plane orbitals in bent NO_2. Further the states of NO_2^+ involving ionization from all orbitals apart from the singly occupied $4a_1$ orbital are split into singlet and triplet states because of the spin interaction of the two singly occupied orbitals, and they appear in the photoelectron spectrum as pairs of similar systems with a 1 : 3 intensity ratio (Brundle et al.).[15]

The $4a_1$ shell is filled in O_3, SO_2, and CF_2 so that in NF_2 the additional electron present in its structure has to go into the $3b_1$ orbital and a new band appears in its photoelectron spectrum. This orbital becomes filled in OF_2. The spectrum of XeF_2, in which another orbital has to be provided for the two additional electrons, is shown as the final spectrum in Fig. 16. The additional band corresponding to these electrons is evident in the spectrum. Electrons in this orbital are clearly responsible for causing a return to linearity. A full discussion of this spectrum is not possible here.

The building-up principle illustrated so far for di- and triatomic molecules has been extended to molecules containing four, five, six, and seven atoms, particularly for molecules of the type AX_n, where X is a halogen atom.[13]

While other techniques have to be adopted for more complicated systems, it is clear that a sound understanding of what happens in the simplest cases is a very desirable preliminary to considerations of the larger molecules.

G. Methods of Orbital Assignment of Bands in Photoelectron Spectra

It would be misleading to infer from the foregoing sections that the methods described therein are generally applicable for the analysis of the spectra of more complicated molecules. This is certainly not the case. It is desirable whenever possible to assign bands from direct experimental data, as is the practice in the spectroscopy of simple molecules. However, because the resolution of photoelectron spectroscopy is not adequate to give rotational fine structural details of bands, even if these were present (which they usually are not because of the short lifetimes of the electronic excited states of polyatomic states), alternative methods must be used to find the nature of the state to be associated with a band in the photoelectron spectrum.

Obviously the vibrational pattern of a band gives a great deal of information about the orbital from which the electron has been removed, and this is greatly assisted if various isotopic species can be studied. The study of groups of compounds in which one atom is successively changed to a heavier atom in the same group of the periodic table is also valuable. Any related set of molecules, such as homologous series in which the orbitals of a chromophoric group are modified in a minor way by a substituent, can obviously help assignment. The variation in the relative intensity of the bands with photon energy, the angular dependence of photoelectron cross-sections, and the possibilities of autoionization are all properties related to the orbital wavefunction, and if given the proper interpretation can facilitate assignment. An extension of the "aufbau" process to build the orbital structures of molecules with more atoms from molecules with fewer atoms (and electrons) is also valuable. Theoretical computations of orbital order using both *ab initio* and semiempirical approaches have been widely used to assign the bands of a photoelectron spectrum, and with improvement in molecular parameters this method has had increasing success. It has been popular with many authors who use it to justify both the assignment and the computational theory. This is obviously a dangerous procedure, and it is clearly desirable as far as possible to make assignments purely from experimental criteria. When this has been done, the predictions can then be compared with the calculations, and, if the comparison is favourable, it can be used to justify the parameters employed in the theory. In this way, the experimental data can be utilized to yield molecular properties not derivable from the spectra directly and the theory can be simultaneously justified.

H. Spectra of Ionic Molecules

The binding in the molecules we have so far considered is of the covalent type. In such binding the valence electrons are approximately equally attracted to either of the fragments into which the united atom has been partitioned. When this is not so as, for example, in the case of LiF for which the united atom is magnesium (configuration $1s^2 2s^2 2p^6 3s^2$) partitioning off a $3+$ fragment will remove first the weakly bound $3s^2$ electrons to form the $1s^2$ shell of Li^+ while the more tightly bound $2p$ electrons will tend to remain in the $2p^6$ shell thus forming an F^- fragment. The field tending to pull over the third electron is clearly dependent on the difference between the ionization potential of the metal atom $I(M)$ and the electron affinity of the halogen atom $E_a(X)$. Values of these differences for the alkali halides range between 2·33 eV for LiF and 0·28 eV for CsCl, whereas for covalent bonds they are much larger, e.g. H_2 or CH_3—Cl where the differences would be about 13 and 6 eV respectively. The $I(M)$—$E_a(X)$ differences affect the bonding through the changes which arise in the short-range repulsive forces associated with the overlap of the closed shells of the ions. A study of the electrons in these closed shells by photoelectron spectroscopy might therefore be expected to give direct information concerning the repulsive factors which largely determine the strength of this type of binding.

Photoelectron spectra of the chlorides, bromides, and iodides of Na, K, Rb, and Cs recorded in the vapour state by a molecular beam technique are shown in Figs 17 and 18 (Potts *et al.*[17]). All the spectra correspond to the removal of an electron from the $X^-(p)^6$ shell of the molecule. This produces an M^+X^0 ion in which the ionic bond has been destroyed leaving only a small residual bonding due to the attraction between the M^+ ion and the induced dipole in the X^0 atom. The photoelectron spectra of the M^+X^- molecules correspond to transitions to the various states of the M^+X^0 ions, the relative separations of which may be deduced by consideration of the polarization of the X^0 atom by the M^+ ion, i.e. by Stark splitting of the ^2P state of the X^0 atom in fields of different strength for each particular halide. On the basis of Mulliken's treatment,[18] we can represent the probable splitting of the X^0 ^2P state by the

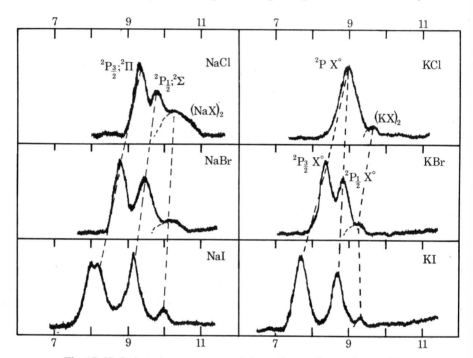

Fig. 17. He I photoelectron spectra of the sodium and potassium halides.

energy-level diagram shown in Fig. 19. The states of the X^0 ^2P atom and their relative energies should correspond to those observed for the M^+X^0 ion in the photoelectron spectra of M^+X^-. Describing the states of M^+X^0 in terms of the states of the X^0 atom, for the "zero field" situation, the spectra should show structure corresponding to $^2P_{\frac{3}{2}}$ and $^2P_{\frac{1}{2}}$ atomic X^0 states, the separation of the bands being just $\frac{3}{2} \times$ the spin–orbit coupling constant ζ for the $p^5, ^2P$ state of the free halogen atom. The spectra obtained for the Cs and Rb halides do in fact correspond to just this situation as might be expected from their low values of $I(M)-E_a(X)$ which are 0·280 and 0·563 eV for CsCl and RbCl respectively, i.e. if any negative charge is placed between M^+ and X^0 the attraction towards M^+ is largely counterbalanced by that due to the electron affinity of X^0. The separation found for the $^2P_{\frac{3}{2}}$ and $^2P_{\frac{1}{2}}$ bands is as predicted although in the case of the chlorides the two bands expected are not resolved.

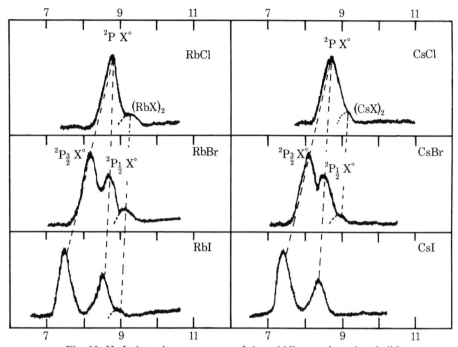

Fig. 18. He I photoelectron spectra of the rubidium and caesium halides.

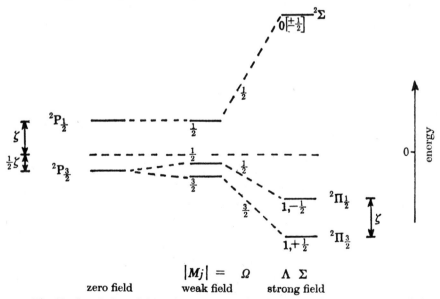

Fig. 19. Correlation of the various energies of ionized states formed by Stark splitting of the X^0 ^2P state.

For the potassium halides (Fig. 8) the spectra are beginning to show evidence of a polarizing field and show greater splittings than are found for free X atoms. The spectra of the sodium halides provide examples of polarization ranging from the weak field to the strong field situation. For NaI the $^2P_{\frac{3}{2}}$ state is split into $M_j = \frac{3}{2}$ and $\frac{1}{2}$ components and the $^2P_{\frac{3}{2}}$–$^2P_{\frac{1}{2}}$ splitting is slightly greater than $\frac{3}{2}\zeta$. In the spectrum of NaBr the $^2P_{\frac{3}{2}}$–$^2P_{\frac{1}{2}}$ splitting is increased to 0·65 eV as compared with a value of 0·46 eV expected for zero electric field, while for NaCl the splitting is increased from 0·11 eV to 0·46 eV. These indications of the presence of large fields agree with the relatively large $I(M)$–$E_a(X)$ differences which are 1·525 eV, 1·775 eV, and 2·075 eV for NaCl, NaBr, and NaI respectively.

Features to the high ionization energy side of the $(p)^{-1}$ bands observed in a number of the spectra are attributed to ionization of the $(MX)_2$ dimer which will be present in increasing amounts for the lighter halides. Comparison with the reliable values for the polymer (solid crystal) which are now available (Poole et al.,[19]) shows that the energies required to ionize the dimers and polymers are greater than those needed for the monomer. The greatest increment occurs in going from monomer to dimer, the subsequent changes being an order of magnitude smaller. For example, the first ionization potential of NaCl is 9·34 eV, that of $(NaCl)_2$ is 0·96 eV higher while that of $(NaCl)_n$ crystal is raised still further by 0·15 eV. These increments are clearly due to increases in the polarization stabilization of the ionized states of Cl^- arising from the increasing population of its environment by polarizable ions. At present we are still at an early stage in the study of ionic binding by photoelectron spectroscopy, but valuable new information is already being accumulated (Berkowitz et al.[20, 21]).

I. Spectra of some simple aromatic molecules

Although this chapter is primarily concerned with relatively small molecules it could hardly omit a brief account of the photoelectron spectrum of benzene and some of its simple derivatives because of the great importance of the electronic arrangement in the aromatic nucleus which is responsible for the special properties of six-membered π ring systems. Such systems are too large to be treated by the united-atom techniques and orbital assignments must be made by comparison of the bands observed in the photoelectron spectrum with the set of molecular orbitals which can be formed by combining atomic orbitals into orthogonal sets appropriate to the molecular geometry. Jonsson and Lindholm[22] made the first assignment of the benzene spectrum based on several different types of information including SCF calculations. More recently further information has been obtained on the nature of the bands from measurements of the relative band intensities in spectra obtained with Ne I, He I, He II,[23] and Mg Kα irradiation,[24] from detailed analyses of the few bands which have resolvable vibrational fine structure,[23] from angular distributions of the photoelectrons,[25] from comparisons with Rydberg series,[26] and from photoionization cross-section mass spectrometric studies.[27] After a certain amount of controversial discussion the Jonsson and Lindholm assignment is now fairly firmly established. It is shown in Fig. 20. Using this assignment the orbital structure of benzene omitting $C(1s)$ combinations can be written in a simplified form as $s^2 s^4 s^{*4} r^2 s^{*2} t^2 r^4 \pi^2 t^4 \pi^4$ where s corresponds

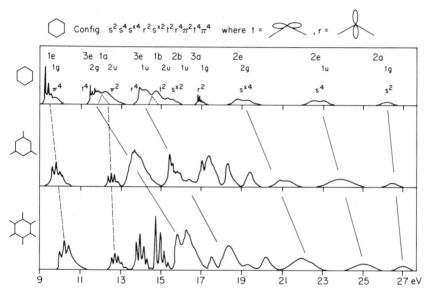

Fig. 20. Photoelectron spectrum of benzene, 1,3,5-trifluorobenzene, and hexafluorobenzene (composite He I and He II spectra).

to orbitals built from $(2s)$ orbitals, t to orbitals built from $C(2p)$ orbitals orientated tangentially to the ring, r orbitals constructed from $C(2p) + H(1s)$ orbitals oriented radially and π orbitals from $C(2p)$ orbitals oriented perpendicular to the molecular plane. The successive orbitals are simply the various linear combinations of the atomic orbitals which have progressively increasing numbers of nodal surfaces. The order in which they are given is that of binding energy decreasing from left to right. Writing these orbitals in terms appropriate to D_{6h} symmetry the ground-state configuration of benzene is $(2a_{1g})^2 (2e_{1u})^4 (2e_{2g})^4 (3a_{1g})^2 (2b_{1u})^2 (3e_{1u})^4 (1a_{2u})^2 (3e_{2g})^4 (1e_{1g})^4$. Molecular orbital calculations can be used to identify some of the bands directly. All calculations indicate that the first ionization potential at 9·25 eV is to be identified with $(\pi, 1e_{1g})^{-1}$ ionization. This is supported by the appearance in its vibrational pattern of vibrations of e_{2g} symmetry in addition to totally symmetrical vibrations. Jahn–Teller splitting would normally be expected for the ionization of the $(1e_{1g})^4$ filled shell but the bonding power of the electrons is so small in this case that the only indication of the degeneracy of the ionic state is the appearance of the e_{2g} vibrations in single quanta.[23]

The next π^{-1} ionization potential $(1a_{2u})^{-1}$ is expected to lie in the 12 eV region but as this is also where the ionization $(t, 3e_{2g})^{-1}$ is expected to occur some uncertainty has existed in the assignment in this region. Various features of the photoelectron spectrum in the region 11·4–13·4 eV have been invoked to try to settle this question but as only a small part of it shows sharp structure a satisfactory understanding of this region has taken some time to achieve. A deeper understanding of the vibrational pattern in terms of the Jahn–Teller effect as well as shifts and splittings brought about by various F substitutions (see later) now appear definitely to favour the assignment shown in Fig. 20 as originally suggested by Johnson and Lindholm.[22, 26] All the other assignments of bands in the photoelectron spectrum to orbitals of the valence electrons are now generally accepted and since for hydrocarbons the bottom of the valence shell can be reached with He II radiation our knowledge of the valence shell orbitals of benzene is complete. Photoelectron spectroscopy has certainly revealed the orbital structure of this 12-atom molecule to a degree which a decade ago seemed beyond our reach.

There are of course many as yet unanswered questions about the photo-electron spectrum. For example, it would be of interest to know why most of the bands corresponding to ionization from the inner orbitals are continuous and do not show vibrational structure. The very short lifetime of the ionized state which this implies may be caused by an electron from an outer orbital quickly falling in to the vacancy resulting from the primary photoionization. Alternatively, the rapid dissociation of the ion into an ionized and a neutral fragment could be the cause of the continuous character of some of the photoelectron bands. It might therefore be asked to what extent is the energy of the ionized state altered by interaction with the repulsive state through which dissociation may occur. No doubt these questions will be answered by further research.

J. The Spectra of Some Fluorobenzenes

Some interesting points on the effect of substituents on group orbital systems are illustrated in the spectra of the fluorine-substituted benzenes shown in Figs. 20, 21, and 22. These have relevance to the orbital assignment of the bands in benzene itself. Figure 20 compares the spectrum of 1,3,5-tri-fluoro- and hexafluorobenzene with that of benzene. The mixing of $2p\pi$ F orbitals into the π^4 and π^2 orbitals is well known to cause little shift of π levels due to the compensation of the blue shifting inductive effect by an opposing mesomeric effect.[28] The in-plane t orbitals in 1,3,5 trifluoro- and

energies. This separates the $1a_{2u}$ clearly from the $3e_{2g}$ orbitals and the spectra of the tri- and hexafluorobenzenes show the $1e_{1g}$ and $1a_{2u}$ bands appearing in clearly recognizable form with an expected intensity ratio of $2/1$.

Tetra - fluoro - benzenes

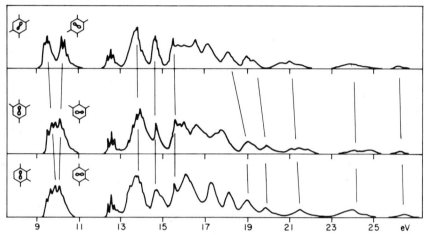

Fig. 21. Photoelectron spectra of benzene, monofluorobenzene, and p-difluorobenzene.

P.E. spectra of C_6H_6 , C_6H_5F and $p - C_6H_4F_2$

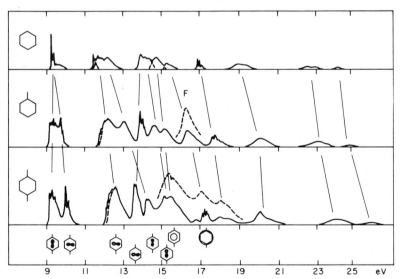

Fig. 22. Photoelectron spectra of the tetrafluorobenzenes.

Figure 21 shows the substitutional splitting of the $(1e_{1g})^4$ orbital as we pass to C_{2v} symmetry in going from benzene to monofluorobenzene and to p-difluorobenzene. The doubly degenerate orbitals $(1e_{1g})^4$ now split into two mutually orthogonal components. One of these is a $2b_1$ orbital, which has a nodal surface perpendicular to the molecular plane and passing through the CF bond. This orbital is therefore unaffected by the substitution. The other is a $1a_2$ orbital, one lobe of which overlaps the $2p\pi$ F orbital and therefore interacts with it (*see* Lindholm[26]). These orbital differences are strikingly confirmed by the vibrational patterns observed in the two photoelectron bands into which the first benzene band divides. The first of these shows a fairly wide vibrational pattern and indicates that it corresponds to ionization from the $1a_2$ orbital which overlaps the $2p\pi$ F orbital. The vibrational pattern of the second band is hardly altered from that of unsubstituted benzene and is thus to be correlated with the $2b_1$ orbital which has a nodal plane passing through the CF bond. The vibrational pattern of the first band is not in fact greatly different from that of the $1a_{2u}$ orbital ($1b_1$ in C_{2v} symmetry) which must also overlap the $2p\pi$ F orbital. Although this cannot be seen in the spectra of Fig. 21 it becomes clear in the spectra of the tetrafluoro benzenes (Fig. 22) where the $(1a_{2u})^{-1}$ band is well separated from the band systems arising from in-plane orbitals.

The orbital interactions arising from the insertion of $(2p\pi_{x,y})^4$ electrons with every F substitution differ somewhat for the in-plane orbitals from those for the π orbitals just discussed. These differences afford valuable criteria for orbital identification. A $2p$ F orbital lying in the molecular plane and perpendicular to a CF bond can interact mesomerically with a σ ring orbital which happens to have a major nodal plane coinciding with the nodal plane of the $2p$ F orbital. A band associated with such an orbital would not be shifted so much to high energies as one associated with another (orthogonal) orbital for which the inductive effect of the fluorine was not compensated by opposing mesomeric repulsion. In the C_{2v} symmetry situation arising in mono and p-difluoro-benzene it is clear from Fig. 21 that the band system assigned as due to $3e_{2g}^{-1}$ ionization is split by the substitution, the lower ionization potential half suffering minor and the higher ionization potential half suffering major shifts to higher energies. Reference to the orbital diagrams shows that such behaviour is to be expected if these orbitals are respectively $7b_2$ and $11a_1$ species in C_{2v} symmetry. The shape of the singly degenerate t^2 orbital $1b_{2u}$ indicates an appreciable mesomeric effect and thus a small shift is to be expected for this band. In 1,3,5-trifluoro and hexafluorobenzene where D_{3h} and D_{6h} symmetries pertain there can be no mesomeric effect for the in-plane orbitals as in C_{2v} symmetries. Consequently there is no splitting of band systems and all orbitals are shifted to higher energies by unopposed inductive effects. These features are indicated by connecting lines in the figures. Similar behaviour is shown by the in-plane r^4 and r^2 systems as shown by the t^4 and t^2 systems with respect to changes in molecular symmetry arising from F substitution. The two sharp band systems in hexafluoro-benzene appearing in the range 13·5–15·5 eV seem to be the π^{*2}, π^{*4} which together with a π^4 and π^2 set at higher energies account for the twelve $2p\pi$ F electrons of the C_6F_6 molecule.

IV. PHYSICAL ASPECTS OF PHOTOELECTRON SPECTROSCOPY

A. The Intensities of Photoelectron Spectra

Apart from the quantitative data concerning orbital energies to be obtained by photoelectron spectroscopy, the intensities of the bands obviously contain much information about the nature and extent of the orbitals themselves. In this connection we must consider the transition moment of an electron from the initial orbital ϕ to the state ε in the ionization continuum into which it is ejected by the incident photon. This is illustrated in Fig. 23 where it can be seen that the integral representing the transition moment depends upon the spread and nodal character of the wavefunction of the initial state and the wavelength and phase of the photoelectron (i.e. the final state). For the purposes of this illustration the wavefunctions are taken as those of a one-dimensional oscillation of an electron in a $V(1/r)$ potential field, that is, pseudo-Rydberg bound states. Plane wavefunctions are used for the continuum states ($\lambda[1 \text{ eV}] \sim 12 \text{ Å}$). The simplest fact that this illustrates is the fall-off of the photoionization cross-section with higher energy (shorter wavelength) photoelectrons. This explains the relative transparency of electrons in valence shells to x-irradiation, where the area of the graph for the transition moment is divided into small compensating positive and negative regions. It can be appreciated that the cross-section appropriate to any orbital will vary with the dimensions and nodal character of this orbital and the wavelength of the photoelectron. As a rough generalization it might be expected that the photoionization cross-section would maximize when $\lambda/4$ of the photoelectron is not less than the orbital dimensions. This would explain experimental observations that near the threshold of ionization the cross-sections of molecular orbitals built from p atomic orbitals are greater than those arising from s orbitals. This situation is reversed for high photoelectron energies, where, because of the smaller radial spread of s orbitals, the transition moment integral maximizes with shorter wavelength photoelectrons. The study of the variation of the relative intensity of the photoelectron spectra of different orbitals with varying irradiating wavelengths is clearly going to yield much information on the nature of orbital wavefunctions. A summary of the information available for gases on this topic has been given by Price *et al.*[29] Photon sources of different energies derived from synchrotron radiation have already been used by Eastman and Grobman[30] to study the band structure of gold and the density of intrinsic surface states

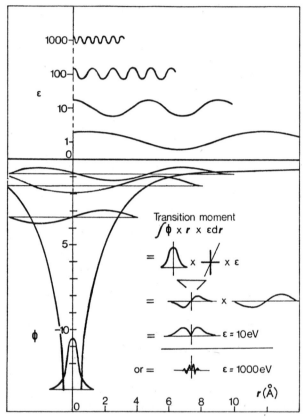

Fig. 23. Diagrammatic illustration of features of importance in the photoionization of an electron from a bound state ϕ to a continuum state ε.

in Si, Ge, and GeAs. The application of this source of photons of variable energy is still little exploited for gases, and it must be stated that the problem of getting adequate energy in a sufficiently narrow band is a very formidable one. It will be particularly important in the study and interpretation of autoionization phenomena and has already been used for this purpose.

B. Selection Rules

The photoionization process in the orbital approximation is a one-electron transition resulting from the interaction of the radiation with the electric dipole of the orbital. The difference of angular momentum between the neutral molecule and the molecular ion is given by

$$\Delta L = 0, \pm 1, ..., \pm l$$

where l is the angular momentum of the ionized electron, i.e. L can change by m_l. The free electron which can be an s, p, d, or f wave requires an angular momentum of $l \pm 1$, and when two different angular momenta are possible by this rule, e.g. $p \to s$ or d, both may be utilized giving an outgoing wave of mixed character. Since one electron with spin $\pm \frac{1}{2}$ is removed the rule for the change of multiplicity is

$$\Delta(2S+1) = \pm 1$$

Two electron processes are weakly allowed only when there is appreciable correlation between the motions of electrons in different orbitals, the intensities of their bands reflecting the extent of the interaction. Because all one-electron photoionizations are allowed the partial photoionization cross-sections, or the probabilities of ionization to different states of the ion, are all of the same order of magnitude, i.e. integrated band intensities per orbital are roughly equal. Orbital and spin degeneracies provide multiplying factors which affect the relative intensities of related bands.

C. Dependence of Cross-Section on Photoelectron Energy

As already indicated, the dependence of the transition moment on the continuum wavefunction implies a variation of the cross-section for photoionization with the wavelength of the photoionizing radiation. The cross-section is generally highest near the threshold after which it decreases with increasing photon energy. When the photon energy is very high compared with the threshold energy as, for example, in the photoionization of valence electrons by soft x-rays, the cross-section should be proportional to ν^3. The cross-sections near threshold exhibit a more complicated behavior, depending on the nature of the overlap between the initial and continuum orbitals involved. Because of experimental difficulties, few absolute measurements are available, but the different ways in which different orbitals behave with increasing photon energy can be judged from the changes in the relative intensities of the bands in spectra obtained first with low- and then with high-energy photons. Figure 24 illustrates the changes in the relative intensities of bands from p- and s-type orbitals with ionizing photons of low and high energies. It can be seen that at low energies the photoelectron bands of p-type orbitals are generally much stronger than those of s bands from the same shell, whereas the reverse is true for their relative intensities in photoelectron spectra obtained with high energy x-radiation. This behavior, which is quite general in a large number of atomic and molecular cases, reflects the greater charge density of s orbitals at smaller radii, since it is at these short distances that the major contribution to the transition moment for short wavelength photoelectrons arise. Many other examples of variation of cross-section with photon energy are given by Price *et al*.[29] Calculations of the relative partial cross-sections of various orbitals from near threshold to high photoelectron energies have been carried out by many authors, the early

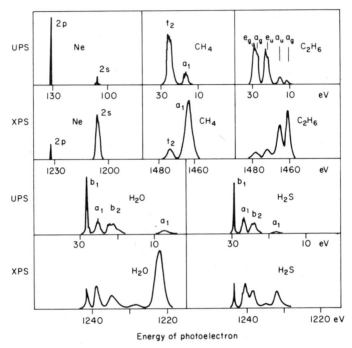

Fig. 24. Examples of changes in relative intensities of bands from *p*- and *s*-type orbitals with ionizing photons of low and high energies.

work being referenced by Marr.[31] More recently Lohr,[32] Gelius,[33] and Iwata and Nagakura[34] have made cross-section calculations using hydrogen-like continuum functions for the ejected electron and SCF wavefunctions for the orbitals of the bound electrons both in the initial and the residual ionized states. They obtained values which agree reasonably well with the experimental results particularly when the photoelectrons have high energies. Schweig and Theil,[35] by dividing the cross-section into one- and two-center terms representing, respectively, contributions from atomic orbitals and their bond combinations as modified by the population coefficients, have obtained reasonable agreement for relative band intensity changes between He I and He II spectra. Their results from threshold up to about 5 eV are not expected to be significant since they use plane wave continuum functions.

D. Autoionization

The ionization continua of the outer orbitals of a system are overlapped on an energy diagram by a manifold of excited bound states associated with electrons from inner orbitals (so-called super-excited states which are usually Rydberg in type). If the symmetry of the product of the wavefunctions of a lower ionized state and its photoelectron ($M^+ \times \varepsilon$) is the same as that of an overlapping superexcited bound state then these states mix. The theory in the case of atomic processes was given by Fano[36] and Fano and Cooper.[37] It has been generalized for more complex systems by Mies[38] and Bardsley.[39] The situation in the case of molecules is complicated relative to that in atoms by vibrational effects which arise from the different shapes of the potential energy curves of the three states involved in the process, viz. the neutral, the super-excited, and the ionic state. In the case of a diatomic molecule, for example, absorption can take place to a vibrational level of the bound state lying vertically above the ground-state minimum followed by a swing of the system to the opposite turning point on its potential energy curve. Auto-ionization is most probable from values of r where ψ_{vib}^2 has maxima, i.e. particularly at the turning points of the vibration. Thus at the far turning point, autoionization can take place strongly to the vibrational state of M^+ vertically below it. Ionization can occur to vibrational levels of the M^+ state which, on Franck–Condon grounds, are not accessible by direct ionization to M^+, and one or more additional maxima may appear in the vibrational pattern. Two factors are therefore important: (1) in addition to the symmetry of the electronic wavefunction the nature of the autoionization is affected by the shape of the potential energy curve of the autoionizing state through its vibrational wavefunction, and (2) information about both can be expected from a study of the phenomenon.

One of the most interesting examples of autoionization is that of oxygen under neon irradiation. Figure 25(a) and (b) show spectra taken with constant energy bandwidth with Ne I and He I radiation respectively. Figure 25(c) shows a part of the spectrum taken with Ne I radiation under much higher resolution. The neon radiation consists of a strong line at 16·85 eV and a much weaker line at 16·67 eV. As can be seen from Fig. 26 the latter does not coincide with an absorption band of O_2. The $(\pi_g)^{-1}$ photoelectron spectrum due to this 16·67 eV photon corresponds to the minor peaks in the 12·2 to 13·2 eV region which have the same intensity distribution as the spectrum obtained with 584 Å irradiation and thus exhibit no autoionization. The major peaks which show a much extended system with an additional maximum are due to the 16·85 eV line which lies on top of a rather broad absorption band as shown in Fig. 26 (Geiger and Schröder[68]). The He I and the Ne I photoelectron spectra are plotted along the ordinate of Fig. 27 which gives also the potential curves of the states of O_2^+. It can be seen by comparison of the two spectra that both the $^2\Pi_g$ and the $^4\Pi_u$ states are affected by autoionization. While in the former an abnormal vibrational intensity pattern is produced, in the latter, as far as can be judged from the

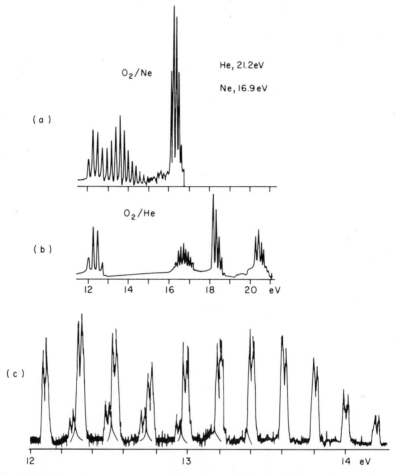

Fig. 25. (a) Photoelectron spectrum of O_2 taken with Ne I radiation. (b) Photoelectron spectrum of O_2 taken with He I radiation. (c) High resolution spectrum of O_2 using Ne I radiation showing autoionization enhancement by the 16·85 eV line (major peaks) but not by the 16·67 eV line (minor peaks).

limited position of the $^4\Pi_u$ spectrum lying above 16 eV available with the neon line, autoionization enhancement occurs from the start of the system. It also appears that the autoionization in the 16 eV region produces a much stronger enhancement than that in the 12 to 15 eV region. The absorption band lying on the 16·85 eV line of neon is undoubtedly a Rydberg band going to the $(\sigma 2p)^{-1}\,^4\Sigma_g^-$ state of O_2^+. The symmetry requirement would be met for the $^4\Pi_u$ autoionization if the bound state was $^4\Sigma_g^-(4p\pi_u)_3 : {}^3\Pi_u$, this symmetry being also matched by the ionized system $^4\Pi_u\,\varepsilon s\sigma_g : {}^3\Pi_g$. (This bound state has been tentatively assigned by Lindholm[41] to a relatively sharp band lying at 16·93 eV, but it does not appear that a state with the

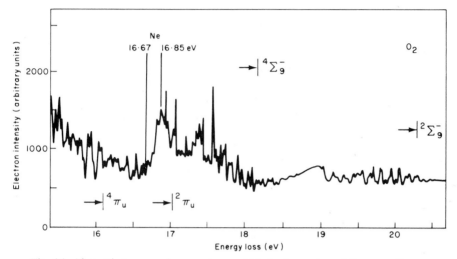

Fig. 26. Absorption energy loss spectrum of O_2 in the region of the neon lines.

right symmetry to autoionize should stay sharp.) The intensities in the vibrational pattern are also compatible with this process for $v = 0$ in the bound state (see Fig. 26). The question now is whether this state could also be responsible for the autoionization of the $X^2\Pi_g$ state (in the 12 to 15 eV range). Continuum states $X^2\Pi_g\ (\varepsilon p\sigma_u) : {}^3\Pi_u$ would give the right symmetry, but according to the Franck–Condon calculations of Kissinger and Taylor,[42] it is necessary to assume that the bound state would have $v = 1$ to get the right intensity distribution in the vibration pattern of the autoionized system. This does not appear to be compatible with the absorption spectrum of O_2, and it may be that another overlying bound state is involved in autoionizing $X^2\Pi_g$ such as a $v = 4$ of a very high Rydberg state going to $A^2\Pi_u$ (note the I_{adiab} of $A^2\Pi_u$ is 17·04,[45] i.e. greater than the 16· eV neon photon).[43] Autoionization through $^3\Sigma_u^-$ as well as $^3\Pi_u$ states of O_2 can occur to both $X^2\Pi_g$ and a $^4\Pi_u$ states of O_2^+.

Studies of autoionization in oxygen have engaged the attention of many workers, because oxygen is one of the major sources of free electrons in the upper atmosphere through photoionization by solar radiation. Also such studies could help in interpreting features of the O_2 absorption spectrum which, because of its diffuseness and complexity, has not yet been fully elucidated. Blake *et al.*[44] observed autoionization at several different irradiating wavelength bands isolated from a helium continuum with a vacuum monochromator. The object was to obtain information about the autoionizing state by irradiating it in its successive vibrational levels, each of which will give rise to a different vibrational intensity distribution in the

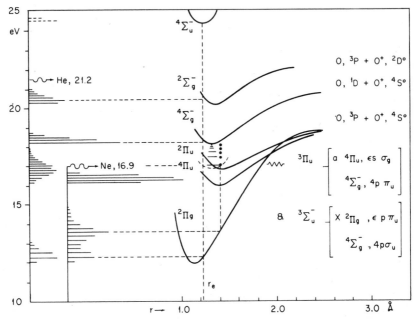

Fig. 27. Potential energy curves of O_2^+ with photoelectron spectrum drawn along the ordinate indicating the nature of the autoionizing processes.

autoionized spectrum. The work has been carried out under higher resolution and with synchrotron radiation by Kissinger and Taylor,[42] who find reasonable agreement with the theory given by Smith.[45] This theory yields the autoionizing linewidth and the Fano line profile index of the bands of the autoionizing state which can be used to determine autoionizing rates and lifetimes of the excited states as well as assigning v' values. Autoionization bands very similar to those in O_2 have been found for NO (Kleimenov *et al.*,[46] Collin and Natalis,[47] Gardner and Samson.[48] It must be said, however, that as yet no case of the autoionization of a diatomic molecule is completely understood, particularly with reference to its effect on the energies of the superexcited state.

Since low-energy photoionizing lines are always likely to coincide with the Rydberg bands of inner orbital electrons, it is not surprising that many intensity anomalies are found in photoelectron spectra excited with hydrogen, argon, neon, and even helium resonance radiation. The spectra taken with 304 Å (He$^+$) radiation are free of these, since few molecules have Rydberg bands of valence electrons at this wavelength. Figure 28 shows certain striking

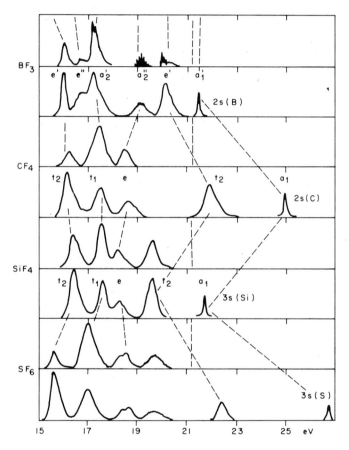

Fig. 28. Spectra of XF$_n$ molecules with 584 Å and 304 Å irradiation indicating intensity changes due to autoionization in the 584 Å systems.

differences in the spectra of some XF$_n$ molecules photoionized by 584 and 304 Å radiation. It is the second and/or the third band in these spectra that are enhanced by autoionization if our explanation is correct. SF$_6$ has, in fact, an enormously strong second band under 584 Å excitation. It is possible that this results from interaction with the dissociative process into $F + FS_5^+$ which occurs within this band. Under 304 Å excitation the intensity pattern is similar to that of UF_6 under 584 Å excitation. (Note: UF_6 has no absorption bands at 584 Å.) The enhancement in these cases is not accompanied by displacement of the maximum of the band envelope as in O_2 and NO, which is reasonable in view of the relatively nonbonding nature of many of the orbitals in these molecules.

A very general type of autoionization is found in the photoelectron spectra of aromatic hydrocarbons. These spectra usually are characterized by an initial set of sharp bands arising from electrons in π orbitals followed by a set of more diffuse bands due to σ-type orbitals at higher energies. The intensities of the σ bands increases substantially relative to the π bands when Ne I radiation is used instead of He I. A similar behavior is observed in angular distribution studies when the photoelectron angle is changed from 90° to 30° with the light beam. Another very interesting case has been found in the photoelectron spectra of C_6H_6 and C_6D_6 under neon (16·86 eV) irradiation (Price[49]). A band system in the region (11·6–13 eV) was found to be selectively enhanced in C_6D_6 relative to C_6H_6. This appears to be due to the fact that one orbital of the benzene molecule has an ionization energy which in C_6H_6 is about 0·01 eV below the photon energy, whereas in C_6D_6, due to the blue zero-point energy shift, its ionization lies 0·01 eV above the energy. In the latter case the excited electron remains attached to the molecule and cannot get away. It ultimately interacts with an electron in a lower orbital thereby increasing the strength of its band by autoionization. In C_6H_6 this electron escapes into the ionization continuum and therefore does not produce autoionization. In a simple way this may be regarded as an internal electron impact phenomenon.

Recent studies by Baer and Tsai[50] have been made of the resonant photoelectron spectra from autoionizing states of CH_3I. It was found that Bardsley's formulation of the configuration interaction theory of autoionization was consistent with the peak intensities of the vibrational bands observed. In his theory the photon energy need not be in exact resonance with a Rydberg state in order for that state to contribute to autoionization, though its contribution is inversely proportional to the energy mismatch.

A further type of autoionization known as vibrational autoionization is possible though it has not yet been found to be very prevalent in photoelectron spectroscopy. The autoionizing state must have an electron in a Rydberg orbital with a high principal quantum number and also enough vibrational ionization to bring the total energy above the minimum molecular ionization energy. Autoionization follows by interaction of the electronic and vibrational motions in which the vibrational energy of the core is converted into electronic energy and the Rydberg electron is ejected. This type of autoionization has been studied theoretically by Berry[51] and experimentally by Berkowitz and Chupka.[52]

E. Angular Distribution of Photoelectrons

The angular distribution of the photoelectrons ejected from a molecule relative to the direction of the incident photon beam is determined by the angular momentum they must carry in order to satisfy the dipole selection rules, i.e. by the s, p, d, or f character of the outgoing electrons. In the ionization of an s electron to a p photoelectron, the axis of the p-wave function is defined by the direction of the electric vector of the light wave. Taking this as the z axis, the angular part of the electron wavefunction is the same as that of an atomic p_z electron and the probability of observing an electron is proportional to its square. Thus ionization of an s electron leads to a $\cos^2 \theta$ distribution about the electric vector.

$$I(\theta) = \text{const } Y_{10}{}^2 = \text{const } \tfrac{3}{4}\pi \cos^2 \theta$$

Ionization from a p orbital leads to s and d outgoing waves. The s waves have a spherical distribution while the d waves are peaked along the direction of the electric vector. Whatever the mixture of s, p, d, etc., waves it turns out that the photoelectron angular distribution can be expressed by the formula

$$I(\theta) = (\sigma/4\pi)[1 + \tfrac{1}{2}\beta(3\cos^2 \theta - 1)]$$

where σ is the total cross-section integrated over all angles and β is an anisotropy parameter. Since in photoelectron spectrometers the light is normally unpolarized, angular variations in intensity must be measured in directions θ' away from the direction of the light beam. The distribution of intensity is then

$$I(\theta') = (\sigma/4\pi)[1 + \tfrac{1}{2}\beta(\tfrac{3}{2}\sin^2 \theta' - 1)]$$

For a pure p wave β has the value 2 thus giving a $\sin^2 \theta'$ variation. β can range from -1 to $+2$. Expressions for β in terms of quantities such as l, the angular momentum of the electron before ionization; σ_{l-1} and σ_{l+1}, the partial cross-sections for production of the $l-1$ and $l+1$ waves; and $(\delta_{l+1} - \delta_{l-1})$, the phase difference between the two waves have been worked out for atoms (Cooper and Zare[53]) and diatomic molecules (Buckingham et al.[54]). However, calculations of β are impracticable for large molecules, so that hopes of assigning a photoelectron band explicitly by a measurement of its asymmetry parameter are not likely to be fulfilled in such cases. Measurements of angular distribution will help, however, to differentiate between photoelectron bands especially when these overlap, and it can be used in a semi-empirical way for band assignment when the behavior of established types of bands are known. For example, experiment shows that $(\pi)^{-1}$ bands in linear or planar molecules are relatively much weaker at low angles with the light beam than $(\sigma)^{-1}$-type bands, e.g. π bands in ethylene, benzene, water, methyl, and iodide show this behavior. The variations in relative intensities for nonplanar polyatomic systems are much less and are frequently hardly within the accuracy with which the measurements can be made.

Carlson and his co-workers (Carlson,[55, 56] Carlson *et al.*,[57] Carlson and McGuire[58]) have been among the more active workers in this field. They obtain good agreement between theory and experiment for the inert gases, H_2 and N_2. In their examination of many polyatomic gases they find in addition to the features already mentioned that the angular parameter depends mostly on the initial orbital from which the electron was ejected and not on the different final states that may arise as a result of Jahn–Teller splitting, spin–orbit splitting, or spin-coupling between two unfilled orbitals. The relative intensities of the vibrational bands of a given electronic system are usually unaltered, but in certain cases irregularities are found connected with either autoionization or some breakdown in the Born–Oppenheimer approximation involving coupling between electronic and vibrational motions.

The experimental arrangement of Carlson and others utilizing a source rotatable with respect to the electron analyzer usually involves considerable loss in photoelectron flux, which results in lower resolution spectra. This can be avoided either by using two slits and two detectors to cover two electron paths in a hemispherical analyzer (Ames *et al.*[59]) or by using two discharge tubes, one aligned parallel to the slit ($\theta' = 90°$) and one perpendicular to it which allows photoelectrons to be accepted by the analyzer at an angle of 30° with the beam (W. C. Price and K. Glenn, unpublished, 1971). Spectra taken by the latter method are shown in Fig. 29.

The theory of the partial cross-sections and angular distributions for individual rotational transitions in the photoionization of diatomic molecules has been worked out by Buckingham *et al.*[59] The only molecule for which resolved rotational photoelectron spectra can be obtained is H_2. Niehaus and Ruf[60] using Ne I radiation measured β for the different rotational transitions, finding for $\Delta J = 0$, $\beta = 1.95$, and for $\Delta J = 2$, $\beta = 0.85$. When $\Delta J = 0$ the value of β is so close to $+2$ that the outgoing electron must be almost a pure p wave. For $\Delta J = 2$, theory indicates it should be a mixture of p and f waves (Sichel,[61]). The experimental value ($\beta = 0.85$) agrees well with a pure f wave for which $\beta = 0.80$. This would correspond to a simple addition of the angular momentum changes in rotational and electronic motion in determining the angular momentum of the outgoing electron. The agreement might be fortuitous since the value $\beta = 0.8$ could result from interference between p and f waves of appropriate phases and amplitude.

F. Electron–Molecule Interactions

Anomalous peaks have been observed in the photoelectron spectra of certain diatomic and triatomic molecules when the pressure in the ionization

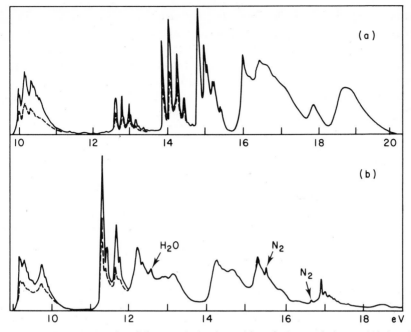

Fig. 29. Spectra showing different relative intensities of photoemission at 90° (———) and 30° (– – – – –) with the photon beam: (a) C_6F_6, (b) C_6H_5Cl.

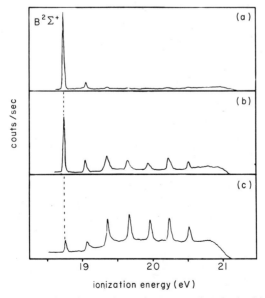

Fig. 30. Photoelectron spectra at elevated pressure of N_2 ionized by helium radiation (21·22 eV): (a) 15 μ (b) 50 μ (c) 110 μ.

chamber is relatively high. These are attributed to collisions between emitted photoelectrons and neutral molecules leading to the formation of temporary negative ion states (Streets *et al.*[62]). Since the negative ion can return to the various vibrational levels of the neutral ground state, the re-emitted electron will then have its initial energy reduced by the appropriate number of vibrational quanta of the ground state. One of the most interesting examples of this occurs in the ionization of N_2 to its $(\sigma_u 2s)^{-1}\ ^2\Sigma_u^+$ state at 18·76 eV

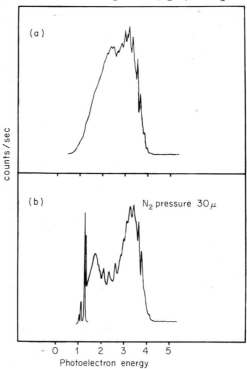

Fig. 31. Resonance absorption spectrum of N_2^- formed by the absorption of photo-electrons emitted from methane ionized by neon radiation: (a) CH_4, (b) $CH_4 + N_2$.

by He I (21·22 eV). The photoelectron spectra obtained at various pressures are shown in Fig. 30. As the pressure is increased it is seen that the photo electrons from the main band are degraded into a series of lower energy bands with an interval which corresponds to the neutral ground-state frequency of N_2.

The resonance absorption spectrum of N_2^- can also be observed and measured with high precision by photoelectron spectroscopy. The upper curve in Fig. 31 shows the spectrum of CH_4 obtained with Ne I radiation. This of course corresponds to a well-defined band of photoelectrons having energies from 0 eV to 4 eV. If nitrogen is now admitted to the ionization chamber, its absorption of the CH_4 photoelectrons can be observed in the photoelectron spectrum, i.e. an electron absorption spectrum can be obtained which gives the vibration pattern appropriate to the ground state of N_2^-. The halfwidths of the bands indicate the lifetimes of the negative ion states. The process is illustrated in Figs. 31 and 32. Many other negative ions can be examined in a similar way.

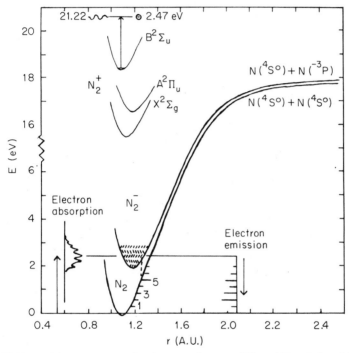

Fig. 32. Potential energy curves for the ground states of N_2 and N_2^- deduced from the excitation and absorption data.

G. Transient Species

If a short-lived radical or free atom can be transported sufficiently quickly into the ion chamber of a photoelectron spectrometer, a spectrum can be obtained from which its orbital structure can be determined. This can be done by fast pumping from the source of the transients through the ion chamber. The unstable species are produced as close as possible to the ion chamber usually by means of a high-frequency discharge, a high temperature furnace, or a gas reaction chamber (Jonathan *et al.*,[63, 64] Cornford *et al.*[65]). Spectra have been obtained for atomic hydrogen, oxygen, nitrogen, and the halogens, vibrationally excited N_2, $O_2(^1\Delta_g)$, $SO(^3\Sigma^-)$, CHO^+. and CS. It is clear that much valuable information will be forthcoming as this technique is developed.

V. CONCLUSION

In the brief period in which it has been developed photoelectron spectroscopy has had phenomenal success in revealing the electronic structure of matter in a particularly direct way. Chemists can see the orbital structure of even fairly large molecules and no longer have to rely on the predictions of theoreticians. Future advances in the study of solids, adsorbed species, etc., can confidently be predicted to yield information on the nature and functions of the electrons involved. It is also clear that the subject will provide a happy hunting ground for the physicist as well as the chemist for many years to come.

REFERENCES

1. W. C. Price, *J. Chem. Phys.* **4**, 147, 539 (1936).
2. K. Watanabe and J. R. Mottl, *J. Chem. Phys.* **26**, 1773 (1957).
3. F. I. Vilesov, B. C. Kurbatov, and A. N. Terenin, *Dokl. Akad. Nauk SSSR* **138**, 1320 (1961); F. I. Vilesov, B. C. Kurbatov, and A. N. Terenin, *Sov. Phys.–Dokl.* **8**, 883 (1962).
4. K. Siegbahn, C. Nordling, A. Fahlman, R. Norberg, K. Hamrin, J. Hedman, G. Johansson, T. Bergmark, S. E. Kartson, I. Lindgren, and B. Lindberg, "ESCA—Atomic, Molecular and Solid State Structure studied by means of Electron Spectroscopy" (North Holland Publ. Co., Amsterdam, 1967).
5. K. Siegbahn, C. Nordling, G. Johanssen, J. Hedman, P. F. Heden, K. Hamrin, U. Gelius, T. Bergmark, L. O. Werme, R. Manne, and Y. Baer, "ESCA—Applied to Free Molecules" (North-Holland Publ. Co., Amsterdam, 1969).
6. N. Jonathan, A. Morris, M. Okuda and K. J. Ross, *Chem. Phys. Letters*, **7**, 497 (1970).
7. N. Jonathan, A. Morris, M. Okuda, K. J. Ross, and D. J. Smith, *Discuss. Faraday Soc.* **54**, 48 (1972).
8. A. W. Potts, H. J. Lempka, and W. C. Price, Unpublished work.
9. A. J. Blake, *Proc. Roy. Soc. Lond.* **A325**, 555 (1971).
10. T. Koopmans, *Physica*, **1**, 104 (1934).
11. A. W. Potts and W. C. Price, *Proc. Roy. Soc. Lond.* **A326**, 165 (1972).
12. A. W. Potts and W. C. Price, *Proc. Roy. Soc. Lond.* **A326**, 181 (1972).
13. A. W. Potts, H. J. Lempka, D. G. Streets, and W. C. Price, *Phil. Trans. Roy. Soc. Lond.* **A268**, 59 (1969/70).
14. N. Jonathan and L. Golob,
15. A. D. Walsh, *J. Chem. Soc. Lond.* pp. 2260 and 2266 (1953).
16. C. R. Brundle, D. Neumann, W. C. Price, D. Evans, A. W. Potts, and D. G. Streets, *J. Chem. Phys.* **53**, 705 (1970).
17. A. W. Potts, T. A. Williams, and W. C. Price, *Proc. Roy. Soc. Lond.* **A341**, 147 (1974).
18. R. S. Mulliken, *Rev. Mod. Phys.* **2**, 60 (1930).

19. R. T. Poole, J. Liesegang, J. G. Jenkin, and R. C. G. Leckey, *Chem. Phys. Lett.* **23**, 194 (1973).
20. J. Berkowitz, *J. Chem. Phys.* **56**, 2766 (1972); **57**, 3194 (1972).
21. J. Berkowitz, J. L. Dehmer and T. E. H. Walker, *J. Chem. Phys.* **59**, 3645 (1973).
22. B. O. Jonnson and E. Lindholm, *Ark. Fysik*, **39**, 65 (1969).
23. A. W. Potts, W. C. Price, D. G. Streets, and T. A. Williams, *Discuss. Faraday Soc.* **54**, 168 (1973).
24. U. Gelius, to be published.
25. T. A. Carlson and C. P. Anderson, *Chem. Phys. Lett.* **10**, 56 (1971).
26. E. Lindholm, *Discuss. Faraday Soc.* **54**, 200 (1973).
27. W. A. Chupka, private communication.
28. R. Bralsford, P. V. Harris, and W. C. Price, *Proc. Roy. Soc. Lond.* **A258**, 459 (1960).
29. W. C. Price, A. W. Potts, and D. G. Streets, in "Electron Spectroscopy", D. A. Shirely, ed. (North Holland Publ. Co., Amsterdam, 1972), pp. 187–198.
30. D. E. Eastman and W. D. Grobman, *Phys. Rev. Lett.* **28**, 1327, 1329 (1972).
31. G. V. Marr, "Photoionization Processes in Gases" (Academic Press, New York, 1967).
32. L. L. Lohr, in: "Electron Spectroscopy", D. A. Shirley, ed. (North Holland Publ. Co., Amsterdam, 1972), p. 245.
33. V. Gelius, in: "Electron Spectroscopy", D. A. Shirley, ed. (North Holland Publ. Co., Amsterdam, 1972), p. 311.
34. S. Iwata and S. Nagakura, *Mol. Phys.* **27**, 245 (1974).
35. A. Schwerg and W. Thiel, *J. Electron Spectrosc.* **2**, 199 (1973).
36. U. Fano, *Phys. Rev.* **124**, 1866 (1971).
37. U. Fano and J. W. Cooper, *Rev. Mod. Phys.* **40**, 441 (1968).
38. F. H. Mies, *Phys. Rev.* **175**, 164 (1968).
39. J. N. Bardsley, *Chem. Phys. Lett.* **2**, 329 (1968).
40. A. Gieger and B. Schröder, *J. Chem. Phys.* **49**, 740 (1968).
41. E. Lindholm, *Ark. Fys.* **40**, 117 (1969).
42. J. A. Kissinger and J. W. Taylor, *Int. J. Mass. Spectrom. Ion Physics* **11**, 461 (1973).
43. O. Edqvist, E. Lindholm, L. E. Selin, and L. Asbrink, *Physica Scripta* **1**, 25 (1970).
44. A. J. Blake, J. L. Bahr, J. H. Carver, and V. Kumar, *Phil. Trans. Roy. Soc. Lond.* **A268**, 159 (1970).
45. A. L. Smith, *Phil. Trans. Roy. Soc. Lond.* **A268**, 159 (1970).
46. V. I. Kleimenov, Yu. V. Chisov, and F. I. Vilesov, *Opt. Spectrosc.* (*USSR*) **32**, 371 (1971).
47. J. E. Collin and P. Natalis, *Chem. Phys. Lett.* **2**, 414 (1968).
48. J. L. Gardiner and J. A. R. Samson, *J. Electron Spectrosc.* **2**, 153 (1973).
49. W. C. Price, *Discuss. Faraday Soc.* **54**, 206 (1972).
50. T. Baer and B. P. Tsai, *J. Electron. Spectrosc.* **2**, 25 (1973).
51. R. S. Berry, *J. Chem. Phys.* **45**, 1228 (1969).
52. J. Berkowitz and W. A. Chupka, *J. Chem. Phys.* **51**, 2341 (1969).
53. J. Cooper and R. N. Zare, *J. Chem. Phys.* **48**, 942 (1968).
54. A. D. Buckingham, B. J. Orr, and J. M. Sichel, *Phil. Trans. Roy. Soc. Lond.* **A268**, 147 (1970).
55. T. A. Carlson, *Chem. Phys. Lett.* **9**, 23 (1971).
56. T. A. Carlson, *J. Chem. Phys.* **55**, 4913 (1971).

57. T. A. Carlson and G. E. McGuire, *J. Electron Spectrosc.* **1**, 209 (1973).
58. T. A. Carlson, G. E. McGuire, A. E. Jonas, K. L. Cheng, C. P. Anderson, C. C. Lu, and B. P. Pullen, in: "Electron Spectroscopy", D. A. Shirley, ed. (North Holland Publ. Co., Amsterdam, 1972), p. 207.
59. D. L. Ames, J. P. Maier, F. Watt, and D. W. Turner, *Discuss. Faraday Soc.* **54**, 277 (1972).
60. A. Niehaus and M. W. Ruf, *Chem. Phys. Lett.* **11**, 55 (1971).
61. J. W. Sichel, *Mol. Phys.* **18**, 95 (1970).
62. D. G. Streets, A. W. Potts and W. C. Price *Int. J. Mass. Spectrom Ion Phys.* **10** 123 (1972–73).
63. N. Jonathan, A. Morris, M. Okuda, and D. J. Smith, in: "Electron Spectroscopy", D. A. Shirley, ed. (North Holland Publ. Co., Amsterdam, 1972), pp. 345–350.
64. N. Jonathan, A. Morris, M. Okuda, K. J. Ross, and D. A. Smith, *Discuss. Faraday Soc.* **54**, 48 (1972).
65. A. B. Cornford, D. C. Frost, F. G. Herring, and C. A. McDowell, *Discuss. Faraday Soc.* **54**, 56 (1972).

ADDITIONAL BIBLIOGRAPHY

A. Books

A. D. Baker and D. Betteridge "Photoelectron Spectroscopy" (Pergamon Press, Oxford, 1972).

A. B. F. Duncan, "Rydberg Series in Atoms and Molecules". (Physical Chem. Monograph 23, Academic Press, New York, 1971).

J. H. D. Eland, "Photoelectron Spectroscopy". (Butterworths, London, 1974).

G. V. Marr, "Photoionization Processes in Gases". (Academic Press, New York, 1967).

J. A. R. Samson, "Techniques of Vacuum Ultraviolet Spectroscopy". (J. Wiley & Sons, New York, 1967).

K. D. Sevier, "Low Energy Electron Spectrometry". (Wiley Interscience, New York, 1969).

K. Siegbahn, C. Nordling, G. Johannsen, J. Hedman, P. F. Heden, K. Hamrin, V. Gelius, T. Bergmark, L. O. Werme, R. Manne, and Y. Baer, "ESCA Applied to Free Molecules". (North Holland Publ. Co., American Elsevier, Amsterdam, New York, 1969).

D. W. Turner, A. D. Baker, C. Baker, and C. R. Brundle, "Molecular Photoelectron Spectroscopy", (Wiley–Interscience, New York, 1970).

B. Conference Reports

"A Discussion on Photoelectron Spectroscopy." A Royal Society discussion organized by W. C. Price and D. W. Turner, *Phil. Trans. Roy. Soc. Lond.* **A268**, 1–175 (1970).

"Electron Spectroscopy." Proceedings of an International Conference held at Asilomar, Calif., Sept. 1971, D. A. Shirley, ed. (North Holland Publ. Co., Amsterdam, 1972).

"The Photoelectron Spectroscopy of Molecules" *Discuss. Faraday Soc.* **54**, 1973.

C. Review Articles

A. D. Baker, Photoelectron spectroscopy, *Accounts Chem. Res.* **3**, 17–25 (1970).

J. Berkowitz, photoionization mass spectrometry and photoelectron spectroscopy in high temperature vapours, *Adv. High. Temp. Chem.*

R. S. Berry, Electronic spectroscopy by electron spectroscopy, *Ann. Rev. Phys. Chem.* **20**, 357–406.

D. Betteridge and M. Thompson, *J. Mol. Struct.* **21**, 341 (1974).

D. C. Frost, Progress in ultraviolet photoelectron spectroscopy, *J. Electron Spectrosc.*, **5**, 99–132 (1974).

A. Hamnett and A. F. Orchard, "Electronic Structure of Inorganic Compounds", Vol. 1, Specialist Reports of Chem. Soc., P. Day, ed. (London, 1972). p. 1.

R. L. De Kock and D. R. Lloyd, Vacuum ultraviolet photoelectron spectroscopy of inorganic molecules, *Advances in Inorganic Chemistry and Radiochemistry*, **16**, 65–107 (1974).

S. Leach, in: "Vacuum Ultraviolet Radiation Physics", N. Damany, B. Vodar, and J. Ramond, eds. (Pergamon Press, Oxford, 1974).

W. C. Price, Photoelectron spectroscopy and the electronic structure of matter, *Phys. Bull.* **23**, 87–92 (1972).

W. C. Price, Photoelectron spectroscopy, *Adv. in At. and Mol. Phys.* **10**, 131–171 (1974).

S. D. Worley, Photoelectron spectroscopy in chemistry, *Chem. Rev.* **71**, 295–314 (1971).

5

Some Aspects of Organic Photoelectron Spectroscopy

E. HEILBRONNER and J. P. MAIER

Physikalisch-Chemisches Institut der Universitat Basel,
4056 Basel, Klingelbergstrasse 80,
Switzerland

I. INTRODUCTION

Photoelectron spectroscopy of larger organic molecules (molecular weight of the order 10^2) suffers from the same handicaps and enjoys the same advantages as the more familiar technique of electronic spectroscopy applied to such molecules. In both fields, the spectra of small molecules[1, 2] yield a lot of detailed information which in many cases allows a unique assignment of the radical cation or electronic excited states, whereas such information is usually lacking in the spectra of the larger systems.[3, 4] This is almost self-evident if one compares the photoelectron spectra of di- or tri-atomic systems (e.g. N_2[2, 5] or CO_2[6]) with those of polyatomic molecules (e.g. 1,4,7-cyclo-nonatriene (**1**)[7] or Dewar-benzene (**2**)[8]), as shown in Fig. 1.

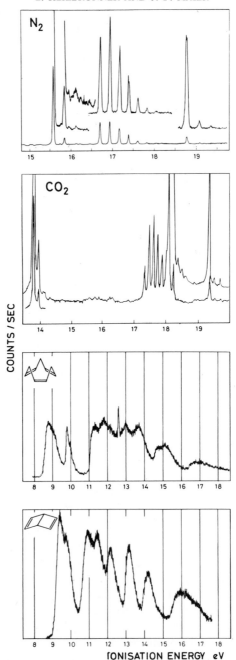

Fig. 1. HeI photoelectron spectra of N_2,[2] CO_2,[6] 1,4,7-cyclononatriene (1),[7] and Dewar-benzene (2).[8]

1   **2**

The disadvantages one has to cope with in the field of organic photo-electron spectroscopy are mainly the following:

(1) Most organic molecules M have no or at best rather low symmetry, i.e. C_1, C_2 or C_s. From this point of view, the two examples **1** and **2** of Fig. 1 are *not* typical. As a consequence the description of the electronic states $\tilde{X}, \tilde{A}, \tilde{B}, \ldots$ of their radical cations M^+ in terms of electronic configurations Ψ'_j is more difficult, because many of the electronic configurations lying in a given energy range will mix, belonging necessarily to one or the other of a small set of irreducible representations. Here and in the following sections the radical cation states $\tilde{X}, \tilde{A}, \tilde{B}, \ldots$ are only those which can be reached from M in its electronic ground state $(^1\Psi_0)$ by a photoionization process, involving only the ejection of a single electron.

(2) Individual bands in the photoelectron spectra of larger organic molecules usually lack resolvable vibrational fine-structure, because too many of the $3n$–6 normal modes of M^+ will be excited during the ionization process. Only if a normal mode Q is heavily concentrated on the same structural feature of M, and thus of M^+, whose bonding characteristics change most on ionization (e.g. the π-bond(s) in alkenes, polyenes,[10] acetylenes,[11] or in 1,4,7-cyclononatriene **(1)** shown in Fig. 1) will a band with a dominating series of vibrational fine-structure components be observed. Thus a detailed analysis of the vibrational modes of the radical cation M^+ becomes impossible.

In the identification of the vibrational modes excited in the radical cations, an advantage rests in that only totally symmetric modes (with respect to the symmetry elements common to the ground $M(^1\Psi_0)$ and cationic $M^+(^2\Psi'_j)$ states) can be excited in any number of quanta in electronic transitions starting from the ground vibrational state.[1] The non-totally symmetric modes can be excited in zero or even number of quanta; however, the 0–0 transition is then by far the most intense.[1] Thus the identification of double quanta excitation is rare, e.g. $^2\Pi_g$ state of diacetylene cation.[2] In instances of vibronic interaction the selection rules are relaxed, albeit the fine structure when discernible tends to be too complicated for confident analysis. The disadvantage encountered then is that, at best, normal coordinate analysis would be restricted to the totally symmetric vibrations.

The linear polyatomics provide the examples where the vibrational fine structure, Franck–Condon profiles and factors, can be discussed within the concepts used for diatomic and triatomics, in respect to the characteristics of

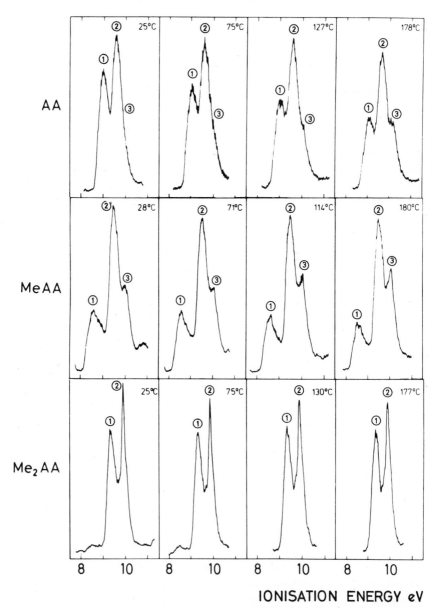

IONISATION ENERGY eV

Fig. 3. HeI photoelectron spectra of acetylacetone (AA), 3-methylacetylacetone (MeAA), and 3,3-dimethylacetone (Me$_2$AA) in the region of the ionization of the oxygen lone pairs, n_- and n_+, for the keto form (cf. Me$_2$AA) and oxygen lone pair and π MO for the enol form, as function of temperature.[13]

the MO associated with the band in the photoelectron spectrum. On the other hand, in most other species the fine-structure may only indicate that it is in accord with the bonding features predicted by the model used for the band assignment. Usually, only a limited amount of information can be deduced from the Franck–Condon envelope of the bands. Nevertheless, the profile of the band is a feature not to be by-passed. In unsaturated hydrocarbons the π-bands are usually well structured and the 0–0 vibrational peak dominant when the skeleton is rigid. When the bands are broad, an ambiguity arises in deciding whether this is attributed to the superposition of the photoelectron spectra of conformers or due to changes of geometry on ionization (e.g. change of dihedral angle between conjugating π bonds), or both. At another extreme, the appearance of the Franck–Condon profile, such as the first band in methane or allene,[2] is suggestive of vibronic complications.

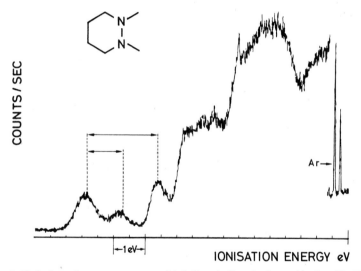

Fig. 2. HeI photoelectron spectrum of 1,2-dimethylhexahydropyridazine (3), (4), and (5). (Redrawn from ref. 12).

(3) In addition to their lack of vibrational fine-structure, many bands in the photoelectron spectrum will overlap strongly, as is evident from the examples shown in Fig. 1 and later on in this chapter. The reason is, that the valence shell of large organic molecules contains by necessity also a large number of valence electrons. This in turn yields a large number of radical cation states $\tilde{X}, \tilde{A}, \tilde{B}, \ldots$ in a restricted range of energies. The overlapping of bands which thus result is often so complete that it becomes almost impossible to evaluate the number of superimposed bands, let alone their individual positions on the ionization energy scale. This restricts the assignment of the photoelectron

spectra to the low ionization energy region, where the bands are either sufficiently separated or the number of overlapping bands can be deduced, by substitution for example. As a consequence, the interpretation of the spectra of the unsaturated hydrocarbons is concerned, by and large, with the ionization energies of the π electrons. The bands are assigned until the onset of the σ-bands is encountered (c.f. Tables I, II) where profuse overlap sets in. In the aromatic hydrocarbons ≈ 11 eV marks the dividing line.

(4) A further complication stems from the fact that many large organic molecules can assume different conformations, so that their photoelectron spectrum is necessarily the superposition of the spectra of the individual conformers, in ratios depending on the conditions prevailing in the target chamber of the spectrometer. The same is true for valence isomers or tautomers.

However, under certain favourable conditions, as in the case of hydrazine derivatives, e.g.

it is possible to deduce the ratio of the conformers such as **3** vs **4** plus **5**, from the photoelectron spectrum, as has been shown by Nelsen and his co-workers[12] (see Fig. 2). Even more remarkable is the result due to Schweig and his group,[13] who have been able not only to show the presence of the tautomers 6, 7 of substituted acetylacetones, but also to derive the value of the keto-enol equilibrium constant $K_T = [7]/[6]$ from the temperature dependence of the band intensity ratios of the relevant bands in the photoelectron spectrum (see Fig. 3).

(5) A final complication consists in the ease with which some organic molecules can isomerize or fragment in the target chamber, prior to the ionization process, thus yielding photoelectron spectra which are those of a molecule of different structure. Typical examples are hexamethyl-prismane (**8**) which yields under certain conditions the spectrum of pure hexamethyl

benzene (**9**) or of difluoromethano[10]annulene (**10**) which, when recorded at higher temperature, yields the spectrum of naphthalene (**11**).[15]

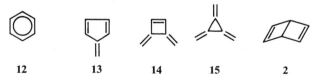

From the above, the reader might conclude that an analysis of the photo-electron spectra of organic compounds should be almost impossible. However, there are also some considerable advantages in working with large molecules, advantages which make "organic photoelectron spectroscopy" a field of its own. Because of their size, it is possible to perturb such molecules $M \rightarrow M'$, $M'', \ldots$ (e.g. by substitution or by minor changes in structure) without altering the build-up of their valence shell in a fundamental way. Comparison of the photoelectron spectra of a set of related molecules $M, M', M'', \ldots$ will therefore provide a rather detailed picture of the sequence and the character of the radical cation states of M^+, reached in a one-electron process especially if the small changes from one system to another can be mirrored in simple quantum mechanical models. The key-words are therefore: "correlation-diagrams" and "perturbation treatments". The type of correlations to be considered are mainly of four types:

(1) In such systems where the energy of the lowest radical cation states $\tilde{X}, \tilde{A}, \tilde{B}, \ldots$ are mainly determined by the topology of the molecular frame-work, e.g. in π-systems, the correlation of the results obtained from a set of isomers is particularly revealing, e.g. of the C_6H_6 isomers presented in Fig. 4.

A further comparison would be provided by investigating molecules contain-ing the same number of π-electrons in their π-system as the molecules **12–15**,

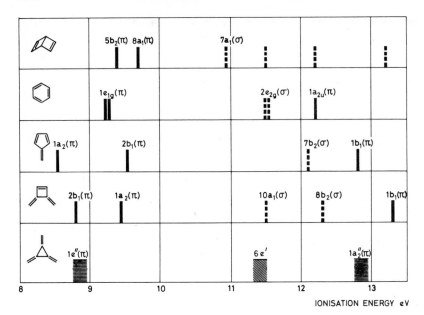

Fig. 4. Comparison of the $I_{v,j}$ values of the C_6H_6 isomers **2**, **12**, **13**, **14**, and **15** and their assignment as deduced from the HeI photoelectron spectra. The $I_{v,j}$ values for **15** represent the predicted photoelectron spectrum, deduced from the $I_{v,j}$ data of the hexamethyl derivative.[146]

16   17   18

such as the C_6H_8 hydrocarbons **16**, **17**, and **18**, or π-systems which derive in a systematic fashion by suitable enlargement of the original system, e.g. the benzologues **19–21** of benzene (**12**) and of fulvene (**13**):

11   19

20   21

(2) Another type of correlation, which has been used with success (also in the field of small molecules[16]), is the comparison of isoelectronic and isosteric

species. However, in large organic molecules the term "isoelectronic" will usually be taken to mean "isoelectronic valence shell", as exemplified by the series **22–25** and **26–31**[17-19] (see Figs 5 and 6).

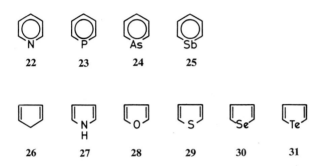

22	**23**	**24**	**25**

26	**27**	**28**	**29**	**30**	**31**

(3) A special type of correlation is obtained if the lowest radical cation states are generated by the ejection of an electron from an orbital of M which is strongly localized in a particular well-delimited region of the molecule, or more precisely if the positive charge in the resulting radical cation M⁺ is localized mainly in this particular place. This will be the case if the organic molecule contains double bonds and/or lone-pairs. Barrelene **(32)**[20] and diazabicyclo[2.2.2]octane **(36)**[21] are such examples. The analysis of the corresponding spectra is greatly facilitated by including those molecules, in the investigation, which differ from the original ones in the number and/or positions of the ionization sensitive regions:

32	**33**	**34**	**35**

36	**37**

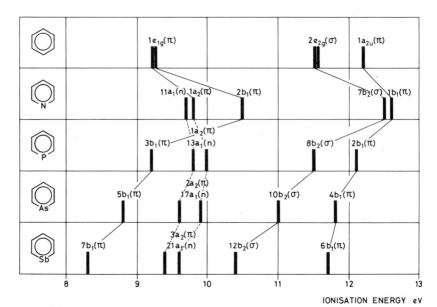

Fig. 5. Correlation diagram of the ionization energies of the "isoelectronic valence shell" series of benzene (**12**), pyridine (**22**), phosphabenzene (**23**), arsabenzene (**24**), and stibabenzene (**25**).[17]

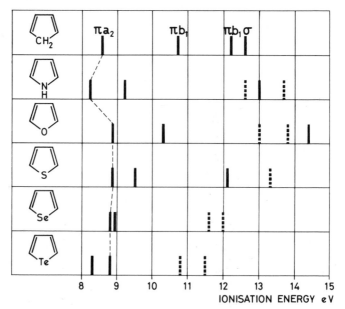

Fig. 6. Correlation diagram of the ionization energies of the "isoelectronic valence shell" series of cyclopentadiene (**26**), pyrrol (**27**), furan (**28**), thiophen (**29**), selenophen (**30**), and tellurophene (**31**).[19] The $I_{v,j}$ values corresponding to ionization from π-orbitals are indicated by solid bars and from σ-orbitals by broken bars.

As a typical example we present in Fig. 7 the results observed for the series bicyclo[2.2.2]octane (**35**) to barrelene (**32**).

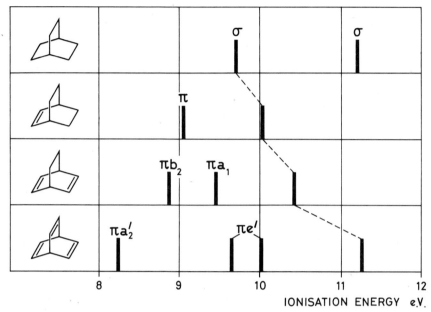

Fig. 7. Correlation diagram of the ionization energies of the series bicyclo[2.2.2]octane (**35**) to barrelene (**32**).

(4) Finally, there is the possibility of perturbing a given system by introducing substituents of various kinds in different positions. Typical examples are the investigation of the series of alkyl-substituted alkenes (**38**)[22] or alkynes (**39**)[23] where the substituents R are changed systematically along the traditional series R = methyl, ethyl, propyl, etc.

$$R_1\diagdown_{}\diagup R_3$$
$$C=C \qquad \textbf{38} \qquad R_1-C\equiv C-R_2 \quad \textbf{39}$$
$$R_2\diagup^{}\diagdown R_4$$

A particularly interesting example of this kind of technique, introduced by Robin and his co-workers,[24] consists in replacing all the hydrogen atoms in a planar π-system M (i.e. **12** or **41**) by fluorine atoms yielding M(perfluoro) (i.e. **40** or **42**). This change leaves the energies of those states of M$^+$(perfluoro) which are antisymmetric with respect to the molecular plane

12 **40** **41** **42**

essentially unaffected relative to those of M^+, whereas the symmetric states are shifted to much higher energies. The spectra of the four compounds **12**, **40**, **41**, and **42** are shown in Fig. 8.

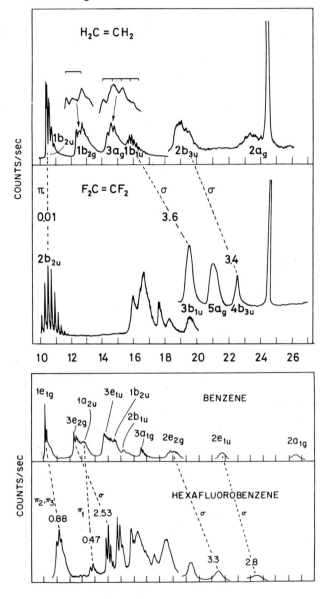

Fig. 8. HeI and HeII photoelectron spectra of ethylene (**41**), tetrafluoroethylene (**42**), benzene (**12**), and hexafluorobenzene (**40**).[24]

In this chapter we shall try to show how one can make use of these different correlation techniques in view of an interpretation of the photoelectron spectrum of an organic compound. In doing this we shall have to concentrate necessarily on a small selection of simple systems, e.g. unsaturated hydrocarbons and some of their derivatives.

To conclude this introduction, attention should be drawn to the rather important role played by theoretical models in this particular field, especially by SCF molecular orbital models.[25] Because of the lack of ancillary, experimental information, these models are often invoked as an important aid for the identification of the electronic states of the radical cation M^+ to be correlated with the individual peaks in the photoelectron spectrum of M. Sometimes, as a last resort, they are even the only basis for a given assignment. Apart from that, some of the more qualitative aspects of these molecular orbital models have to be used as guide-lines for the correlation procedures discussed above. For this reason a few comments on the underlying theoretical assumptions seem to be in order.

II. MOLECULAR ORBITAL MODELS

In the wake of the Woodward–Hoffmann rules,[26] organic chemists have become increasingly "orbital-conscious" and this is presumably the origin of the enthusiasm with which the results of uv-photoelectron spectroscopic investigations have been accepted and interpreted in terms of molecular orbital models.[27] More often than not these interpretations invoke Koopmans' approximation.[28] In view of the widespread misconceptions which are found in the literature and because of the special role molecular orbital models play in organic photoelectron spectroscopy, it seems worthwhile to recapitulate some of the fundamental facts, although this may duplicate part of previous chapters.

The starting point for the discussion are the many-electron SCF Hartree–Fock calculations of the Roothaan-type[29] in either *ab initio* or one of the various semiempirical formulations. (We shall not concern ourselves with other methods, e.g. the $X\alpha$-muffin-tin model,[30] although they may become more important in the future.) These treatments are applied to the primary process

$$M(^1\Psi_0) + h\nu \rightarrow M^+(^2\tilde{\Psi}_j) + e^-(T_j) \tag{1}$$

where M is an organic molecule in its electronic singlet ground state $^1\Psi_0$, and $^2\tilde{\Psi}_j$ is a doublet state of the radical cation M^+. For the time being we neglect changes in vibrational and/or rotational energies and we assume that M^+ has exactly the same structure, i.e. the same internal coordinates $R_{\mu\nu}$ and $\phi_{\kappa\mu\nu}$ as M. The ground state $^1\Psi_0$ of the closed shell molecule M will be represented by a single determinantal wave function ($2N$ = number of electrons)

$$^1\Psi_0 = \| \psi_1 \bar{\psi}_1 \cdots \psi_j \bar{\psi}_j \cdots \psi_N \bar{\psi}_N \| \tag{2}$$

where the ψ_j are SCF orbitals obtained by minimizing the total energy $\mathscr{E}(^1\Psi_0)$ of M, that is by solving the Hartree–Fock equations

$$\mathscr{F}\psi_j = \sum_i \lambda_{ij}\psi_i \tag{3}$$

Applying the analogous procedure to the open shell system M^+ one obtains for the doublet state

$$^2\tilde{\Psi}'_j \begin{cases} \tilde{\Psi}'^\alpha_j = \| \tilde{\psi}_1 \bar{\tilde{\psi}}_1 \cdots \tilde{\psi}_j \cdots \tilde{\psi}_N \bar{\tilde{\psi}}_N \| \\ \tilde{\Psi}'^\beta_j = \| \tilde{\psi}_1 \bar{\tilde{\psi}}_1 \cdots \bar{\tilde{\psi}}_j \cdots \tilde{\psi}_N \bar{\tilde{\psi}}_N \| \end{cases} \tag{4}$$

The SCF orbitals $\tilde{\psi}_j$ referred to in (4) necessarily differ from the ψ_j used in (2). In (2) and (4) the same spatial functions ψ_j or $\tilde{\psi}_j$ have been used for electrons with spin α or β: this is the restricted Hartree–Fock approximation. Apart from the limitations inherent in the model approximations (e.g. basis set, adjustable parameters in the semiempirical models) the only effect not taken care of is that due to electron–electron correlation.[31] On the other hand, this approach does include electron rearrangement in the radical cation.

Although this is well known, it must be emphasized that there is a great ambiguity of orbitals. In particular any unitary transformation of a set of SCF orbitals $\{\psi_j\}$ or $\{\tilde{\psi}_j\}$, satisfying (3), into a new set

$$\{\psi'_j\} = \mathbf{U}\{\psi_j\} \tag{5}$$

will leave all predictions of true observables derivable from the SCF model invariant, in particular the energies of the states $\mathscr{E}(^1\Psi_0)$ or $\mathscr{E}(^2\tilde{\Psi}'_j)$ of M and M^+ respectively. In contrast the orbitals ψ_j and the matrix elements λ_{ij} defined in (3) do *not* belong to this class of observable quantities, depending, as they do, on the arbitrary choice of a unitary transformation (5).

Chemically speaking the primary process (1) can be regarded as the "reaction" of M in its closed shell ground state with a photon $h\nu$ to yield as products the radical cation M^+ in the state $^2\tilde{\Psi}'_j$ and an electron of kinetic energy T_j. The latter quantity is the one measured by the photoelectron spectrometer. Because of energy and momentum conservation we have, in view of $m(M^+) \gg m(e^-)$,

$$\mathscr{E}(^2\tilde{\Psi}'_j) + T_j - \mathscr{E}(^1\Psi_0) - h\nu = 0 \tag{6}$$

and thus, for the ionization energy

$$I_{vj} \equiv \mathscr{E}(^2\tilde{\Psi}'_j) - \mathscr{E}(^1\Psi_0) = h\nu - T_j \tag{7}$$

It is obvious that $I_{v,j}$ depends on the electronic structure of both the "educt" M and of the "product" M^+, a point of some importance with respect to the simplification to be now introduced.

Among all the sets $\{\psi'\}$ or orbitals that can be obtained according to (5), there is one of particular interest, namely the set $\{\varphi_j\}$ of canonical orbitals which is chosen in such a way that the matrix (λ_{ij}) becomes diagonal. Under these conditions (3) becomes

$$\mathscr{F}\varphi_j = \varepsilon_j \varphi_j \qquad (8)$$

where the ε_j are the "orbital energies" of the canonical orbitals φ_j.

Koopmans has shown[28] that if one expands the $2N\text{-}1$ molecular spin orbitals of the ground state $^2\tilde{\Psi}_0$ of M^+ in terms of the $2N$ spin orbitals of $^1\Psi_0$ of M, then the $2N\text{-}1$ lowest canonical spin orbitals φ_j, $\bar{\varphi}_j$ with φ_N (or $\bar{\varphi}_N$) singly occupied yield the optimum description of ground state M^+. This result can be formulated in the following more practical form: If we make use of the simplifying assumption $\tilde{\psi}_j \equiv \varphi_j$ in (4), then it can be shown that $I_{v,j}$ as defined in (7) satisfies the simple relationship

$$I_{v,j} = -\varepsilon_j \qquad (9)$$

usually referred to as Koopmans' Theorem. It must be emphasized again that (9) is only valid for a model under the arbitrary assumption that $\tilde{\psi} \equiv \varphi_j$, i.e. that the orbitals of M^+ do not change relative to the canonical orbitals of M, even though an electron has been removed from one of these orbitals.

Looked at from this point of view, the set of canonical orbitals of M is nothing but a convenient basis for a simplified description of M^+ in one of the doublet states $^2\tilde{\Psi}'_j$. This is closely analogous to the fact that the four sp^3 hybrid atomic orbitals of carbon constitute a more adequate basis for the description of methane, rather than the completely equivalent set $2s$, $2p_x$, $2p_y$, and $2p_z$. In other words, the set $\{\varphi_j\}$ of canonical orbitals is the "coordinate system" of choice for a convenient and heuristically useful rationalization of photoelectron–spectroscopic data, in the same sense that a cartesian co-ordinate system is ideal for the description of the rectilinear movement of a free falling particle, in contrast to a system of polar coordinates. However, the observation of the motion of such a particle will not allow us to conclude that a cartesian coordinate system "exists" and, for the same reasons, it will be impossible to draw conclusions concerning the "existence" of canonical orbitals φ_j and their orbital energies ε_j from photoelectron spectroscopic data. These constructs are *not* observables.

So far the structure of the radical cation M^+ has been assumed to be identical to that of the parent molecule M. However, if after the vertical ejection process the radical cation M^+ in the state $^2\tilde{\Psi}'_j$ could lose its excess vibrational energy it would relax to the vibrational ground state corresponding to its new equilibrium structure

$$M^+(^2\tilde{\Psi}'_j) \to M^+_{\text{equ.}}(^2\tilde{\Psi}'_j) \qquad (10)$$

the internal coordinates $R_{\mu\nu}{}^+, \phi_{\kappa\mu\nu}{}^+$ of which depend on the electronic state $^2\tilde{\Psi}'_j$. Necessarily $M^+_{\text{equ.}}$ lies below M^+ in energy by $-\Delta E_{\text{relax}}$. The ionization energy associated with the process $M(^1\Psi'_0) \to M^+_{\text{equ.}}(^2\tilde{\Psi}'_j)$ is called the adiabatic ionization energy

$$I_{a,j} = I_{v,j} - \Delta E_{\text{relax}} \qquad (11)$$

In the preceding argument we have assumed that M^+ in its electronic state $^2\tilde{\Psi}'_j$ is a stable radical cation, i.e. that it has no (or little) tendency to fragment in its vibrational ground state. If, however, the energy hypersurface of $M^+(^2\tilde{\Psi}'_j)$ has no minimum, then the adiabatic ionization energy can not be defined. In such a case the position of the band-onset in the photoelectron spectrum is better labelled $I_{\text{onset},j}$ rather than $I_{a,j}$.

Comparison of the equilibrium structure of M^+ in the state $^2\tilde{\Psi}'_j$ with the structure of M, i.e. of the internal coordinates $R_{\mu\nu}{}^+, \phi_{\kappa\mu\nu}{}^+$ with $R_{\mu\nu}, \phi_{\kappa\mu\nu}$ will allow the assessment of the normal modes excited in the radical cation M^+ by the ejection process (1).[32] Depending on the Franck–Condon factors one or more of these modes may dominate and determine the vibrational fine structure of the corresponding band.

The "true" assignment of the photoelectron spectrum of an organic compound (and indeed of any compound) would imply that for each band in the spectrum the following information were available both from the experimental and theoretical side:

(a) The state energy differences (7) and (11), i.e. the ionization energies $I_{v,j}$ and $I_{a,j}$.

(b) The irreducible representations of the symmetry group $\mathscr{G}$ of M to which the individual states $^2\tilde{\Psi}'_j$ of M^+ belong.

(c) The symmetry group $\mathscr{G}^+$ and the internal coordinates of $M^+_{\text{equ.}}(^2\tilde{\Psi}'_j)$, i.e. of the radical cation after the relaxation process (10).

(d) The transition probability (i.e. the experimental integrated band intensity corrected for the instrumental transmission function and the angular dependence of the ejected photoelectrons) for the process (1) and the Franck–Condon factors for the different normal modes excited in M^+.

(e) The frequencies and the normal coordinates of these normal modes, i.e. the normal modes of $M^+_{\text{equ.}}(^2\tilde{\Psi}'_j)$.

As mentioned in the introduction, it is not possible, except in rare cases, to derive this information from the experimental data available for larger organic molecules. On the other hand, a complete *a priori* calculation of the $(3n-6)$-dimensional energy profiles for M in its different electronic states $^2\tilde{\Psi}'_j$ is out of the question. As a consequence of this sad situation, what is called an "assignment" of the photoelectron spectrum of an organic molecule will be by necessity an assignment with respect to an arbitrarily chosen, simplified theoretical model and thus depend heavily on the particular assumptions implied in this model. It must be emphasized that for this reason all assignments are part experiment and part model. In particular the choice of a different reference model may thus lead to a different assignment. It is obvious, that of two different assignments derived on the basis of two different theoretical models, one is not necessarily "right" and the other "wrong", but rather, that this unavoidable situation raises the question which of the models available yields the "best" assignment. The appraisal and also the criticism of an assignment will hinge among other things on the answers to the following questions:

(A) Is the band assignment obtained on the basis of our model consistent with all the information that can be deduced within the inherent limits of error from the particular model chosen (e.g. band intensities, bandshapes, spin-orbit splittings, vibrational fine structure)?

(B) Will the model, perhaps after suitable calibration on a small set of molecules, allow reasonably correct predictions for the photoelectron spectra of other, similar compounds?

(C) Is the electronic structure of M and/or M^+ implied by a model, which would be acceptable according to the criteria (A) and (B), consistent with other physico-chemical properties of M and/or M^+?

(D) Is the model "economical"? By this we mean that from two models satisfying the conditions (A) to (C) we shall retain the simpler one, i.e. the one implying fewer *ad hoc* assumptions and/or adjustable parameters.

(E) Is the model heuristically useful?

Although these points may seem trivial, they must be taken into consideration when qualitative or semi-quantitative models are used for the rationalization of a proposed assignment.

Semiempirical procedures are usually calibrated to fit a particular property, e.g. CNDO/2[33] to mimic *ab initio* results, MINDO/2[34] and MINDO/3[35] to yield enthalpies of formation, or SPINDO[36] and HAM[37] to predict photoelectron band positions. Because of the large number of parameters involved it is rather difficult to assess how a particular set of values for these parameters reflects in the orbital energies ε_j and thus in the predicted ionization energies $I_{v,j}$, assuming Koopmans' Theorem. To compare the different models in a chemically and heuristically useful way, one makes use of the unitary transformation (5), choosing a matrix $U \equiv L$ which transforms the canonical orbitals φ_j into localized orbitals λ_j, i.e.

$$\{\lambda_j\} = L\{\varphi_j\} \tag{12}$$

according to a preselected localization criterion, e.g. the one proposed by

Edmiston and Ruedenberg:[38]

$$\sum_j \left\langle \lambda_j(1)\, \lambda_j(2) \left| \frac{e^2}{r_{12}} \right| \lambda_j(1)\, \lambda_j(2) \right\rangle = \text{Maximum} \tag{13}$$

The matrix elements $F_{\lambda,ij}$ of the transformed Hartree–Fock matrix

$$\mathbf{F}_\lambda = (F_{\lambda,ij}) = \mathbf{L} \mathbf{F}_\varphi \mathbf{L}^+ \tag{14}$$

show a high degree of transferability from one compound to another and their configurational and/or conformational dependence is similar within a given semiempirical model.[39] In contrast, the absolute values of the different $F_{\lambda,ij}$ differ considerably from one semiempirical procedure to another.

The matrix elements $F_{\lambda,ii} \equiv A_i$ are called the self-energies of the localized orbitals λ_i and the $F_{\lambda,ij} = B_{ij}$ $(i \neq j)$, the localized orbital interaction terms. The λ_i can now be transferred to, and used as a basis for, the description of another molecule in the framework of an essentially Hückel-type approximation:

$$\varphi = \sum_j c_j \lambda_j \;\therefore\; \langle \lambda_i | \mathscr{H} | \lambda_i \rangle = A_i; \quad \langle \lambda_i | \mathscr{H} | \lambda_j \rangle = B_{ij} \tag{15}$$

This LCBO-scheme originally introduced by Hall[40] (LCBO = Linear Combination of Bond Orbitals) has been elaborated and used quite successfully in the past.[41] In particular it yields a justification for using simple, calibrated Hückel molecular orbital models for the rationalization and systen.atization of photoelectron spectroscopic data and we shall therefore rely heavily on such models. Together with the correlation techniques described in the introduction, they are the method of choice to deal with large organic molecules.

III. π-ELECTRON SYSTEMS

Not unexpectedly, the simplest photoelectron spectra of organic compounds of medium size are those of the unsubstituted, planar unsaturated molecules for which the traditional σ/π separation underlying the simple Hückel[42] or the Pariser–Parr–Pople[43] treatment is a good approximation. The highest occupied orbitals are π-orbitals, although for larger systems the π- and σ-manifolds do overlap in the region of 10 eV to 13 eV approximately. As had been found for many other physico-chemical properties of such systems,[42] the positions of the bands due to electron ejection from a π-orbital depend also primarily on the topology of the π-system, i.e. on the eigenvalues x_j of the bond matrix $\mathbf{B} = (B_{\mu\nu})$, with $B_{\mu\nu} = 1$ if μ is connected to ν, and $B_{\mu\nu} = 0$, otherwise. Originally this had been shown for the first ionization energies by Streitwieser and Nair[44] in pre-photoelectron-spectroscopic times, and first been extrapolated to the higher ionization energies by Eland and Danby.[45] To a first, crude approximation the vertical π-ionization energies can be parameterized in the form

$$I_{v,j}{}^{\mathrm{HMO}} = -(\alpha + x_j \beta) \tag{16}$$

as long as there is no strong first-order bond localization, which implies that the simplifying assumption $\beta_{\mu\nu} = \beta$ for all π-bonds μ, ν is acceptable.

However, this latter assumption is no longer valid if the system contains strongly localized double bonds (e.g. polyenes) or if it is rather large.[46] In these cases, formula (16) can be improved upon by including first-order corrections for the differences in π-bond lengths $R_{\mu\nu}$ of M and for the changes $\Delta R_{\mu\nu,j} = R_{\mu\nu,j}{}^+ - R_{\mu\nu}$ which accompany the relaxation process (10). Here $R_{\mu\nu,j}{}^+$ and $R_{\mu\nu}$ are the lengths of the bond μ, ν in $M_{\mathrm{equ.}}^+({}^2\tilde{\Psi}_j)$ and $M({}^1\Psi_0)$ respectively. The necessary correction $\delta I_{v,j}$, that has to be added to $I_{v,j}{}^{\mathrm{HMO}}$ of (16), can be easily computed in the framework of the model from the changes in bond order $p_{\mu\nu}$ which accompany the removal of an electron from the orbital $\psi_j{}^{\mathrm{HMO}}$ by a simple first-order perturbation treatment:[4]

$$I'_{v,j} = I_{v,j}{}^{\mathrm{HMO}} - \underbrace{b \sum_{\mu\nu} (p_{\mu\nu,j}{}^+ - p_{\mu\nu})(p_0 - p_{\mu\nu})}_{y_j} \tag{17}$$

Here $p_{\mu\nu}$ is the bond order of the bond $\mu\nu$ in M, $p_{\mu\nu,j}{}^+$ in $M_{\mathrm{equ.}}^+({}^2\tilde{\Psi}_j)$, and $p_0 = 2/3$ the bond order for a π-bond of standard length $R_0 = 1 \cdot 39$ Å, i.e. in benzene. The constant b depends on the force constants of a sp^2–sp^2 single bond and of a double bond of length R_0, as well as on the derivative $d\beta/dR$ of the resonance integral. In practice b is handled as an adjustable parameter:

$$I'_{v,j} = -\varepsilon_j = -(\alpha + \beta x_j + b y_j) \tag{18}$$

The degree of improvement obtainable by the treatment embodied in (18) over the simple HMO model (16) is obvious from Fig. 9 where both models have been applied to the following mixed sample of π-systems:

46   47   48

49   50

Formulae (17) and (18) have been applied successfully to the benzenoid $C_{18}H_{12}$ hydrocarbons **46**, **51–54**,[48] the series of the acenes **12**, **11**, **45**, **46**, **55**[49] (see Fig. 10), and especially by Schmidt and his co-workers to a large set of benzenoid hydrocarbons[50] and to the [n]helicenes[51] with n = 6 to 14 (e.g. [6]helicene **56**). The correlations between calculated and observed ionization energies are in all these cases excellent.

51   52   53

54   55   56

57

From these results the optimal parameters α and β to be used in conjunction with (16) have been deduced by least squares techniques for the particular sample used (* = no error quoted, † = 90 percent confidence limits,

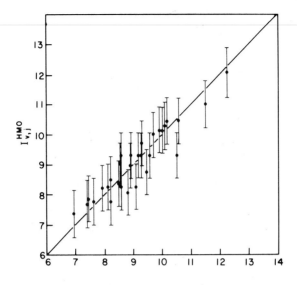

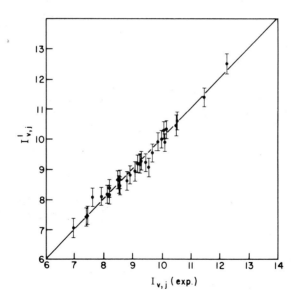

Fig. 9. Correlation of HMO orbital energies $-\varepsilon_j = I_{v,j}^{\text{HMO}}$ (formula 16), and $-\varepsilon'_j = I'_{v,j}$ obtained by the perturbation treatment (formula (17)), with the experimental ionization energies $I_{v,j}$(exp.)[47]. The confidence limits attached to each point refer to 90% security.

§ = standard error):

$-\alpha(eV)$	$-\beta(eV)$	Ref.	
6·47*	2·70*	45	
6·55 ± 0·34†	2·73 ± 0·33†	47	
6·05 ± 0·11†	2·87 ± 0·10†	48	(19)
6·07 ± 0·23†	2·91 ± 0·21†	49	
5·93 ± 0·05§	2·88 ± 0·05§	50	

For the perturbation treatment embodied in formula (18) one finds:

$-\alpha(eV)$	$-\beta(eV)$	$-b(eV)$	Ref.	
5·85 ± 0·16	3·33 ± 0·15	7·7 ± 1·0	47	
5·78 ± 0·11	3·20 ± 0·12	5·1 ± 1·6	48	(20)
5·86 ± 0·11	3·20 ± 0·11	7·9 ± 1·5	49	

The acenes **12**, **11**, **45**, **46**, **55**, and **57** provide a nice example for the correlation technique discussed in section I. Coulson has shown[52] that in the framework of standard Hückel-MO theory the $4N+2$ π-orbital energies of an N-acene A(N) can be obtained in closed form, i.e.

$$\varepsilon = \alpha+\beta; \quad \alpha-\beta; \quad \alpha+\frac{\beta}{2}\left\{\pm 1\pm\left[9+8\cos\left(\frac{\pi J}{N+1}\right)\right]^{\frac{1}{2}}\right\} \qquad (21)$$

where $J = 1, 2, ..., N$. Using the abbreviations $\varepsilon = \alpha+x\beta$ and $r_J = [9+8\cos(\pi J/(N+1))]^{\frac{1}{2}}$ the $2N+1$ bonding orbitals have energies determined by the coefficients

$$\varepsilon_j = \alpha+x_j\beta \quad \begin{cases} x^0 = 1 \\ x_J^+ = (r_J+1)/2 \\ x_J^- = (r_J-1)/2 \end{cases} \qquad (22)$$

The corresponding linear combinations ψ^{HMO} belong to the following irreducible representations of D_{2h}, if the axes are chosen as indicated in the diagram of formula (21):

$$
\begin{array}{lll}
N & : \text{even} & \text{odd} \\
\psi^{\text{HMO},0} & : B_{1u} & B_{3g} \\
J & : \text{even} & \text{odd} \\
\psi_J^{\text{HMO}+} & : B_{3g} & B_{1u} \\
\psi_J^{\text{HMO}-} & : A_u & B_{2g}
\end{array}
\tag{23}
$$

Making use of these results it is easy to construct the correlation diagram shown in Fig. 11. Note that the deviations observed for the individual ionization energies relative to the predictions (22), if the parameters (19) are used, are fully accounted for if the perturbation treatment (17), (18) is applied.[49]

It is of interest to investigate how an SCF calculation, which takes the electron–electron interaction explicitly into account would perform for such systems. To this end we apply the Pariser–Parr–Pople treatment[43] to the set of acenes **12, 11, 45, 46, 55** assuming all bonds of equal length, i.e. $R_{\mu\nu} = 1{\cdot}397$ Å for all bonded pairs $\mu\nu$ and all angles 120°. Using the parameters $\beta_{\mu\nu} = -2{\cdot}371$ eV, $\gamma_{\mu\mu} = 10{\cdot}959$ eV, $\gamma_{\mu\nu}$ (bonded) $= 6{\cdot}783$ eV, $\gamma_{\mu\nu}(R \leqslant 6$ Å; not bonded$) = (328{\cdot}77 + R_{\mu\nu})/(30{\cdot}0 + 12{\cdot}341R_{\mu\nu} + R_{\mu\nu}^2)$ eV, $\gamma_{\mu\nu}(R > 6$ Å$) = 14{\cdot}395/R_{\mu\nu}$ eV and scaling the eigenvalues $\varepsilon_j^{\text{SCF}}$ so obtained according to

$$
I_{v,j} = A + B\varepsilon_j^{\text{SCF}}
\tag{24}
$$

by a least squares fit ($A = 9{\cdot}816 \pm 0{\cdot}078$ eV, $B = -0{\cdot}984 \pm 0{\cdot}054$) the results summarized in Table I are calculated. Note that the residual variance is significantly greater than the one found by the Hückel perturbation treatment, which takes bond length changes $\Delta R_{\mu\nu}$ into account. As shown by Schmidt[50, 51] this result is typical for all benzenoid hydrocarbons.

Polyenes[10] and polyines,[11] as well as other hydrocarbons containing conjugated or cumulated double and triple bonds,[53, 54] are characterized by strong first-order bond alternation (see Table II for a selection of photoelectron spectroscopic data of such unsubstituted systems). For this reason a description in terms of localized two-centre π-orbitals is the model of choice, i.e. expressions (15) with $\lambda_j \equiv \pi_j$. A simple illustration is provided by the two series ethylene (**41**), butadiene (**43**), hexatriene (**16**) and acetylene (**58**), diacetylene (**59**), triacetylene (**60**), for which the correlation diagrams are

HC≡CH   HC≡C−C≡CH   HC≡C−C≡C−C≡CH

58   **59**   **60**

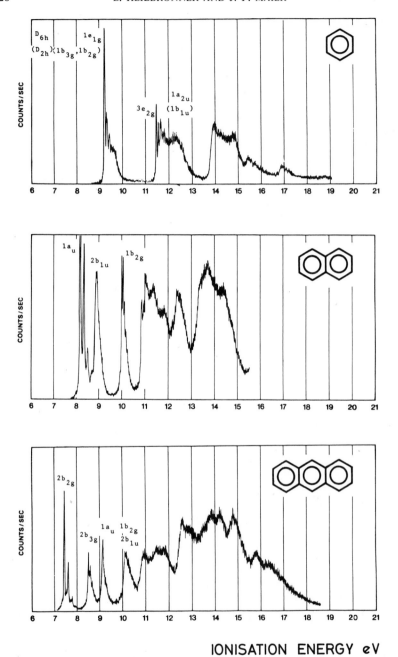

Fig. 10(a). HeI photoelectron spectra of the acenes: benzene (12), naphthalene (11), and anthracene (45).[49]

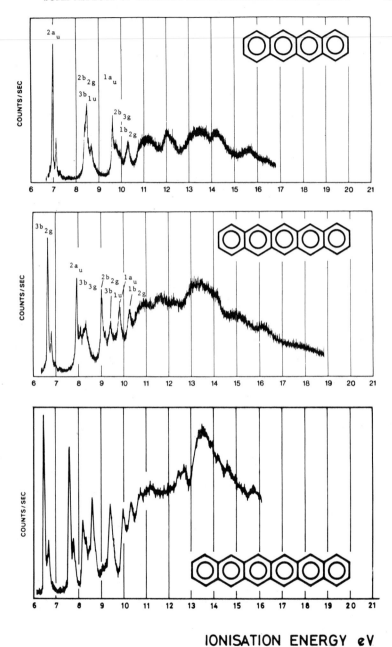

Fig. 10(b). HeI photoelectron spectra of the acenes: tetracene (**46**), pentacene (**55**), and hexacene (**57**).[49, 50]

TABLE I

Vertical π-Ionization Energies ($I_{v,j}(\pi)$) of Aromatic Hydrocarbons

(A) CATACONDENSED SYSTEMS

(a) Acenes

Compound	IE 1	IE 2	IE 3	IE 4	IE 5	IE 6	IE 7	IE 8	IE 9
Benzene[a] (D_{6h})	9·25 $1e_{1g}$	12·2 $1a_{2u}$							
Naphthalene[b] (D_{2h})	8·15 $1a_u$	8·88 $2b_{1u}$	10·08 $1b_{2g}$	10·85 $1b_{3g}$					
Anthracene[c] (D_{2h})	7·47 $2b_{2g}$	8·57 $2b_{3g}$	9·23 $1a_u$	10·26 $1b_{2g}$	10·40 $2b_{1u}$				
Tetracene (D_{2h})	7·04 $2a_u$	8·44 $3b_{1u}$	8·63 $2b_{2g}$	9·60 $1a_u$	9·75 $2b_{3g}$	10·26 $1b_{2g}$			
Pentacene[c] (D_{2h})	6·74 $3b_{2g}$	8·03 $2a_u$	8·40 $3b_{3g}$	9·09 $2b_{2g}$	9·49 $3b_{1u}$	9·88 $1a_u$	10·33 $1b_{2g}$		
Hexacene (D_{2h})	6·44 $3a_u$	7·55 $3b_{2g}$	8·14 $4b_{1u}$	8·56 $2a_u$	9·36 $3b_{3g}$	9·36 $2b_{2g}$	9·95 $1a_u$	9·95 $3b_{1u}$	10·30 $1b_{2g}$

(b) Phenes

Compound	IE 1	IE 2	IE 3	IE 4	IE 5	IE 6	IE 7	IE 8
Phenanthrene[c] (C_{2v})	7·86 $4b_1$	8·15 $3a_2$	9·28 $2a_2$	9·89 $3b_1$	10·59 $2b_1$			
Tetraphene[c] (C_s)	7·47 $9a''$	8·05 $8a''$	8·86 $7a''$	9·39 $6a''$	9·95 $5a''$	10·41 $4a''$		
Pentaphene (C_{2v})	7·34 $6b_1$	7·47 $5a_2$	8·60 $4a_2$	8·93 $5b_1$	9·57 $4b_1$	9·89 $3a_2$	10·35 $3b_1$	
Hexaphene (C_s)	7·02 $13a''$	7·47 $12a''$	8·31 $11a''$	8·60 $10a''$	9·16 $9a''$	9·49 $8a''$	9·90 $7a''$	10·25 $6a''$

TABLE I (cont.)

(c) Other $C_{18}H_{12}$ systems (4 benzene rings)

Compound							
Chrysene[c] (C_{2h})	7·60 $5a_u$	8·10 $4a_u$	8·68 $4b_g$	9·46 $3b_g$	9·76 $3a_u$	10·25 $2b_g$	10·6
3,4-Benzophenanthrene	7·62 $5b_1$	8·00 $4a_2$	8·96 $3a_2$	9·1 $4b_1$	9·95 $3b_1$	10·26 $2a_2$	
Triphenylene (D_{3h})	7·89 $3e''$	8·66 $1a_1''$	9·68 $2e''$	10·06 $2a_2''$			10·7

(d) Other $C_{22}H_{14}$ systems (5 benzene rings)

Compound									
Picene (C_{2v})	7·54 $6b_1$	7·67 $5a_2$	8·36 $4a_2$	9·06 $5b_1$	9·28 $4b_1$	9·92 $3a_2$	10·55 $2a_2$		
3,4,5,6-Dibenzophenanthrene (C_{2v})	7·47 $5a_2$	7·7 $6b_1$	8·29 $5b_1$	8·81 $4a_2$	9·48 $3a_2$	9·77 $4b_1$	9·97 $3b_1$	10·63 $2a_2$	
3,4-Benzotetraphene (C_s)	7·14 $11a''$	7·88 $10a''$	8·18 $9a''$	9·01 $8a''$	9·29 $7a''$	9·90 $6a''$	10·3 $5a''$		10·4
1,2,3,4-Dibenzanthracene (C_{2v})	7·44 $5a_2$	7·92 $6b_1$	8·30 $4a_2$	9·16 $3a_2'$	9·40 $5b_1$	9·97 $4b_1$	9·97 $2a_2$		
1,2,5,6-Dibenzanthracene (C_{2h})	7·38 $6b_g$	7·82 $7b_g$	8·43 $5a_u$	9·02 $4a_u$	9·26 $4b_g$	10·04 $3b_g$	10·04 $3a_u$	10·73 $2b_g$	10·6
1,2,7,8-Dibenzanthracene (C_{2v})	7·39 $6b_1$	7·80 $5a_2$	8·62 $4a_2$	8·82 $5b_1$	9·61 $4b_1$	9·61 $3a_2$	10·4 $3b_1$		10·7

TABLE I (cont.)

(B) PERICONDENSED SYSTEMS

Compound								
Perylene[c] (D_{2h})	$7{\cdot}00$ $2a_u$	$8{\cdot}55$ $3b_{2g}$	$8{\cdot}68$ $3b_{1u}$	$8{\cdot}90$ $2b_{3g}$	$9{\cdot}34$ $2b_{2g}$	$10{\cdot}4$ $2b_{1u}$		
1·12-Benzoperylene[c] (C_{2v})	$7{\cdot}19$ $5a_2$	$7{\cdot}86$ $6b_1$	$8{\cdot}70$ $4a_2$	$8{\cdot}85$ $5b_1$	$9{\cdot}05$ $3a_2$	$9{\cdot}88$ $4b_1$	$10{\cdot}85$ $3b_1$	
Coronene[d] (D_{6h})	$7{\cdot}36$ $2e_{2u}$	$8{\cdot}65$ $2e_{1g}$	$9{\cdot}19$ $1b_{1g}$	$9{\cdot}19$ $1b_{2g}$	$10{\cdot}4$ $2a_{2u}$			
Ovalene[c] (D_{2h})	$6{\cdot}86$ $4b_{2g}$	$7{\cdot}46$ $4b_{3g}$	$8{\cdot}08$ $3a_u$	$8{\cdot}30$ $5b_{1u}$	$8{\cdot}74$ $2a_u$	$8{\cdot}74$ $3b_{2g}$	$9{\cdot}00$ $4b_{1u}$	$9{\cdot}75$ $3b_{3g}$
Pyrene[d] (D_{2h})	$7{\cdot}41$ $2b_{3g}$	$8{\cdot}26$ $2b_{2g}$	$9{\cdot}00$ $3b_{1u}$	$9{\cdot}29$ $1a_u$	$9{\cdot}96$ $2b_u$	$10{\cdot}5$		
Anthanthrene (C_{2h})	$6{\cdot}92$ $6a_u$	$8{\cdot}08$ $5a_u$	$8{\cdot}22$ $5b_g$	$8{\cdot}7$ $4b_g$	$9{\cdot}42$ $4a_u$	$9{\cdot}42$ $3b_g$	$10{\cdot}34$ $3a_u$	

TABLE I (cont.)

(C) NON-BENZENOID SYSTEMS

	IP₁	IP₂	IP₃	IP₄	IP₅	σ-onset
Azulene (C_{2v})	7·43 $2a_2$	8·50 $3b_1$	10·07 $1a_2$	10·85 $2b_1$		11·0
Acenaphthylene (C_{2v})	8·22 $4b_1$	8·39 $2a_2$	8·99 $3b_1$	10·87 $2b_1$		10·7
Fluranthene (C_{2v})	7·95 $3a_2$	8·1 $5b_1$	8·87 $4b_1$	9·50 $2a_2$	10·39 $3b_1$	10·8
Acepleiadylene (C_{2v})	7·13 $3a_2$	7·83 $5b_1$	8·77 $4b_1$	9·51 $2a_2$	9·87 $3b_1$	10·5
Biphenylene (D_{2h})	7·61 $2b_{3g}$	8·90 $1a_u$	9·68 $1b_{2g}$	10·08 $2b_{1u}$		10·7

All values (in eV) are taken from the work of Schmidt et al.[50, 57] except those referenced. The values are uncertain to ±0·04 eV. The last column lists the onset of the σ-band system.

[a] Refs. 2, 36.   [b] Ref. 45.   [c] Refs. 47, 49, 50.   [d] Ref. 50

TABLE II

Vertical π-Ionization Energies ($I_v(\pi)$) of Unsaturated Hydrocarbons

41	$H_2C=CH_2$	10·51 $1b_{1u}$					12·38	2
43		9·03 $1b_{1g}$	11·46 $1a_u$				12·2	36
16		8·29 $2a_u$	10·26 $1b_g$	11·9 $1a_u$			11·6	10
17		8·32 $2b_1$	10·27 $1a_2$	11·9 $1b_1$			11·5	10
58	$HC{\equiv}CH$	11·40 $1e_u$					16·7	2
59	$\equiv-\equiv$	10·17 $1e_g$	12·62 $1e_u$				17·0	2
60	$\equiv-\equiv-\equiv$	9·50 $2e_u$	11·55 $1e_g$	12·89 $1e_u$			16·8	142
61[a]		9·63 $2a''$	10·61 a'	12·01 $1a''$			13·2	143
62[a]		8·72 $2a_u$	9·97 b_u	11·00 $1b_g$	12·01 $1a_u$		12·95	143
63[a]		9·07 $2a_u$	10·55 a_g	10·85 b_u	11·18 $1b_g$	12·43 $1a_u$	13·75	143
64[a]		9·10 $2b_1$	10·54 a_1	10·78 b_2	11·10 $1a_2$	12·39 $1b_1$	13·74	143

TABLE II (cont.)

65[b]	$H_2C{=}C{=}CH_2$	10·04 $1e'$				15·0	2
66[b]	$H_2C{=}C{=}C{=}CH_2$	9·30 $1b_{3g}$	9·63[c]	9·98 $2b_{3u}$			54
13		8·55 $1a_2$	9·54 $2b_1$	12·8 $1b_1$		12·1	144
14		8·80 $2b_1$	9·44 $1a_2$	13·3 $1b_1$		11·5	144
67		8·09 $4a''$	8·69 $3a''$	10·80 $2a''$		11·6	145
68		7·69 $3b_1$	10·22 $1a_2$	11·24 $2b_1$		11·5	145
69		7·40 $4b_1$	7·6 $2a_2$	9·8 $1a_2$	10·16 $3b_1$	11·25	145

All values given in eV. For determination, assignment, and uncertainty of the given values, the original works (referenced in the last column of the table) should be consulted. The penultimate column lists the first σ-band.

[a] For the species 61, 62, 63, and 64, the in-plane acetylenic π-ionization energies are labelled by their symmetry alone.

[b] The first $I_v(\pi)$ values given for 65, and 66, are taken from the first maximum of the bands, which have a complex profile due to a presumed Jahn–Teller distortion.

[c] This ionization process is attributed to simultaneous excitation of a π electron accompanying the π electron ejection.

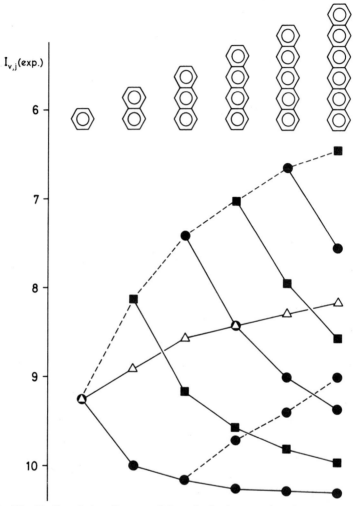

Fig. 11. Correlation diagram of the π ionization energies of the acenes.

shown in Fig. 12. For both sets the most simple approach would be the following LCBO model:

$$\phi = \sum_{j=1}^{n} c_j \pi_j \quad \begin{cases} \langle \pi_j | \mathscr{H} | \pi_j \rangle = A \\ \langle \pi_j | \mathscr{H} | \pi_{j\pm1} \rangle = B \end{cases} \tag{25}$$

which yields:

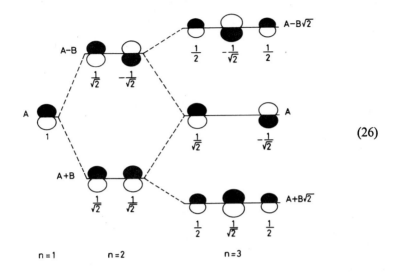

(26)

A comparison with Fig. 12 shows that (26) is a rather good approximation of what is observed. Taking butadiene (43) and diacetylene (59) as a reference, the calibration of (26) yields (in eV):

$$A = -10 \cdot 25 \qquad A = -11 \cdot 40$$
$$B = - \ 1 \cdot 22 \qquad B = - 1 \cdot 23$$

(27)

from which one would predict for the corresponding chain of three π-orbitals

	calc.	obs.	calc.	obs.
$I_1(\pi)$	8·52	8·29	9·66	9·50
$I_2(\pi)$	10·25	10·26	11·40	11·55
$I_3(\pi)$	11·98	11·9	13·14	12·89

(28)

in satisfactory agreement with observation. However, a more detailed investigation of such systems,[55] especially of substituted ones, leads to the conclusions that a more realistic description of such molecules demands the inclusion of antibonding localized π-orbitals $\lambda_j^* \equiv \pi_j^*$ into the linear

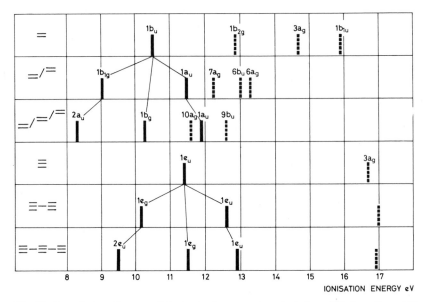

Fig. 12. Correlation diagram of the ionization energies of ethylene (**41**), butadiene (**43**), hexatriene (**16**), and acetylene (**58**), diacetylene (**59**), and triacetylene (**60**).

combination φ, as indicated qualitatively in the following self-explanatory diagram:

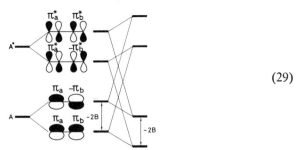

(29)

As can be seen, admixture of antibonding orbitals lowers the orbital-energies of the bonding π-orbitals. In addition, they tend to polarize the localized bond orbitals. It should be noted that use of a complete set of bonding and anti-bonding two-centre basis orbitals is of course equivalent to using the set of individual atomic $2p$-orbitals as a basis.

IV. SUBSTITUTED π-SYSTEMS

"Organic chemists have tended to be hypnotized by the kind of approach embodied in the Hammett equation; they have consequently been dominated by a lemming-like urge to study systems amenable to $\rho\sigma$-plots." (M. J. S. Dewar[56])

There is hardly any property of molecules which, for better or for worse, has not been subjected to an LFER-analysis (LFER = Linear Free Energy Relationship) based on one or the other of the different sets of Hammett-type σ-values.[57] Ionization potentials are no exception as exemplified by the work of Watanabe et al.,[58] Kaufman and Koski,[59] Turner,[60] Van Cauwelaert,[61] and the references given therein. These earlier investigations were concerned only with first ionization potentials, i.e. with a process (1), with $^2\tilde{\Psi}_j \equiv {}^2\tilde{\Psi}_0$, in which an electron vacates the highest occupied molecular orbital ψ_{HOMO}. With the advent of photoelectron spectroscopy all ionization energies $I_{v,j}$ or $I_{a,j}$, corresponding to ejection of an electron from any one of the occupied molecular orbitals ψ_j of M, have become available. In the following we shall concern ourselves again only with such cases where the ψ_j are π-orbitals. Other systems are quoted only for the sake of comparison.

Due to the rather extensive number of data now available for unsubstituted and the corresponding substituted π-systems, the parameterization of π-ionization potentials in terms of polar and/or orbital interaction effects due to the substituting groups, has gained in interest. Indeed a set of empirical parameters characterizing the influence of the substituent(s) if possible in the framework of a simple molecular orbital model, would be of considerable help in the analysis of PE spectra of unsaturated molecules.

Of particular interest are alkyl substituents R. Apart from the important role they play in the majority of LFER investigations, they have from our point of view the added advantage that their basis orbitals $\phi(R)$ lie at much lower energies than most of the π-orbitals. Consequently the π-bands in the photoelectron spectrum of an alkyl substituted derivative M' of the parent molecule M are not overlapped by the σ-bands of R and are thus easily correlated with those of M. The changes brought about by alkyl substitution can then serve to characterize the π-orbital ψ_j which has lost the photoelectron in the process (1).

Although the limitations of Koopmans' approximation (9) are well known, it is usually assumed that it will at least apply in the restricted form

$$\Delta I_{v,j} = -\Delta\varepsilon_j \tag{30}$$

i.e. for the prediction of changes $\Delta I_{v,j}$ in the ionization potential $I_{v,j}$ from changes $\Delta\varepsilon_j$ of the orbital energy of ψ_j, calculated for a given small perturbation (e.g. alkyl substitution) by a first- or second-order perturbation treatment.

It is noteworthy that even (30) suffers from severe limitations, as has been shown for example in the case of substituted fulvenes.[62] (See also section VII.)

Nevertheless, it will be found that a rather satisfactory and simple LFER-type relationship between $\Delta I_{v,1}$ and the type and number of substituting groups R is observed in many cases, although such a correlation is the resultant of a rather complex electronic mechanism for the interaction of R with the parent system. This is in agreement with recent experimental and theoretical results concerning the structure of carbonium-[63] and carbenium-ions[67] and with the gas-phase acidities and basicities of alkyl-substituted systems.[64]

An LFER treatment of photoelectron spectroscopically determined ionization potentials has first been investigated in some detail by Cocksey et al.[65] who showed that the first adiabatic ionization potential $I_{a,1}(XR)$ of a molecule XR(X = HCO, R'CO, OH, I, and OR'; R' = alkyl) can be fitted to a Hammett-type relationship

$$I_{a,1}(XR) = I_{a,1}(XMe) + \chi_X \mu_R \tag{31}$$

where μ_R is a parameter characteristic for the alkyl group R and χ_X for the moiety X containing the orbital ψ_j from which the electron is ejected. By convention the parameter μ_R is defined as

$$\mu_R = I_{a,1}(IR) - I_{a,1}(IMe) \tag{32}$$

i.e. applying (31) with $\chi_I = 1 \cdot 00$ to the alkyl iodides IR. Within the limits of error of the method, relationship (31) will also apply to the vertical ionization potentials $I_{v,j}$ of molecules XR. (For a summary see ref. 66.)

In Fig. 13 the first vertical ionization potentials $I_{v,1}(RX)$ as obtained by photoelectron spectroscopy for X = Br,[66, 67] OH[66, 68], $CH_2 = CH$ (vinyl),[69] $HC \equiv C$ (ethinyl),[70] C_6H_5 (phenyl),[71] and C_6H_5(6-fulvenyl)[62] have been plotted vs

$$\overline{I_{v,1}(IR)} = (I_{v,1}(IR; {}^2\Pi_{\frac{3}{2}}) + I_{v,1}(IR; {}^2\Pi_{\frac{1}{2}}))/2 \tag{33}$$

i.e. the mean of the ionization energies corresponding to the ejection of an electron from one of the pairs of 5p-orbitals of iodine in the corresponding alkyl iodides. The split between these ionization potentials is due to spin-orbit coupling[72] which would leave a radical cation IR+, with an axis of order $g \geqslant 3$ (e.g. IH, IMe, ItBu), in a ${}^2\Pi_{\frac{3}{2}}$ or ${}^2\Pi_{\frac{1}{2}}$ state. (For details see ref. 73). As can be seen from Fig. 13 the linear regression expected according to (31) and (32) is fulfilled within reasonable limits of error.

Cocksey et al.[65] have derived sets of μ_R-parameters for (31) using either definition (32) with $I_a(IR) = I_v(IR, {}^2\Pi_{\frac{3}{2}})$, or by subjecting the ionization potentials of fifteen sets of compounds XR (with χ_X ranging from 0·5 to 2·0) to a least squares analysis. The latter parameters have been labeled μ'_R in Table III, where they are compared to the original μ_R, to μ''_R derived by a

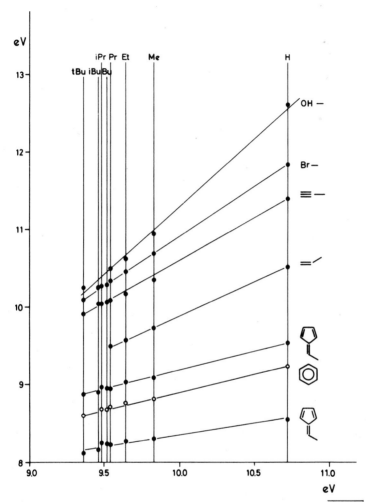

Fig. 13. Plot of the first ionization energies $I_{v,1}(\text{RX})$ (ordinate) *vs* the $\overline{I_{v,1}(\text{IR})}$ values (abscissa) (evaluated according to formula 33). The substituents X are shown to the right of the diagram and the alkyl groups R are correlated by the solid vertical lines.

least-squares analysis of $I_{v,1}(\text{XR})$ *vs* $I_{v,1}(\text{IR})$ (see formula (33)) and to Taft's σ^* values.[57] The general agreement is rather satisfactory, with the notable exception that μ_H'' is over twice as large as μ_H. However, the value $\mu_\text{H}'' = 0\cdot89$ fits perfectly on a linear regression based on the standard σ^* values as an independent variable

$$\mu_\text{R}'' = (-0\cdot029 \pm 0\cdot026) + (1\cdot822 \pm 0\cdot115)\,\sigma_\text{R}^* \qquad (34)$$

$$\text{corr. coeff.} = 0\cdot9883$$

Note that the regression (34) does not miss the origin ($\mu_R'' = \sigma_R^* = 0$) significantly and that the regression μ_R vs σ_R^* yields a correlation coefficient of only 0·970.

If the moiety X is susceptible to multiple substitution in symmetry equivalent positions ρ one might expect that in analogy to the behaviour of the traditional Hammett $\rho\sigma$-rule[57] the relation (31) can be generalized to yield

$$I_v(XR_1 R_2 \ldots) = I_v(XMe_n) + \chi_X \sum_{R_\rho} \mu_{R\rho}'' \tag{35}$$

where $I_v(XMe_n)$ is the vertical ionization potential of the fully methylated reference compound. For practical reasons it is more convenient to rewrite (35) in the following form:

$$I_v(XR_1 R_2 \ldots) = I_v(XH_n) + \chi_X \sum_{R_\rho} (\mu_{R\rho}'' - \mu_H'')$$
$$= I_v(XH_n) + \chi_X \sum_{R_\rho} \mu_{R_\rho}{}^H \tag{36}$$

Here the $\mu_{R_\rho}{}^H = \mu_{R\rho}'' - \mu_H''$, given in Table III, are now similar to the classical σ-constants in Hammett's original treatment, i.e. $\mu_H{}^H = 0$ (rather then $\mu_{Me}'' = 0$).

We shall now investigate how well the additivity rule (36) is obeyed in practice, by considering two examples.

In Table IV are given the π-ionization potentials $I_{v,1}(\pi)$ for a series of alkyl substituted ethylenes 38[22] and the corresponding sum over the substituent parameter $\mu_R{}^H$ which has to be inserted into (36). According to the latter formula we expect a linear dependence of $I_{v,1}(\pi; RR'C=CR''R)$ on the sum of the $\mu_R{}^H$. As shown in Fig. 14 the regression is not strictly linear. The quadratic contribution is significant, but the departure from linearity is not severe.

TABLE III

Comparison of the Alkyl Group Parameters μ_R and μ_R' with Taft's σ^* Values

| | Ref. 65 | | This work | This work | Ref. 57 |
	μ_R	μ_R'	μ_R''	$\mu_K{}^H$	σ^*
H	0·41		0·89	0·00	0·49
Me	0·00	0·00	0·00	−0·89	0·00
Et	−0·20	−0·19	−0·19	−1·08	−0·10
Pr	−0·29	−0·28	−0·29	−1·18	−0·115
iPr	−0·36	−0·36	−0·35	−1·24	−0·19
Bu	−0·34	−0·37	−0·31	−1·20	−0·13
iBu	−0·38	−0·38	−0·37	−1·26	−0·125
tBu	−0·52	−0·50	−0·47	−1·36	−0·30

TABLE IV

π-Ionization Potentials of Alkyl Substituted Ethylenes $RR'C=CR''R'''$

	$I_v(\pi)^a$	$R\mu_R^H$
$H_2C=CH_2$	10·52	0·00
$H_2C=CHMe$	9·73	−0·89
$H_2C=CMe_2$	9·23	−1·78
MeHC=CHMe (cis)	9·13	−1·78
MeHC=CHMe (trans)	9·13	−1·78
MeHC=CMe$_2$	8·68	−2·67
Me$_2$C=CMe$_2$	8·30	−3·56
$H_2C=CHEt$	9·58	−1·08
$H_2C=CHPr$	9·50	−1·18
$H_2C=CHBu$	9·46	−1·20
EtHC=CHMe (cis)	9·11	−1·97
EtHC=CHMe (trans)	9·06	−1·97
PrHC=CHMe (trans)	9·06	−2·07
EtHC=CHEt (trans)	9·02	−2·16
$H_2C=CEt_2$	9·11	−2·16

All values in eV. The parameters μ_R^H used in the summation given in the last column are defined as $\mu_R^H = \mu_R - \mu_H$ relative to the constants given in Table III.

^{a} $I_v(\pi)$ values tabulated as given in ref. 15.

In the second example we discuss the dependence of the first vertical π-ionization potential of benzene on the number of substituting methyl groups R = Me.[71,74] Benzene (D_{6h}) is a particular case, because its two highest occupied π-orbitals ($1e_{1g}$) are degenerate. In all methyl substituted derivatives, with the exception of 1,3,5-trimethylbenzene (D_{3h}) and of hexamethylbenzene (D_{6h}), this degeneracy is lifted, leading to two molecular orbitals $\pi'(A)$, $\pi'(S)$ of differing orbital energies $\varepsilon(\pi'(A)) \neq \varepsilon(\pi'(S))$. Assuming that the perturbation due to the methyl-substituents is small, the two orbitals $\pi'(S)$ and $\pi'(A)$ will not differ significantly from the real representations $\pi(S)$ and $\pi(A)$ of the unperturbed $1e_{1g}$ pair (in ZDO approximation):

$$\left.\begin{aligned}
\pi(S) &= \sum_\mu c_{S\mu}\phi_\mu = \frac{1}{\sqrt{12}}(2\phi_1+\phi_2-\phi_3-2\phi_4-\phi_5+\phi_6) \\
\pi(A) &= \sum_\mu c_{A\mu}\phi_\mu = \tfrac{1}{2}(\phi_2+\phi_3-\phi_5-\phi_6)
\end{aligned}\right\} \tag{37}$$

Because of $c_{S\mu}^2 + c_{A\mu}^2 = \frac{1}{6}$ for all μ it is to be expected that the mean ionization energy

$$\overline{I_v(\pi')} = \tfrac{1}{2}[I_v(\pi'(A)) + I_v(\pi'(S))] \tag{38}$$

of the two bands corresponding to ejection from $\pi'(S)$ and $\pi'(A)$ depends linearly on the number N of substituting methyl groups. The data listed in Table V and displayed in Fig. 15 show that this is indeed the case.

In the general case of substitution in positions ρ which are *not* related by symmetry and for the *j*th ionization energy, formula (36) has to be modified to

$$I_{v,j}(XR_1R_2\ldots) = I_{v,j}(XH_n) + \sum_{R_\rho} \chi_{X,j,\rho}\mu_{R_\rho}{}^H \tag{39}$$

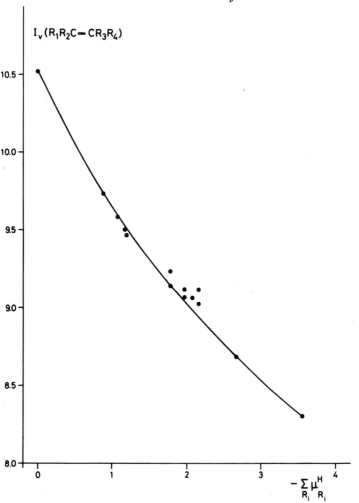

Fig. 14. Dependence of first ionization energies $I_{v,1}(\pi)$ of alkyl-substituted ethylenes[69] on the substitutent parameters $\mu_{R_i}{}^H$.

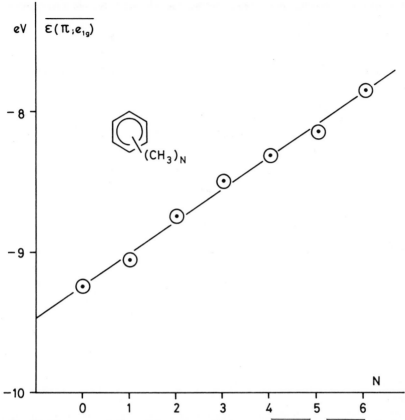

Fig. 15. Linear correlation of the mean orbital energy $\overline{\varepsilon(\pi, e_{1g})} = \overline{-I_v(\pi')}$, obtained according to formula (38), on the number N of substituting methyl groups in the benzene ring.

TABLE V

Mean π-Ionization Potentials $\overline{I_v(\pi)}$ of Methylsubstituted Benzenes

N	$\overline{I_v(\pi)}^a$		
0	9·25		
1	9·06		
2	8·75 (1, 2);	8·75 (1, 3);	8·73 (1, 4)
3	8·48 (1, 2, 3);	8·61 (1, 2, 4);	8·40 (1, 3, 5)
4	8·32 (1, 2, 3, 5);		
5	8·15		
6	7·86		

All values in eV. N = number of methyl-groups. Numbers in parentheses = positions of substitution.

^{a} Calculated according to ref. 38 from $I_v(\pi'(A))$ and $I_v(\pi'(s))$.[74]

where $\chi_{X,j,\rho}$ depends on the moiety X, the vacated orbital ψ_j, and the position ρ which carries the substituent R_ρ.

The examples given above suggest that the restricted form (30) of Koopmans' Theorem should be an acceptable approximation. If this is true, then the LFER-type dependence of $I_{v,j}$ on alkyl substitution should be amenable to a simple parameterization in the framework of an independent electron molecular orbital treatment. To begin with, we shall again use the standard Hückel model for π-systems.

Let $\mathscr{H}$ be the HMO hamiltonian for the parent compound X and for the separated alkyl groups $R_1 R', \ldots$. The normalized π-orbitals of X are $\psi_j = \sum_\mu c_{j\mu} \phi_\mu$ and those of the alkyl groups $\phi(R)$. These orbitals are assumed to be normalized. The corresponding orbital energies are

$$\varepsilon_j = \langle \psi_j | \mathscr{H} | \psi_j \rangle$$
$$A_R = \langle \phi(R) | \mathscr{H} | \phi(R) \rangle \tag{40}$$

Substituting X in positions $\rho, \rho', \ldots$ by alkyl groups $R, R', \ldots$ to yield $XRR' \ldots$ will lead to a new HMO hamiltonian $\mathscr{H}' = \mathscr{H} + h$, where h is the perturbation operator into which we absorb all those effects we wish to take into consideration. It is thus defined by a list of ad hoc matrix elements:

(a) Inductive effect. The "classical" treatment[42] of the inductive effect is to postulate that the atomic Coulomb term $\alpha_\rho = \langle \phi_\rho | \mathscr{H} | \phi_\rho \rangle$ in the moiety X is changed by

$$\langle \phi_\rho | h | \phi_\rho \rangle = \delta\alpha_\rho(R) \tag{41}$$

due to substitution of X in position ρ by R. For alkyl groups $\delta\alpha_\rho(R)$ is positive.

(b) Hyperconjugative effect. According to Mulliken[75] hyperconjugation is described by a resonance integral

$$\langle \phi_\rho | h | \phi(R) \rangle = \beta_{\rho R} \tag{42}$$

which links the orbital $\phi(R)$ of the substituting alkyl group to the atomic orbital ϕ_ρ of X.

(c) Inductive transmission effect. This is a generalization of the effect (a), introduced originally by Pauling and Wheland,[76] which takes into consideration that substitution of position ρ in X may in addition to (41) also change the Coulomb terms in positions τ and τ', ortho to ρ, albeit to a smaller extent, i.e.

$$\langle \phi_\tau | h | \phi_\tau \rangle = \langle \phi_{\tau'} | h | \phi_{\tau'} \rangle = m\delta\alpha_\mu(R) \tag{43}$$

where m is the transmission coefficient ($m < 1$).

(d) Non-bonded through-space interaction. This effect relates to hyperconjugation (b) in the same sense as inductive transmission (c) to the inductive effect (a). Hoffmann and Olofson[77] have suggested that the stability of conformers of extended π-systems depends on the through-space interaction of non-bonded atomic orbitals ϕ_μ. If this concept is applied to alkyl substituted π-systems XRR'... we might expect that through-space interaction of $\phi(R)$ with atomic orbitals ϕ_μ other than ϕ_ρ is not negligible. To account for this effect the necessary matrix element would be

$$\langle \phi_\mu | h | \phi(R_\rho) \rangle = \beta'_{\mu R}; \quad \mu \neq \rho \tag{44}$$

(e) Through-space interaction between substituting alkyl groups R and R'. This type of interaction will presumably be of interest only if both R and R' are substituents of the same centre $\rho = \rho'$, e.g. in $H_2C = CRR'$. The corresponding matrix element is

$$\langle \phi(R) | h | \phi(R') \rangle = \beta_{RR'} \tag{45}$$

The relevance of such long-range interactions between alkyl groups has been discussed recently by Hoffmann et al.[78]

In the traditional application of the HMO model, the influence of an alkyl group R on a π-system is described only in terms of (41) and (42). Assuming for the moment that $\varepsilon_j - A_R$, as defined in (40), is large with respect to the expected shifts $\Delta\varepsilon_j$ to be inserted into formula (30), a perturbation treatment will yield

$$\Delta\varepsilon_j = \langle \psi_j | h | \psi_j \rangle = \sum_\rho c_{j\rho}{}^2 \, \delta\alpha_\rho(R) + \sum_\rho \frac{c_{j\rho}{}^2 \beta_{\rho R}{}^2}{\varepsilon_j - A_R} \tag{46}$$

Note, that because of the correspondence between R and ρ, R' and ρ', etc., the summation index can be either R or ρ. The first summation in (46) contains the contributions of the inductive effects (a) to $\Delta\varepsilon_j$, the second summation those due to hyperconjugation (b). Contraction of the two terms yields

$$\Delta\varepsilon_j = \sum_\rho c_{j\rho}{}^2 \left[\delta\alpha_\rho(R) + \frac{\beta_{\rho R}{}^2}{\varepsilon_j - A_R} \right] = \sum_\rho c_{j\rho}{}^2 \, \delta\alpha'_\rho(R) \tag{47}$$

where $\delta\alpha'_\rho$ stands for the square bracket. From (30) we obtain

$$\Delta I_{v,j} = -\sum_\rho c_{j\rho}{}^2 \, \delta\alpha'_\rho(R)$$

$$\delta\alpha'_\rho(R) = \delta\alpha_\rho(R) + \beta_{\rho R}{}^2/(\varepsilon_j - A_R) \tag{48}$$

which should be compared to (39) in the form

$$\Delta I_{v,j} = \sum_\rho \chi_{X,j,\rho} \mu_R{}^H \tag{49}$$

where $\Delta I_{v,j}$ stands for $I_{v,j}(XR_1 R_2 ...) - I_{v,j}(XH_n)$.

Formulae (48) and (49) lead to some important conclusions, concerning the use of a simple Hückel model and the application of a LFER treatment:

(1) According to (48) $\delta\alpha'_\rho(R)$ can be obtained directly from the $\Delta I_{v,j}$ values observed for a series of alkyl-substituted derivatives XR, if the coefficients $c_{j\rho}{}^2$ are known. On the other hand, it is not possible to differentiate in the framework of our model between an inductive (a) or hyperconjugative (b) effect as long as the assumption $|\varepsilon_j - A_R| \gg |\Delta\varepsilon_j|$ is valid. A given $\delta\alpha'_\rho(R)$ could be interpreted as purely inductive $(\delta\alpha_\rho'(R) = \delta\alpha_\rho(R); \beta_{\rho R} = 0)$ or of purely hyperconjugative origin $(\delta\alpha'_\rho(R) = \beta_\rho{}^2{}_R/(\varepsilon_j - A_R); \delta\alpha_\rho(R) = 0)$ or by a judicious mixture of both effects.

It is useful to discuss briefly the magnitude of the effects involved, taking R = Me as an example. The PE spectra of the alkanes[79] suggest that the basis orbital energy of $\phi(Me)$ is approximately $A_{Me} = -14$ eV. From the regression shown in Fig. 13 and 15 (see Tables IV and V) we obtain (all values in eV)

$$
\begin{array}{cccc}
X & \varepsilon_{HOMO} & \Delta\varepsilon_{HOMO} & \varepsilon_{HOMO} - A_{Me} \\
\text{phenyl} & -9.2_5 & 0.3 & \sim 5 \\
\text{vinyl} & -10.5 & 0.8 & \sim 3.5
\end{array}
\tag{50}
$$

Thus the condition $\varepsilon_j - A_R \gg \Delta\varepsilon_j$ is well fulfilled in the case X = phenyl. The case X = vinyl may well be at the limit where the model underlying (46) to (48) may be used with confidence, if hyperconjugation is a major contribution to $\Delta\varepsilon_j$. The latter is strongly suggested by comparing the ratios $\Delta\varepsilon_{HOMO}(vinyl)/\Delta\varepsilon_{HOMO}(phenyl) = 2.5$ to the ratio of the coefficients $c^2_{HOMO,\rho}(vinyl)/c^2_{HOMO,\rho}(phenyl) = (1/\sqrt{2})^2/(1/\sqrt{3})^2 = 1.5$. This conclusion would seem to be in agreement with the observation that for a given alkyl group R the perturbation $\Delta\varepsilon_j$ seems to increase with decreasing gap $\varepsilon_j - A_R$, as suggested by the regression lines in Fig. 13.

V. "THROUGH-SPACE" AND "THROUGH-BOND" INTERACTIONS

The concepts of "through-space" and "through-bond" interactions were originally introduced by Hoffmann[80] in order to provide some heuristically useful guidelines for the discussion of physico-chemical properties of molecules. Especially when the intramolecular interactions between localised[38] or semi-localized orbitals become significant, these notions lead to an attractive rationalization and have, since then, been widely used either qualitatively or in the framework of independent electron treatments; i.e. the Hückel[42] or the Extended Hückel[81] methods. With the advent of the photoelectron spectroscopic studies of organic molecules, encompassing among others, the model systems previously discussed by Hoffmann, it became apparent that the consequences of these interactions are often clearly represented in the photoelectron spectra. The semilocalized basis molecular orbitals most frequently studied comprise those of π electrons[82] of un-saturated systems and of lone pairs, e.g. N,[83] O,[84-86] S[84,87]. The bands which have to be correlated with orbitals derived mainly from such a semi-localized basis are amenable for study and assignment, as they occur in the low ionization energy regions of the spectra. Thus, the ionization energies are expected to be a source of quantitative data of "through-space" and "through-bond" interactions.

However, in order to decompose the experimental data into individual contributions assigned to the "through-space" and the "through-bond" mechanism, models have to be introduced. In this section we shall concern ourselves with the studies of these concepts, based on the data provided by the photoelectron spectra, firstly in the terminology of an independent electron treatment and then within many-electron models.

The first task which has to be faced involves the assignment of the photo-electron spectrum. This requires:

(a) the identification of the bands in the photoelectron spectrum which are correlated with those molecular orbitals which are derived from the basis orbitals in question, and

(b) the determination of the symmetry of the states of the radical cation to which the bands in photoelectron spectrum correspond.

Although the solution of problem (a) is often straightforward, this is by no means always the case. An example is provided by p-benzoquinone **70**[85] for which the true sequence of the lone pair bands may only be established by applying a correlation technique to the set **70**, **71** ($= 1,4$-naphthoquinone), **72** ($= 9,10$-anthraquinone) as has been shown by Schweig *et al.*[86]

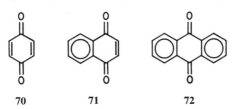

70   71   72

The electronic states $^2\Psi_j(\tilde{X}, \tilde{A}, \tilde{B}, ...)$ of the radical cation M$^+$ belong to the same irreducible representation of M as the vacated canonical orbitals φ_j, provided that φ_j and $^2\Psi_j$ are classified according to the symmetry elements common to M and M$^+$ (the latter in each of its states $\tilde{X}, \tilde{A}, \tilde{B}, ...$). Implicit in this approach is the embracement of Koopmans' Theorem to relate the ionization energies to the corresponding molecular orbitals and their orbital energies. The second task entails the application of a particular model to allow a partitioning of the "observed" orbital energy shifts, relative to assumed basis energies, into "through-space" and "through-bond" contribution. Here it must be emphasized that any such rationalization is strongly dependent on the particular model chosen and therefore the starting point must always be the definition of the theoretical model concepts to be used.

In a Hückel-type independent electron model scheme (e.g. the standard HMO treatment[42]), the "through-space" interaction between two basis orbitals χ_a and χ_b is represented by their interaction matrix element $\langle \chi_a | H | \chi_b \rangle$, i.e. the resonance integral B_{ab}, H being the Hückel hamiltonian. In a first approximation B_{ab} is proportional to the overlap integral $S_{ab} = \langle \chi_a | \chi_b \rangle$ in size, but with opposite sign, i.e.

$$B_{ab} = -kS_{ab} = -k\langle \chi_a | \chi_b \rangle \tag{51}$$

where k is a positive calibration constant. The self energies of χ_a and χ_b are

$$A_a = \langle \chi_a | H | \chi_a \rangle; \quad A_b = \langle \chi_b | H | \chi_b \rangle \tag{52}$$

Note that the basis orbitals χ_a, χ_b have not yet been defined. They can be atomic orbitals ($\chi_i \equiv \phi_i$), group orbitals or localized orbitals, similar to those used in (15) within the framework of a many-electron treatment. With (51) and (52) we obtain the following linear combinations and associated orbital energies, assuming ZDO approximation

$$\chi_+ = a\chi_a + b\chi_b; \quad \varepsilon_+ = \bar{A} + B_{ab}\left[\left(\frac{\Delta A}{2B_{ab}}\right)^2 + 1\right]^{\frac{1}{2}}$$
$$\chi_- = b\chi_a - a\chi_b; \quad \varepsilon_- = \bar{A} - B_{ab}\left[\left(\frac{\Delta A}{2B_{ab}}\right)^2 + 1\right]^{\frac{1}{2}} \tag{53}$$

where $\bar{A} = (A_a + A_b)/2$ and $\Delta A = A_a - A_b$.

From (53) follows that for positive S_{ab} (negative B_{ab}) the orbital sequence is χ_+ *below* χ_- in energy. We define this as the "natural sequence", i.e. the one where the orbital at higher energy possesses a node. If χ_a and χ_b are related by symmetry, then (53) simplifies because of $A_a = A_b = A$ to

$$\chi_+ \equiv \chi_S = (\chi_a + \chi_b)/\sqrt{2}; \quad \varepsilon_+ \equiv \varepsilon_S = A + B_{ab}$$
$$\chi_- \equiv \chi_A = (\chi_a - \chi_b)/\sqrt{2}; \quad \varepsilon_- \equiv \varepsilon_A = A - B_{ab} \tag{54}$$

The lower indices refer to the symmetric (S) and antisymmetric (A) behaviour of the linear combinations relative to a symmetry element of order two.

It should be noted that formulae (53) and (54) have been derived under the ZDO approximation *after* B_{ab} has been calibrated according to (51). Therefore the assumption $B_{ab} = -kS_{ab}$ entails that the difference in ionization energies $\varepsilon_A - \varepsilon_S$, corresponding to electron ejection from either χ_- or χ_+, does depend on S_{ab}. As an example of pure "through-space" interaction, dictated by symmetry (cf. (54)), the energy difference of $\varepsilon_A - \varepsilon_S = 2.5$ eV is observed for the two linear combinations χ_A, χ_S in butadiene (**43**)[10] or di-acetylene (**59**)[11] (cf. Fig. 12 and (27)), where S_{ab} is of the order of 0.1.

The "through-space" interaction between non-conjugated (two-centre) π-orbitals, e.g. in 1,4,7-cyclononatriene (1), Dewar-benzene (2), barrelene (32), or in 1,4-cyclohexadiene (73) is usually called "homoconjugation" if the π-orbitals are separated by one sp^3 centre. The size of this particular "through-space" interaction depends both on the distance and on the mutual orientation of the basis orbitals, as shown in Fig. 16 for a model of 73[88] assuming different dihedral angles.

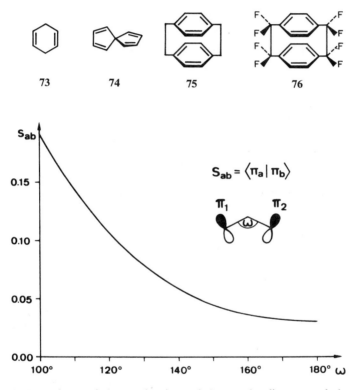

Fig. 16. Dependence of the overlap integral S_{ab} on the distance and the mutual orientation of the basis orbitals, π_a, π_b, defined by the dihedral angle ω.

A special case of homoconjugation is encountered in spiro-connected molecules such as spiro[4.4]nonatetraene (74).[89] Due to the rather special local symmetry (D_{2d}) at the spiro-center in such systems, almost pure "through-space" interaction occurs between the highest occupied π-orbitals of the spiro-connected π-systems, leading to very simple rules for the prediction of the size of the interaction. An example, due to Schweig $et\ al.$[90] is given in Fig. 17.

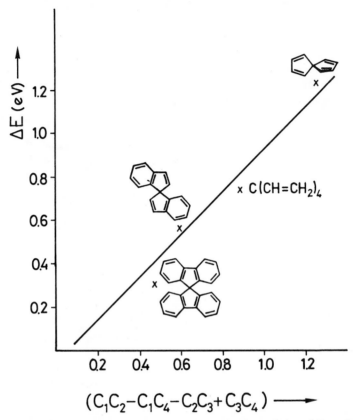

Fig. 17. Correlation of the experimentally measured spiro splitting, ΔE, *vs* the values derived from the expression $(C_1C_2 - C_1C_4 - C_2C_3 + C_3C_4)$. The latter four coefficients of the atomic orbitals of the spiro-conjugated atoms are taken from MINDO/2 calculations.[90] (Redrawn from ref. 90).

Another example where "through-space" interaction is expected to play an important role is provided by the paracyclophanes (e.g. [2,2]paracyclophane **75**[91] and related molecules (e.g. [2.2.2](1,3,5)cyclophane **77**[92] and anti-[2.2]paracyclonaphthane **78**[93]).

77   **78**

This mechanism has been discussed by Schmidt *et al.*[91, 92] However, in agreement with a theoretical prediction by Gleiter[94] it has been shown[95] that "through-bond" interaction, in which the two CC-σ-orbitals of the ethylidene bridge act as relay orbitals, is significant for the $b_{3u}(\pi)$ orbital of **75**. Indeed if the basis orbital energy of the CC-σ-orbitals is strongly lowered by fluorination, to yield the octafluoroderivative **76**, the "through-bond" interaction drops considerably and the observed spectrum reflects (almost) pure "through-space" interaction, as shown in Fig. 18. This will be discussed in more detail below. The linear combinations $b_{2u}(\pi)$ and $b_{3g}(\pi)$, which are

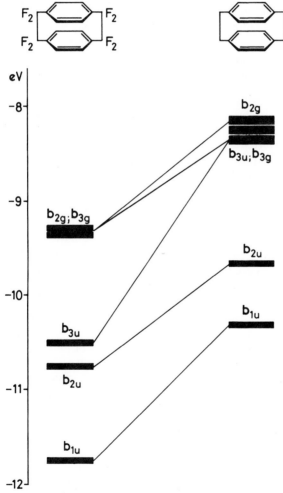

Fig. 18. Correlation diagram of the assignment of the photoelectron spectra of [2.2]-paracyclophane (**75**) and 1,1,2,2,9,9,10,10-octafluoro[2.2]-paracyclophane (**76**).[95]

antisymmetric with respect to the symmetry plane separating the two benzene moieties of **75** and **76**, cannot interact with the ethylidene CC-σ-orbitals. Their splitting $\varepsilon(b_{3g}(\pi)) - \varepsilon(b_{2u}(\pi)) \approx 1{\cdot}5$ eV is thus due to the pure "through-space" mechanism, and again follows the expected dependence on overlap.

In the case of pure "through-bond" interactions, the basis orbitals, χ_a and χ_b, have a negligible overlap $S_{ab} \approx 0$. Then the linear combinations generated, which may be formed irrespective of symmetry,

$$\chi_+ = (\chi_a + \chi_b)/\sqrt{2}; \quad \chi_- = (\chi_a - \chi_b)/\sqrt{2} \tag{55}$$

have necessarily equal expectation values $\langle \chi_+ | H | \chi_+ \rangle = \langle \chi_- | H | \chi_- \rangle$. As mentioned, χ_a and χ_b may, or not, be symmetry equivalent. When there are other bonding or antibonding orbitals, χ_i, belonging to the same irreducible representations as χ_+ (or χ_-), the resulting interaction matrix elements $B_{i+} B_{i-}$ (resonance integrals) are again assumed to be proportional to overlap integrals within an independent electron treatment, i.e.

$$B_{i+} = \langle \chi_i | H | \chi_+ \rangle = (B_{ia} + B_{ib})/\sqrt{(2)} \propto (S_{ia} + S_{ib})/\sqrt{2}$$
$$B_{i-} = \langle \chi_i | H | \chi_- \rangle = (B_{ia} - B_{ib})/\sqrt{(2)} \propto (S_{ia} - S_{ib})/\sqrt{2} \tag{56}$$

Thus if χ_a, χ_b overlap significantly with at least one other basis orbital χ_i, the degeneracy of χ_+ and χ_- may be lifted. The "through-bond" induced shifts $\tau_+ \tau_-$, of the orbital energies of

$$\varepsilon_+ = \langle \chi_+ | H | \chi_+ \rangle; \quad \varepsilon_- = \langle \chi_- | H | \chi_- \rangle \tag{57}$$

can then be evaluated. One should note that both situations, $\tau_+ < \tau_-$, and versa, $\tau_+ > \tau_-$, are possible, leading to either the natural, or the inverted, order of the perturbed orbitals, respectively.[96] Examples of pure "through-bond" interactions are provided by the difference in ionization energies of the nitrogen lone pairs in cyanogen ($\sim$0·4 eV),[97] pyrazine (1·72 eV),[98] and the axial halogen lone pairs of dihaloacetylenes ($\sim$0·6 eV).[99]

However, "through-space" and "through-bond" interactions will in most cases occur simultaneously, and in view of the possibility, outlined above, that the last-named mechanisms can produce a natural or inverted sequence if acting by itself, it follows that reinforcement, cancellation, or over-compensation can be the final result, if both effects are operative. Within the independent electron model outlined, the two effects will yield the MOs, ψ_+, ψ_-

$$\psi_+ = c_+ \chi_+ + \sum_i c_{i+} \chi_i; \quad c_+^2 \gg c_{i+}^2$$
$$\psi_- = c_- \chi_- + \sum_i c_{i-} \chi_i; \quad c_-^2 \gg c_{i-}^2 \tag{58}$$

when the "through-space" contribution is supplemented with some admixture of "through-bond" interactions via σ, or σ^* relay orbitals (χ_i). The resulting orbital energies, $\varepsilon(\psi_+)$ and $\varepsilon(\psi_-)$, differ from those of $\varepsilon(\chi_+)$ and $\varepsilon(\chi_-)$ by τ_+ and τ_-:

$$\varepsilon(\psi_+) = \varepsilon(\chi_+) + \tau_+$$
$$\varepsilon(\psi_-) = \varepsilon(\chi_-) + \tau_-$$

(59)

An instance where this sort of approach has shown to be enlightening is that of the [2,2]paracyclophanes **75**, **76** mentioned above. In contrast to the interaction of the benzene $e_{1g}(\pi)_A$ orbitals resulting in $b_{2u}(\pi)$ and $b_{3g}(\pi)$ orbitals (*vide supra*), the "through-bond" interaction of the benzene $e_{1g}(\pi)_S$ orbitals via the CC-σ-orbitals is important. The consequence of this is that the $b_{3u}(\pi)$ orbital is shifted to higher energy and the corresponding band in the photoelectron spectrum coalesces with two other bands due to electron ejection from $b_{2g}(\pi)$ and $b_{3g}(\pi)$ orbitals, lying in this region. However, fluorination of the methylene groups considerably lowers the basis energy of the CC-σ-orbitals. The resultant effect is a diminution of τ_+ in **76**. In the correlation diagram of Fig. 18 it can be seen that now the $b_{3u}(\pi)$ band lies where expected, for small "through-bond" interaction.

We now examine the situation prevailing in the bicyclic dienes **79**(n), i.e. Dewar-benzene **79**(0)≡**2**,[8] norbornadiene **79**(1),[100] bicyclo[2.2.2]octadiene **79**(2) **33**,[20] bicyclo[3.2.2]nona-6,8-diene **79**(3)[101] and bicyclo[4.2.2]-deca-7,9-diene **79**(4)[96] to probe the interactions of the two bis-homoconjugated two-centre π-orbitals $\chi_a \equiv \pi_a$ and $\chi_b \equiv \pi_b$.

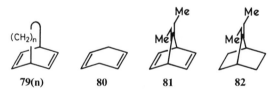

$(CH_2)_n$ 79(n)   80   81   82

In Table 6 are given the π-ionization energies $I_{v,1}$ and $I_{v,2}$ of the bicyclic dienes **79**(n) with $n = 0$–4 and the values for the hypothetical hydrocarbon **79**(∞) which have been extrapolated from those of cyclohexa-1,4-diene **80**[102] by assuming a correction of ~ -0.2 eV for the effect of the infinitely long methylene chain.

With reference to the dependence of the overlap integral $S_{ab} = \langle \pi_a | \pi_b \rangle$ on the dihedral angle ω, shown in Fig. 16, the split ΔI_v would be expected to decrease monotonically, at least from **79**(1) to **79**(∞); the case of Dewar-benzene **79**(0) being a rather special one, due to the presence of two four-membered rings with high-lying Walsh-orbitals and of an extremely long CC single bond possessing a correspondingly high-lying CC-σ-orbital (c.f. ref. 8). However, the ΔI_v's go through a minimum for **79**(3) and then increases again to ~ 1 eV for **79**(∞). These data suggest a compensating interplay of "through-bond" and "through-space" interactions. The primary step must therefore be

TABLE VI

π-Ionization Energies (in eV) of the Bicyclic Dienes 79(n) with n = 0, 1, 2, 3, 4 and of cyclohexa-1,4-diene 80

	79(0)	79(1)	79(2)	79(3)	79(4) ... 75(∞)[a]	80
PE Band 1: $I_{v,1}$	$9 \cdot 4_0$	$8 \cdot 7_0$	$8 \cdot 8_5$	$9 \cdot 0_0$	$8 \cdot 9_5$ ... $8 \cdot 6_0$	$8 \cdot 8_0$
PE Band 2: $I_{v,2}$	$9 \cdot 7_0$	$9 \cdot 5_5$	$9 \cdot 4_5$	$9 \cdot 2_0$	$9 \cdot 3_0$ ... $9 \cdot 6_0$	$9 \cdot 8_0$
$\bar{I}_v{}^c$	$9 \cdot 5_5$	$9 \cdot 1_0$	$9 \cdot 1_5$	$9 \cdot 1_0$	$9 \cdot 1_0$ ... $9 \cdot 1_0$	$9 \cdot 3_0$
$\Delta I_v{}^d$	$0 \cdot 3_0$	$0 \cdot 8_5$	$0 \cdot 6_5$	$0 \cdot 2_0$	$0 \cdot 3_5$ ... $1 \cdot 0_0$	$1 \cdot 0_0$
Ref.	8	100	20	101	102	103

[a] These values have been extrapolated from those of 80 by correcting for the inductive influence of an infinitely long polymethylene chain.

[c] $\bar{I}_v = (I_{v,1} + I_{v,2})/2$.

[d] $\Delta I_v = (I_{v,2} - I_{v,1})/2$.

to establish a definite experimental assessment of the symmetries of the two lowest states of the radical cations at both ends of the series or, in other words, the assignment of the two highest occupied orbitals of the hydro-carbons to definite irreducible representations of the symmetry group C_{2v}.

As a preliminary illustration of the type of problem often encountered, one may cite the case diazabicyclo[2.2.2]octane (36) where the separation of the first two bands associated with the nitrogen lone-pair combinations is $\sim 2 \cdot 5$ eV. In view of the large spatial separation of the two localized lone-pair orbitals, and thus of their small overlap, the magnitude of the separation of the bands leads one necessarily to postulate that the interactions involved are "through-bond" dominated. Nevertheless, from the photoelectron-spectro-scopic data alone one can not further conclude whether a small "through-space" interaction is reinforced by the dominant "through-bond" interaction to yield the natural order ($^2A_1'$ ground state of M$^+$), or whether the latter is so strong as to yield the inverted sequence ($^2A_2''$ ground state of M$^+$). The calculations based on the Extended Hückel method,[103] as well as many-electron treatments[104] predict the inverted sequence and the analysis of the vibrational fine structure discernible on the first two bands of concern[21] is also in harmony with this assignment. On the other hand, the presence of a threefold axis in barrelene (32) and 1,4,7-cyclononatriene (1) results in π-orbitals of e', a_2' and e, a_1 symmetry respectively. The band in these photoelectron spectra, corresponding to the electron ejection from the degenerate orbitals, can be assigned from a Jahn–Teller distortion of the Franck–Condon profile, and to some extent from the integrated intensity of the bands[105] (c.f. Fig. 1). In these cases the order of the states of M$^+$ is ascertained and the natural sequence (i.e. ground states of the cations are ^2E and $^2A_2'$ for 1 and 32 respectively) is established in both instances.

Returning to the problem of the hydrocarbons listed in Table 6, the large split $\Delta I_v = 1 \cdot 0$ eV observed in the spectrum of cyclohexa-1,4-diene **80** has been ascribed to a strong hyperconjugative "through-bond" interaction via the two methylene groups in positions 3 and 4. For an assumed D_{2h} symmetry[106] only the b_{1u} combination of the semi-localized π-orbitals, i.e. $\chi_+ = (\pi_a + \pi_b)/\sqrt{2}$ is of correct symmetry to interact with the σ and/or σ^* methylene orbitals as illustrated in Fig. 19. This leads one to presume that an

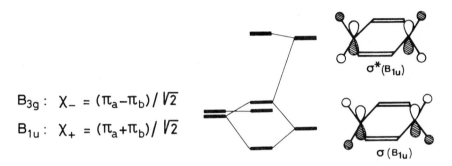

$$B_{3g}: \quad \chi_- = (\pi_a - \pi_b)/\sqrt{2}$$

$$B_{1u}: \quad \chi_+ = (\pi_a + \pi_b)/\sqrt{2}$$

Fig. 19. Interaction of the semi-localized π-orbitals with the σ and σ^* methylene orbitals. The irreducible representations refer to an assumed planar structure of D_{2h} symmetry.

inverted sequence is attained ($^2B_{1u}$ ground state of M$^+$). At the other end of the series, the sequence has been experimentally established for **79**(1) and **79**(2). In the latter species the natural sequence has been deduced by comparison with the data on barrelene (**32**) where the Jahn–Teller contour was strongly suggestive of the natural sequence. The assignment in norbornadiene **79**(1) has been confirmed by the introduction of an additional semi-localized π-orbital π_c, of known symmetry behaviour, in position 7, e.g. in 7-isopropylidene-norbornadiene **81** with the explicit assumption that to a first-order approximation, the overall features of the interactions between the two π-orbitals π_a and π_b are not altered. The comparison of the ionization energy data of **81** and 7-isopropylidene-norbornane **82** with **79**(1) allows an unambiguous assignment to be derived. The situation is illustrated by the left side (a) of Fig. 20, which refers to the set of molecules norbornene (**83**), **79**(1), **81**, and **82**. In view of the fact that the exocyclic π-orbital π_c is necessarily antisymmetric with respect to the plane containing the CC-bonds 1,7 and 4,7 and can thus interact only with the linear combination of π_a and π_b having the same symmetry, the result is self-explanatory. It shows that the sequence of states of the radical cation of **79**(1), or in terms of Koopmans' approximation of the orbitals, is the natural one, i.e. a $^2B_2$-radical cation ground state. This result receives strong support from the data observed for

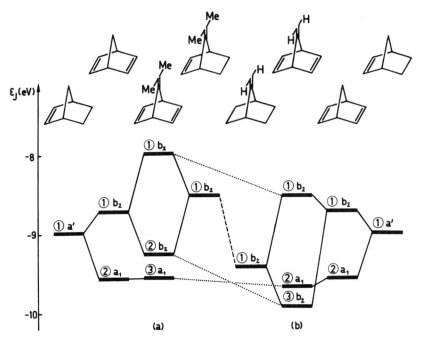

Fig. 20. Comparison of the assignment of the photoelectron spectra of the shown hydrocarbons, which allows an unambiguous assignment for norbornadiene **79(1)** to be derived.

the corresponding non-methylated hydrocarbons 7-methylene-norbornadiene (**84**) and 7-methylene-norbornane (**85**)[107] included on the right (b) of Fig. 20.

Similar arguments can be advanced for the assessment of the orbital sequences in **79(3)**[101] and **79(4)**.[102] In both cases an inverted sequence of the two highest occupied, π-dominated orbitals is deduced.

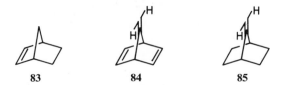

As a result, the data for **79(1)** and **(79(2)** are consistent with a "through-space" dominated interaction between the π-orbitals π_a and π_b, whereas in the systems **79(3)**, **79(4)**, and **79(∞)** "through-bond" interaction overcompensates the "through-space effect. This is shown in the diagram of Fig. 21 which, incidentally, is based entirely on experimental data. Note that the cross-over of the orbitals occurs between **79(2)** and **79(3)**.

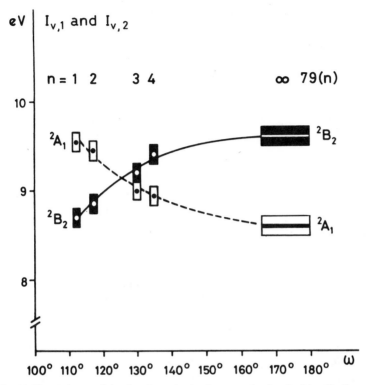

Fig. 21. Dependence of the $I_{v,1}$, $I_{v,2}$ π-ionization energies for the bicyclic dienes **79(n)** on the dihedral angle ω.

The preceding examples exemplify some of the ploys employed in order to determine whether the natural or inverted order of states (or orbitals) is attained. However, further answers are often sought to the following questions:

(a) how do "through-space" and "through-bond" interactions between the orbitals π_a and π_b compare in their relative importance in determining the orbital energies of the CMOs $a_1(\pi)$ and $b_2(\pi)$;

(b) which of the σ-orbital(s) is (are) the important relay-orbital(s) for "through-bond" interaction between π_a and π_b;

(c) how does the choice of a particular semi-empirical SCF model (SPINDO,[36] MINDO/2,[34] CNDO/2[33]) influence the conclusions drawn concerning the question (a) and (b).

In order to achieve these objectives, a more quantitative assessment in the framework of many-electron is needed. One such approach has been described in detail in ref. 108 and is only outlined here to enable the reader to follow the subsequent arguments. In this connection it should again be remembered that the separation of the observed effects into "through-space" and "through-bond" interactions is artificial and closely tied to the model used.

The canonical molecular orbitals (CMOs), φ_j, which are obtained by solving the Hartree–Fock Eq. (3) can be transformed by a unitary transformation (5) into localized molecular orbitals (LMOs), λ_j (12), e.g. according to the criterion given in (13).[38] The off-diagonal elements $F_{\lambda,ab}$ (15) in the corresponding Hartree–Fock matrix $\mathbf{F}_\lambda$, linking two LMOs' λ_a and λ_b, are taken to represent the "through-space" interactions, whereas the diagonal elements $F_{\lambda,aa} \equiv A_a$, $F_{\lambda,bb} \equiv A_b$ (see (15)) are the self-energies of λ_a, λ_b. If the molecules possesses symmetry, one can form symmetry-adapted semi-localized molecular orbitals (SLMOs) ρ_j, e.g.

$$\rho_+ = (\lambda_a + \lambda_b)/\sqrt{2},$$
$$\rho_- = (\lambda_a - \lambda_b)/\sqrt{2} \tag{60}$$

if λ_a and λ_b are related by a symmetry operation of order two. The corresponding Hartree–Fock matrix $\mathbf{F}_\rho$ is blocked out in submatrices $\mathbf{F}_\rho^{(r)}$ belonging to the different irreducible representations $\Gamma^{(r)}$ of the symmetry group. In order to obtain the "through-bond" interaction terms within this framework, from the SLMO matrix $\mathbf{F}_\rho$ (or more precisely from the submatrices $\mathbf{F}_\rho^{(r)}$), all off-diagonal elements linking the SLMOs based on λ_a and λ_b (e.g. ρ_+ or ρ_- of (60)) to other SLMOs ρ_j are removed and the remaining matrix then diagonalized. This yields the so-called precanonical orbitals (PCMO) ψ_j, which do not contain any contribution from λ_a or λ_b. We are now in a position to calculate the cross-terms between the SLMOs given in (60) and the PCMOs ψ_j of appropriate symmetry. These off-diagonal elements (e.g. $F_{\psi,j+}$ or $F_{\psi,j-}$ with reference to (60)) then define the "through-bond" interactions. From them we can assess (a) which of the PCMOs ψ_j are the dominant relay orbitals and (b) the sign and the size of the "through-bond" induced orbital energy shifts (τ_+ and τ_-).

For norbornadiene (**79**(1)) such an approach yields the correlation diagram shown in Fig. 22, starting with the CMOs φ_j obtained by the SPINDO, MINDO/2, and CNDO/2 procedures. The "through-space" contributions are responsible for the separation of the self-energies of the SLMOs ρ_+ and ρ_-, and the total "through-bond" interactions, i.e. the sum of the individual interactions of ρ_+ and ρ_- with the PCMOs ψ_j, for the step from the SLMOs ρ_+ and ρ_- to the CMOs $a_1(\pi)$ and $b_2(\pi)$ respectively. It should be noted that the three semiempirical SCF treatments differ considerably with regard to the size of the individual contributions. In particular the CNDO/2 model predicts a final order in contradiction with the photoelectron spectroscopic results, and with MINDO/2 also, the π-orbital sequence is critically sensitive to small changes in geometry. Table 7 summarizes the results obtained.

TABLE VII

Numerical Results for the "Through-space" and "Through-bond" Interactions between the LMOs $\pi_a \equiv \lambda_a$ and $\pi_b \equiv \lambda_b$ of Norbornadiene 79(1), according to the SPINDO,[36] MINDO/2,[34] and CNDO/2 [33] Treatments.

a	SPINDO	MINDO/2	CNDO/2
$F_{\lambda,aa} = F_{\lambda,bb}$	− 10·44	10·70	− 16·41
$F_{\lambda,ab}$	− 0·54	− 0·78	− 2·09
$F_{\rho+}$	− 10·98	− 11·47	− 18·50
$F_{\rho-}$	− 9·90	− 9·92	− 14·32
$A_1 : \tau_+$	0·85	2·07	6·46
$B_2 : \tau_-$	0·34	0·66	1·99
$\tau_+ - \tau_-$	0·51	1·41	4·50

All values in eV.

a Subscripts a and b refer to the LMOs $\lambda_a \equiv \pi_a$ and $\lambda_b \equiv \pi_b$; Subscripts + and − refer to the SLMOs given in (60).

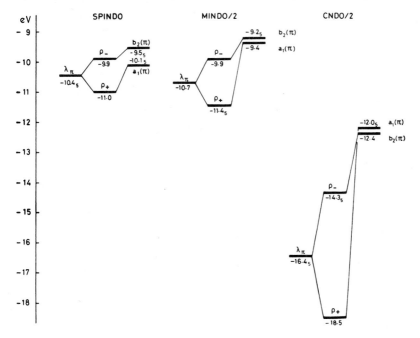

Fig. 22. Correlation diagram for "through-space" and "through-bond" interaction of π_a- and π_b-orbitals in norbornadiene 79(1)[108]

A significant point of interest arises from the data of Table VII in that the "through-bond" contribution τ_- to the b_2 combination ρ_- of the π-orbitals is not negligible. A similar result is also found by such an analysis for diazabicyclo[2.2.2]octane (36),[108] where the MINDO/2 procedure yields $\tau_+ = 5.52$ eV and $\tau_- = 4.0$ eV and negligible "through-space" contribution of 0.14 eV. The σ-orbitals which are found to be important for the "through-bond" interactions in norbornadiene are the CH-σ-orbitals of the methylene group, and the sp^3–sp^2 CC-σ-orbitals of the six-membered ring, for both the b_2 and a_1 combinations of the π-orbitals.

The dependence of the "through-space" and "through-bond" interactions on the dihedral angle ω of a cyclohexa-1,4-diene-system (80) bent along an axis passing through the centres 3 and 6 can also be used to account for the trend in the ionization energies of the bridged hydrocarbons 79(1)–79(∞) shown in Fig. 21. The results of such a treatment are reproduced in Fig. 23 and in Table VIII.

TABLE VIII

Dependence of the "Through-space" and "Through-bond" Contributions to the Splitting of the π-Orbitals of a Hypothetical Cyclohexa-1,4-diene (80) with Variable Dihedral Angle ω

		180°	150°	120°
SPINDO[36]	Through-space-split	0.33	0.50	1.33
	τ_+	1.04	0.97	0.91
	τ_-	0.00	0.07	0.27
MINDO/2[34]	Through-space-split	0.66	0.94	2.01
	τ_+	1.86	1.81	1.75
	τ_-	0.00	0.21	0.73
CNDO/2[33]	Through-space-split	2.50	2.99	4.04
	τ_+	5.80	5.72	6.09
	τ_-	0.00	0.87	4.50

All values in eV.

On examination of the data in Table VIII and Fig. 23, it is seen that the "through-space" interaction between the LMOs $\lambda_a \equiv \pi_a$ and $\lambda_b \equiv \pi_b$, decreases with increasing dihedral angle ω, proportional to the overlap integral $S_{ab} = \langle \pi_a | \pi_b \rangle$, whereas the "through-bond" shift τ_+ is almost independent of ω. These calculations suggest that the major reason for the experimentally observed cross-over near $\omega \approx 120°$–130° (see Fig. 21) is the large increase of τ_- as ω decreases. The τ_- growth results from the increasing interaction of the LCMOs $\lambda_a \equiv \pi_a$ and $\lambda_b \equiv \pi_b$, with the σ-LMO of the two

sp^2-sp^2 CC single bonds on the other side the six-membered ring of the molecules **79**(n) and, in the framework of these models, it is this contribution which causes the crossing.

The approach described above allows further insight to the questions (a)–(c) asked above and also underlines the care that has to be exercised in relying exclusively on semi-empirical SCF treatments for the interpretation of photoelectron spectra.

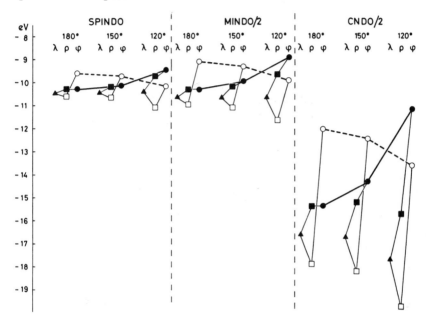

Fig. 23. Dependence of "through-space" and "through-bond" interactions on the dihedral angle ω in a cyclohexa-1,4-diene system (**80**) and on the semiempirical SCF procedure.[108] The open circles and squares refer to orbitals of A_1 symmetry, whereas the full circles and squares to orbitals of B_2 symmetry. The triangles represent the LMO energies, the squares the SLMO energies, and the circles the CMO energies.

VI. TWISTED AND BENT π-SYSTEMS

Applying the localization procedure (12)–(14) to semi-empirical many-electron treatments yields the result[109] that the cross-term $F_{\lambda,ij} = B_{ij}$ between two σ-linked localized orbitals λ_i, λ_j, e.g.

is a function of the twist angle θ

$$B_{ij}(\theta) = B_{ij}(0)\cos(\theta) + C \qquad (61)$$

$B_{ij}(0)$ and C depending on the nature of λ_i and λ_j. Apart from the constant C, this is exactly what one assumes in naive independent electron treatments, e.g. in the Hückel model, where $\beta(\theta) = \beta \cos \theta$ is used for the resonance integral between two neighbouring 2p-AOs, twisted by an angle θ with respect to each other. As shown in Fig. 24, the linear combination of the

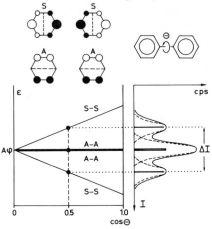

Fig. 24. The combination of the symmetric (S) and antisymmetric (A) molecular orbitals of the two phenyl moieties in diphenyl (**48**) and the dependence of the orbital energies on the twist angle θ according to formula (63). The expected band pattern in the photoelectron spectrum of diphenyl is shown alongside.

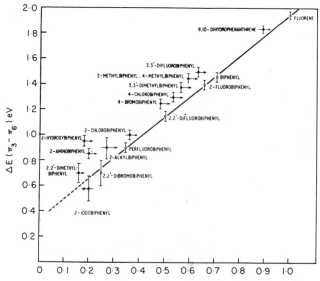

Fig. 25. Plot of $I_{v,4} - I_{v,1} = \Delta E$ values vs the known twist angle θ (from electron diffraction data) for a series of diphenyls (open circles).[110] The full circles represent the experimental ΔE values of the indicated diphenyls.

symmetric (S) and antisymmetric (A) molecular orbitals of the two phenyl moieties in diphenyl (**48**) leads to the symmetry correct orbitals

$$\psi_{A\pm} = (\phi_{A1} \pm \phi_{A2})/\sqrt{2}; \quad \psi_{S\pm} = (\phi_{S1} + \phi_{S2})/\sqrt{2} \tag{62}$$

abbreviated as $A \pm A$ and $S \pm S$. Whereas the orbital energy of $\psi_{A\pm}$ remains unaffected by a twist θ, the energies of ψ_{S+} and ψ_{S-} are

$$\varepsilon_{S+} = A_\phi + (\beta/3)\cos\theta; \quad \varepsilon_{S-} = A_\phi - (\beta/3)\cos\theta \tag{63}$$

respectively, with $A_\phi = \alpha + \beta$. The expected band shape is shown in Fig. 24. The comparison of the values of the ionization energy difference ΔI measured for a series of diphenyls[110] with known twist angle θ results in the expected linear dependence on $\cos\theta$ suggested by

$$\Delta I = -(\varepsilon_{S-} - \varepsilon_{S+}) = -\tfrac{2}{3}\beta\cos\theta \tag{64}$$

as can be seen from Fig. 25. Another rather striking example has been provided by Brundle and Robin[111] who analyzed the photoelectron spectra of butadiene (**43**), 1,1,4,4-tetrafluorobutadiene (**86**), and hexafluorobutadiene (**87**) reproduced in Fig. 26. The conclusion derived from the reduced ΔI value

86   **87**

observed in the latter compound **87** was that this molecule is not planar but twisted by $\theta \approx \pi/4$, a conclusion that has been verified by electron diffraction.[112]

Such results seem to suggest that photoelectron spectroscopy is the method of choice for the investigation of twisted and bent π-bonds and π-systems. Thus naive Hückel theory predicts that twisting the two $2p$-atomic orbitals of a two-centre orbital $\pi = (\phi_a + \phi_b)/\sqrt{2}$ by an angle θ, should yield a shift of $\Delta\varepsilon = -\beta(1 - \cos\theta)$, i.e. with $\beta \approx -3\cdot5$ eV, $\Delta\varepsilon \approx 0\cdot5$ if $\theta = 30°$. However, this is not really the case. Whereas cis- (**88**) and trans-cyclooctene (**89**) *do* differ by $\Delta I = 0\cdot3$ eV in their π-ionization energies ($I_v(\pi) = 8\cdot98$ eV in (**88**), $8\cdot69$ eV in (**89**))

88   **89**

no noticeable shift of the ionization energies $I_v(\pi)$ can be observed in bicyclic hydrocarbons violating Bredt's rule, relative to the expectation value derived for a similarly alkyl-substituted flat double bond. An example is provided by the pair of isomeric "anti-Bredt" hydrocarbons bicyclo[4.2.1]-non-1(2)ene (**90**) and bicyclo[4.2.1]non-1(8)ene (**91**). Energy calculations with an optimized

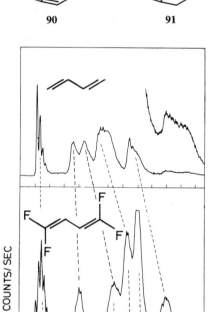

Fig. 26. HeI photoelectron spectra of butadiene (**43**), 1,1,4,4-tetrafluorobutadiene (**86**), and hexafluorobutadiene (**87**)[111]. The decrease in the separation of the first two bands (π) in the spectrum of **87** reflects the non-planarity, $\theta \approx \pi/4$.

valence force field yield twist angles for the double bonds of 19·4° (90) and 16·6° (91), if the hydrocarbons assume conformations of lowest energy. However, the measured ionization energies $I_v(\pi) = 8.45$ eV and $I_v(\pi) = 8.37$ eV[113] for 90 and 91 respectively, coincide within experimental error with the values that would be expected, on the basis of known data for unsaturated hydrocarbons, for a corresponding substituted and fully planar π-bond.

The cycloalkynes are the ideal molecules for the investigation of the influence of bending on π-ionization energies. Figure 27 shows the photoelectron spectrum of cyclooctyne (92)[113] and of 3,3,7,7-tetramethylcycloheptyne (93),[114] both of which are considerably strained hydrocarbons. One

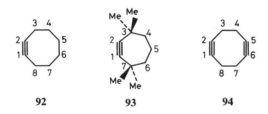

might have expected that in 92 the bending of the CC-single bonds 2, 3 and 8, 1 away from the triple bond axis 1, 2 by 21·5°[115] would lift the degeneracy of the in-plane and out-of-plane π-orbitals. However, as can be seen from Fig. 27, this is not the case.[113] Only the much larger bending angle in the tetramethylderivative (93) of cycloheptyne leads to a clear split between the two π-bands.[114]

A rather more complicated situation prevails in the case of 1,5-cyclooctadiyne (94) which has D_{2h} symmetry.[116] Here the self-energies of the "in-plane" and "out-of plane" two centre π-orbitals of the two triple bonds are presumed to be the same, on the basis of the above information. However, in view of the close proximity of the two triple bonds, which are separated by only 2·57 Å, there are strong "through-space" interactions between the "in-plane" and the "out-of-plane" π-orbitals respectively. This is shown qualitatively in the orbital diagram of Fig. 28. In addition the symmetric "in-plane" combination $\pi_{z,-}$, i.e. the one in which the phases of the two π_z orbitals are opposed to each other, exhibits a dominating CC-σ-orbitals in positions 3, 4 and 7, 8. As a consequence the $7a_g$-orbital has an energy similar to the two anti-symmetric orbitals $2b_{2g}$ and $6b_{3u}$ which are only affected to a small degree, or not at all, by hyperconjugative "through-bond" destabilizations, as indicated in Fig. 28.[117] Over-all the situation is rather similar to the one encountered in the para-cyclophane series discussed above.

In this connection it should be mentioned that the bending of the benzene moieties in para-cyclophane (75)[118] does not reflect in the photoelectron

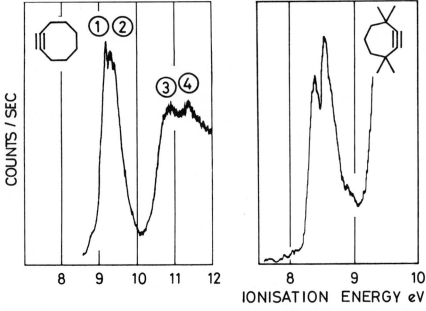

Fig. 27. HeI photoelectron spectra of cyclooctyne (**92**)[113] and 3,3,7,7-tetramethyl-cycloheptyne (**93**)[114] in the π-ionization energy region.

spectrum, the structure of which is completely dominated by the strong "through-space" and "through-bond" interactions of the two phenylene π-systems (cf. Fig. 18). Again, bending does not seem to influence the basis orbital energies by more than 0·2 eV, in close analogy to the situation prevailing in the hydrocarbons **92**, **93**, and **94**.

This perhaps surprising result can be rationalized in terms of an electronic mechanism originally proposed by Hoffmann *et al.*[119] and more recently complemented by Haselbach.[120] Their arguments can be summarized qualitatively as shown in the following diagram which refers to the effect of bending on the π-orbitals of acetylene:

$$
\begin{array}{ccc}
\overset{\displaystyle H}{\underset{\displaystyle H}{C\equiv C}} & \xleftarrow{C_{2h}} \quad H{-}C\equiv C{-}H \quad \xrightarrow{C_{2v}} & \underset{H \quad H}{C\equiv C} \\
 & D_{\infty h} &
\end{array}
\tag{65}
$$

$$
\left.\begin{array}{l} 1a_u(\pi) \\ 3a_g(CC\text{-}\sigma) \end{array}\right\} \qquad 1e_{1u}(\pi) \qquad \left\{\begin{array}{l} 1b_1(\pi) \\ 4a_1(CC\text{-}\sigma) \end{array}\right.
$$

$$
3b_u(CH\text{-}\sigma) \qquad\qquad 3a_{1g}(\sigma) \qquad\qquad 3a_1(CH\text{-}\sigma)
$$

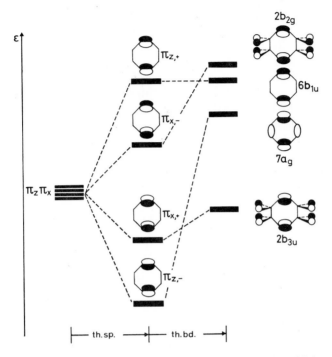

Fig. 28. Qualitative diagram for the interplay of "through-space" and "through-bond" interactions of the basis π-orbitals of 1,5-cyclooctadiyne (**94**).[116] The symmetry designations refer to D_{2h}.

In (65) we are using e_{1u} and a_{1g} instead of π_u and σ_g^+ for the orbitals of acetylene ($D_{\infty h}$) to avoid confusion with the usual chemical σ,π-terminology. The rationalization suggested is that a trans-bend ($D_{\infty h} \rightarrow C_{2h}$) will lead to a lowering of the in-plane component of $1e_u(\pi)$, which for an angle $\theta = 90°$ will end up as the out-of-phase linear combination $3b_u(CH\text{-}\sigma)$ of the $CH\text{-}\sigma$-orbitals, whereas the $3a_g(\sigma)$ orbital of acetylene becomes the $3a_g(CC\text{-}\sigma)$ of the bent model system. The two orbitals $3b_u(CH\text{-}\sigma)$ and $3a_g(CC\text{-}\sigma)$, being respectively u and g relative to inversion, are allowed to cross for an intermediate angle θ as indicated in (65). On the other hand, under a cis-type deformation ($D_{\infty h} \rightarrow C_{2v}$) the corresponding orbitals $4a_1(CC\text{-}\sigma)$ and $3a_1(CC\text{-}\sigma)$ emanating from $1e_u(\pi)$ and $3a_g(\sigma)$ respectively, belong to the same irreducible representation A_1 of C_{2v}. Consequently, the non-crossing rule applies and the corresponding orbital energies change only little in function of θ, as indicated on the right-hand side of (65). In particular for small values of θ the two orbitals dominantly π in character remain close in energy, which suggests that we should expect only a small split between the two photoelectron bands due to electron-ejection from these two highest occupied orbitals. An important assumption underlying the above argument is that the orbitals $1a_u(\pi)$, $3a_g(CC\text{-}\sigma)$ for the C_{2h} system and $1b_1(\pi)$, $4a_1(CC\text{-}\sigma)$ for the C_{2v} system are almost accidentally degenerate for $\theta = 90°$, as suggested by the theoretical results obtained by Haselbach.[120] It should be mentioned that the argument embodied in (65) is similar to the one given by Gimarc[121] for the dependence on of the orbital energies of diimide. It is also in agreement with the *ab initio* calculations carried out by Millie *et al.*[122] Concerning the effect of twisting a double-bond, the reader is referred to the work of Mock *et al.*[123]

A rather special case of distorted π-systems is that of the bridged annulenes, to be discussed briefly in the next section. Of these non-benzenoid "aromatic" systems we mention at this point the hydrocarbon 1,6-methano-[10]annulene

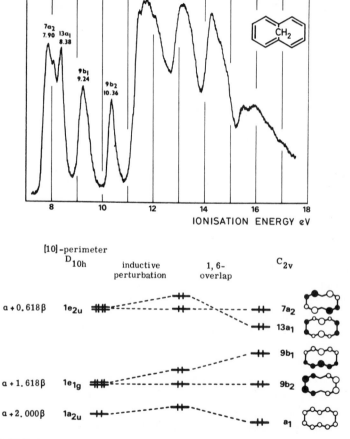

Fig. 29. HeI photoelectron spectrum of 1,6-methano-[10]annulene (95) and the model used to account for the π-band positions.[124]

(95), which has been investigated by Schmidt *et al.*[124] Their results are summarized in Fig. 29. As can be seen the observed π-band positions can be faithfully accounted for by assuming an inductive destabilization, due to the

95

bridging methylene group, of those orbitals having finite orbital coefficients in positions 1 and 6 ($\delta\alpha(-CH_2-) = 0\cdot8$ eV) and a homoconjugative inter-action between the $2p$-atomic orbitals ϕ_1 and ϕ_6, characterized by $\beta_{16} = -2\cdot0$ eV. The success of this simple parametrization is rather surprising, because it involves the complete neglect of effects due to the non-planarity of the peripheral π-system ($\theta_{1,2} = 34\cdot0°$; $\theta_{2,3} = 19\cdot7°$; $\theta_{3,4} = 0°$ [125]). If it is assumed that the individual resonance integrals $\beta_{\mu\nu}$ between bonded centres μ, ν depend on the twist angles $\theta_{\mu\nu}$ according to $\beta_{\mu\nu} = \beta\cos\theta_{\mu\nu}$, where β is the standard resonance integral for a planar π-bond, then the resulting pertur-bation should be

$$\delta\varepsilon_j{}^{\text{twist}} = \left(2\sum_{\mu\nu} c_{j\mu}c_{j\nu}(\cos\theta_{\mu\nu}-1)\right)\beta \qquad (66)$$

where summation extends over all bonds. However, it has been found that the corrections $\delta\varepsilon_j{}^{\text{twist}}$ calculated according to (66) are much too large and not in agreement with experimentally observed effects.[126] Thus, improving the model by including the expected effects of non-planarity which must be present in view of previous experience with other unsaturated systems, carries the model beyond the "Pauling point", i.e. the point at which increased sophistica-tion leads to a worse agreement with experiment.

However, if the model embodied in formula (66) is taken at face value, there is a perfectly good reason for this observation. Assuming that the observed twist angles $\theta_{\mu\nu}$[125] are those to be used, formula (66) is probably wrong. It implies that the $2p$-AOs at centres μ, ν are strictly perpendicular to the local σ-plane, a hypothesis which seems to be supported by the photoelectron spectroscopic results of loosely coupled π-systems, such as those reported in Figs. 25 and 26. In all these cases the change in π-orbital interaction "measured" by photoelectron spectroscopy is well represented by $\beta' = \beta\cos\theta$, where θ is the twist angle between the planes of the two partial systems connected by a sp^2–sp^2 single bond. However, in these molecules the $2p$-AOs at the linked centres are strongly coupled to other $2p$-AOs within each partial system and are thus locked in orientations perpendicular to the planes of these systems. On the other hand, this is no longer true in molecules such as **95** and its higher analogues where the neighbouring $2p$-AOs of the twisted bond are strongly coupled. They will polarize in such a way as to yield optimum overlap within the constraints imposed by the σ-frame. The resulting π-ribbon stretches "elastically" around the periphery and will not necessarily be locally perpendicular to the σ-bonds, as long as such deviations optimize the total energy of the system. As mentioned before, a similar type of adjustment has been predicted by Mock et al.[123] for the distorted π-orbital of a deformed ethylene. Therefore it is reasonable to assume that the observed twist angles $\theta_{\mu\nu}$ exaggerate the true angles between linked atomic orbitals $2p_\mu, 2p_\nu$ participating in the perimeter of **95** or other bridged [n]annulenes, which are thus smaller than the twist angles $\theta_{\mu\nu}$ determined by X-ray analysis. In other words, the π-ribbon adjusts elastically and is no longer locally orthogonal to the σ-frame.

VII. NON-BENZENOID "AROMATIC" HYDROCARBONS

Although in some instances, e.g. in the case of substituted cyclobutadienes (**96**), the photoelectron spectrum may yield information concerning the presence of a cyclically delocalized π-electron system **96(a)** as contrasted to a system consisting of two strongly localized double bonds **96(b)** [$R_1, R_2 = R_3$, $R_4 = -C(Me_2)CH_2SCH_2C(Me_2)-$;[127] $R_1 = R_2 = R_3 = CMe_3$, $R_4 = H$[128]] it

96(a)   96(b)

is fortunate that the photoelectron spectroscopic data known to date for benzenoid hydrocarbons, linear polyenes, azulenes, bridged annulenes, and the systems presented in this section, do not allow the formulation of yet another "aromaticity" criterion.

The type of difficulty involved can be demonstrated through the application of a naive LCBO model to "the" aromatic molecule benzene **12**. It is an experimental fact that the successive introduction of double bonds in a saturated hydrocarbon lowers the self-energies A_j of a two-centre π-orbital π_j by about -0.5 eV per double bond.[129] Consequently the self-energies $A_a = -9.0$ eV of cyclohexene (**97**), $A'_a = A'_b = -9.5$ eV of cyclohexa-1,3-diene (**98**) extrapolate to $A''_a = A''_b = A''_c = -10.0$ eV for a hypothetical cyclohexa-1,3,5-triene (**99**) as shown at the top of Fig. 30.

97   98   99

The resonance integral $B_{ab} = \langle \pi_a | \mathscr{H} | \pi_b \rangle = -1\cdot2$ eV for two conjugated π-orbitals derived from **98** and various other conjugated dienes should also be valid for the triene **99**. With these parameters the prediction of the orbital energies ε_j and thus of the ionization energies $I_{v,j}$ of **99** are those indicated in Fig. 30. As can be seen, those values are close to those observed for the real, delocalized benzene molecule. Obviously minor adjustments of the parameters A_j and B_{ab} would result in complete agreement between calculated and observed band positions. This means, that even from the observed ionization energies of benzene it is impossible to decide whether this prototype of an "aromatic" molecule exhibits strong first order bond localization, i.e. corresponds to the hypothetical structure **99**, or possesses a completely delocalized π-system with equilibrated bond lengths, as it obviously does.

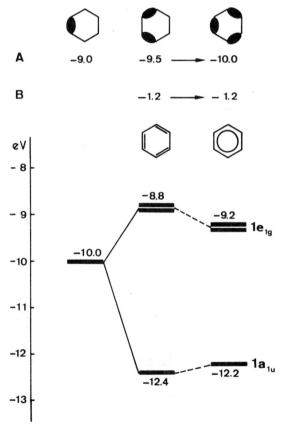

Fig. 30. Correlation diagram for the series cyclohexene (**97**), cyclohexa-1,3-diene (**98**), and cyclohexa-1,3,5-triene (**99**) based on a LCBO model with the self-energies (*A*), and the resonance integrals (*B*), of the π-orbitals as indicated.

Obviously the situation becomes even more ambiguous in most non-alternant π-systems, in which a strong double bond fixation is expected and/or observed.

Extreme examples for the latter type of molecules are the cross-conjugated, non-alternant π-systems such as **13**, **14**, and **15** mentioned in the introduction or 6-vinyl-fulvene (**67**), heptafulvene (**68**), and sesquifulvalene (**69**) (cf. Table II).[74, 130] The photoelectron spectra of the latter three molecules are shown in Fig. 31.

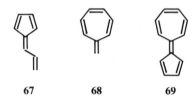

67   68   69

Whereas the simple Hückel treatment, i.e. the application of formula (16) with the parameters given in (19) yields rather respectable predictions for the ionization energies of "well-behaved" π-systems, as can be seen from the regression given on top of Fig. 9, molecules such as fulvene (**13**), 6-vinyl-fulvene (**67**), or sesquifulvalene (**69**) do not at all follow this pattern. This is evident from Fig. 32, in which the first ionization energy $I_{v,1}$ corresponding to the ejection of an electron from the highest occupied π-orbital ψ_{HOMO} has been plotted *vs* the orbital energy coefficient x_{HOMO} for a series of π-systems exhibiting strong bond localization. For topological reasons x_{HOMO} is the same for butadiene (**41**), (**13**), and (**67**) ($x_{HOMO} = 0.618$) and the same value is obtained for the second highest occupied π-orbital of **69**, for which $x_{HOMO} = 0.570$ which explains the occurrence of a first double band at 7.5 eV in the photoelectron spectrum of **69** shown in Fig. 31. Thus we would expect that the positions of the first bands in the photoelectron spectra of **41**, **13**, **67**, and **69** are similar. As can be seen from Fig. 32 this is not at all the case. The reason for the failure of a simple molecular orbital calculation, relying on the validity of Koopmans' approximation, is that in the molecules **13**, **67**, and **69**, the vacated orbital is localized completely on the five-membered ring, more specifically on the four centres not carrying a side-chain. Consequently the positive charge of the resulting radical cation is also strictly localized in this part of the molecule after the "Koopmans' ejection" and, according to Pauling's electroneutrality rule, an influx of electrons into the five-membered ring is induced. This electron rearrangement, which stabilizes the radical cation relative to the corresponding "Koopmans' state", becomes more important the larger the side-chain attached to the five-membered ring. This rationalization explains nicely the trend observed in the data collected in Fig. 32.

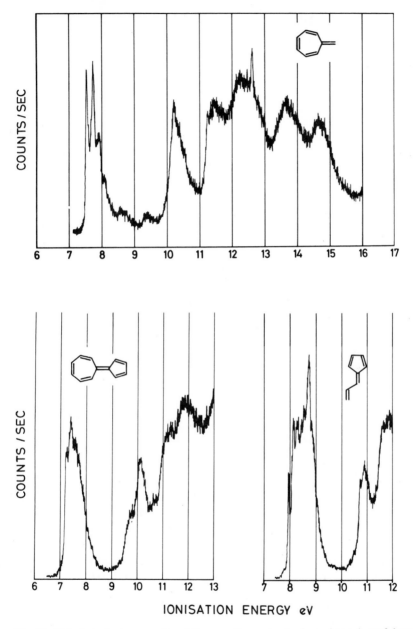

Fig. 31. HeI photoelectron spectra of the non-alternant π-hydrocarbons, heptafulvene (68), sesquifulvene (69), and 6-vinyl-fulvene (67).[145]

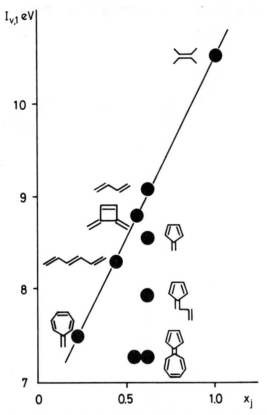

Fig. 32. Correlation of the first ionization energy $I_{v,1}(\pi)$ with the orbital energy co-efficient x_{HOMO}, for the series of π-systems shown.

As a final example of non-benzenoid "aromatic" compounds we mention the photoelectron spectroscopic data of the bridged [14]annulenes ($\mathbf{100}(n)$) and of dicyclohepta[cd. gh]pentalene ($\mathbf{101}$). The π-ionization energies of the 1,6;8,13-alkanediylidene-[14]annulene ($\mathbf{100}(n)$) with $n = 2$, 3, 4 show a dependence on the size n of the bridging groups which is the resultant of a complicated and not uniquely definable interplay of inductive, conjugative effects as indicated in the previous section for 1,6-methano[10]annulene ($\mathbf{95}$). A tentative assignment is shown in Fig. 33.[126] In contrast to the photoelectron

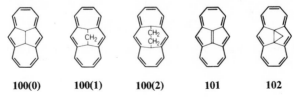

spectra of **100**(n) that of **101** can be rationalized by straightforward molecular orbital calculations. Finally, the molecule 1,6;8,13-cyclopropanediylidene[14]-annulene (**102**) occupies a situation intermediate between **100**(2) and **101** because of the presence of a three-membered ring, the Walsh orbitals of which play a similar role as the π-orbital of the central double bond in **101**.

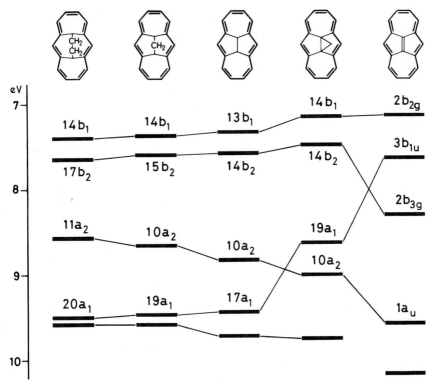

Fig. 33. Correlation diagram of the ionization energies of the bridged [14]annulenes (**100**(n)), n = 0, 1, 2, dicycloheptal [cd.gh]pentalene (**101**) ,and 1,6:8,13-cyclopropanedi-ylidene[14] annulene (**102**).[126]

VIII. THE C2s VALENCE BANDS OF HYDROCARBONS

In this section we consider the data available for an ionization process (1) in which the vacated canonical molecular orbital φ_j belongs to the $C2s$ inner valence shell of a hydrocarbon. For these compounds the corresponding bands in the photoelectron spectrum occupy the region of ionization energies extending typically from 15 eV to 26 eV.

In order to gain access to the whole energy region encompassing these processes, HeIIα ($^2P_{\frac{1}{2}}(2p) \rightarrow {}^2S_{\frac{1}{2}}(1s)$) (40·80 eV) photon radiation has proved to be the most convenient energy source. As most of the *C2s* bands occur in the region below ≈ 25 eV (with the exception of the deepest band), the accompanying HeIα (21·22 eV) photon radiation and higher lines of this series $^1P_1$ $(1s\ np) \rightarrow {}^1S_0$ $(1s^2)$ converging at the ionization energy of He at 24·58 eV, do not usually interfere, for hydrocarbons with first ionization energies $I_{v,1} > \approx 8$ eV.

The inner valence bands of simple saturated and of some unsaturated hydrocarbons have been studied by Price and his pupils[131–134] in the 15 eV to 26 eV region by the HeII radiation, which in some cases was filtered to remove the HeI lines. Some data are also available from monochromatized X-ray excited photoelectron spectra in this region.[135]

In Fig. 34 are reproduced the HeIIα photoelectron spectra of the series methane to neopentane, taken from the work of Potts *et al.*[132] Even from a superficial observation of the pattern of the spectra it is immediately apparent that there is a natural separation on the ionization energy scale of the two

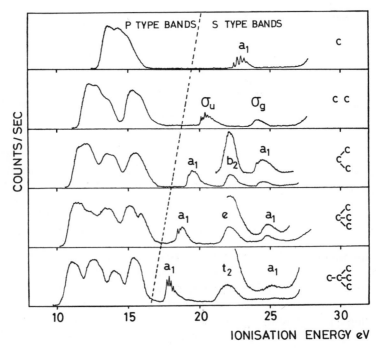

Fig. 34. HeII photoelectron spectra of the methyl methanes (103).[132]

groups of bands labelled as p-type and s-type. This is especially prominent among the smaller members of the homologous series. As the molecular size increases, the boundary moves to lower ionization energy and the common extremities of the two regions come closer together. Whereas there is profuse overlap of the bands in the p-type region, the s-type bands are well separated. Consequently, it is rather more difficult to estimate the $I_{m,j}$ values of all the bands in the p-region. In the s-region in case of a reasonable number of carbon atoms (up to 12 or so), the situation is more clear-cut. The correspondence of the number of bands (including degeneracies or overlap) to the number of carbon atoms is evident. Another feature is the lower intensity of the s-type bands compared to the p-type and also significant is the decrease in relative intensity of the s-bands as one proceeds to higher ionization energy (cf. Fig. 33). The wavelength of the photoelectron, for these low photoelectron kinetic energies, is comparable to molecular dimensions and as a result the photoionization transition moment is very much dependent on the orbital characteristics, i.e. size, symmetry, and nodal properties. Thus the bands are more intense as the number of nodal surfaces in the vacated molecular orbital increases, i.e. the lower the ionization energy. These relative band intensities and their dependence on the energy of the outgoing photoelectron (e.g. by comparing spectra recorded with HeI and HeII radiation) have been discussed qualitatively by Price et al.[131]

The low photoionization cross-section for the deepest $C2s$ bands (c.f. the band near 25 eV in the spectrum of neo-pentane shown in Fig. 33) can sometimes lead to difficulties in locating the band maximum and in the definitive attribution of the band to a s-type orbital. The latter obstacle may arise due to the presence of weak band(s) which represent processes involving simultaneous excitation of a valence electron concomitant to photoionization of another electron. These transitions represent formally dipole forbidden processes, but have, nevertheless, been detected as weak bands in the majority of diatomics and triatomics in the energy regions of the s-bands.[136] None the less, in the examples cited here these complications seem not to have arisen and even in the absence of data for the bands at highest ionization energies, the pattern and the band positions provide a sufficiently secure basis to allow a rationalization in terms of molecular orbital models, in particular of the Hückel model. Note that the set of data represent the antibonding as well as the bonding orbitals (relative to the $2s$ basis atomic orbitals) in contrast to the p-type region where only the filled (bonding) orbitals are amenable to the photoelectron spectroscopic measurements.

The ionization energies of the $2s$-type bands of open-chain n-alkanes (**103**), the cycloalkanes (**104**), cycloalkenes (**105**), and of some other unsaturated hydrocarbons are collected in Table IX. By and large, the data are taken from the work of Potts and Streets[133, 134] and supplemented with some data

$$H_3C(CH_2)_{n-2}CH_3$$

103

$$\begin{matrix} H_2C \\ | \\ H_2C \end{matrix} (CH_2)_n$$

104

$$\begin{matrix} HC \\ \| \\ HC \end{matrix} (CH_2)_n$$

105

from the authors' laboratory.[74, 137] In Fig. 35 are reproduced the photoelectron spectra of some cycloalkanes,[133] where the p- and s-regions are again defined. In the larger members of the unsaturated hydrocarbons the two regions actually overlap. Benzene provides such a case and the lowest ionization energy $C2s$ band (15·45 eV) lies below the $3a_{1g}p\sigma$ band (16·84 eV).[138] The HeII excited spectrum is shown in Fig. 35.

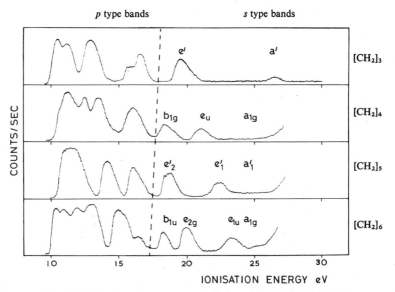

Fig. 35. HeII photoelectron spectra of the cycloalkanes (**104**).[133]

The simplest interpretation of the observed data is within the framework of the Hückel model as applied traditionally to the π-systems of unsaturated hydrocarbons. This was the approach adopted by Potts and Streets[133, 134] and we shall outline the results of such a treatment.

The orbital energies (ε_j) corresponding to the linear combinations ψ_j of the $C2s$ atomic orbital $\phi_\mu \equiv 2s_\mu$ in the traditional Hückel approximation are $\varepsilon_j = \alpha + x_j\beta$, the symbols having their usual meaning, albeit with reference to a $2s$ basis. In this approach it is assumed that the $C2s$ molecular orbitals can be treated independently of the molecular orbitals with which the p-type

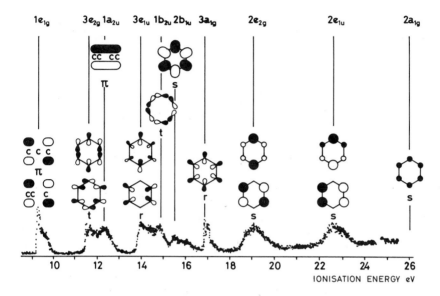

Fig. 36. HeII photoelectron spectrum of benzene and the assignment of the bands to molecular orbitals (in D_{6h} symmetry) from which the electrons are considered to be photoionized.[138]

region is associated. The simplest relationship that may be expected is a linear dependence of the "observed" orbital energies, derived from the $I_{m,j} \approx I_{v,j}$ values by means of Koopmans' Theorem on x_j. For the set of molecules of the type 103, 104, or 105 this does not yield an especially good correlation. However, a considerably better correlation is obtained when different values of α are given to carbon atoms which differ in the number of attached hydrogens. In other words, the well-used practice of calibrating the parameters α_μ and $\beta_{\mu\nu}$ on the photoelectron spectral data to suit the model is adopted. This implies the assumption that the $2s$-linear combinations are not independent of the p-type orbitals, but that the influence of the latter can be treated as a perturbation which is incorporated into the Coulomb integrals α_μ and the resonance integrals $\beta_{\mu\nu}$ to be used.

Based on the ionization energies of the $2s$-bands of the saturated hydrocarbons 103 and 104 the following α_μ values have been obtained: $\alpha_{CH_4} = -22 \cdot 9$ eV, $\alpha_{CH_3} = -22 \cdot 0$ eV, $\alpha_{CH_2} = -21 \cdot 6$ eV, $\alpha_{CH} = -21 \cdot 3$ eV, and $\alpha_C = -21 \cdot 0$ eV.[133, 134] These α_μ-values are derived from the set of linear equations

$$\sum_\mu \alpha_\mu(R) = -\sum_j I_{v,j}(R) \qquad (67)$$

which can be set up for a series of hydrocarbons R and which are then solved for the $\alpha_\mu(R)$ with $\mu = CH_4, CH_3, CH_2, CH,$ and C under the assumption

TABLE IX

Ionization Energies of the C2s-type Orbitals of some Alkanes and Unsaturated Hydrocarbons

						Assumed symmetry	Ref.
CH_4	22·91 $2a_1$					T_d	133
CH_3CH_3	23·9 $2\sigma_g$	20·42 $2\sigma_u$				$D_{\infty h}$	133
$CH_3CH_2CH_3$	24·5 $3a_1$	22·1 $2b_2$	19·15 $4a_1$			C_{2v}	133
$CH_3(CH_2)_2CH_3$	24·7 $3a_g$	23·0 $3b_u$	20·7 $4a_g$	18·80 $4b_u$		C_{2h}	133
$CH_3(CH_2)_3CH_3$	24·8 $4a_1$	23·7 $3b_2$	21·7 $5a_1$	19·9 $4b_2$	18·74 $6a_1$	C_{2v}	133
$(CH_3)_2CHCH_2CH_3$	24·9 $4a'$	23·3 $3a''$	22·0 $4a''$	20·4 $5a'$	18·34 $6a'$	C_s	133
$(CH_3)_3CH$	24·8 $3a_1$	21·9 $2e$	18·37 $4a_1$			C_{3v}	133
$(CH_3)_4C$	25·1 $3a_1$	21·9 $2t_2$	17·81 $4a_1$			T_d	133
$(CH_2)_3$	26·5 $2a_1'$	19·5 $2e'$				D_{3h}	133
$(CH_2)_4$	— $2a_{1g}$	21·0 $2e_u$	18·25 $2b_{1g}$			D_{4h}	133

Molecule							Sym.	Ref.
$(CH_2)_5$	— $2a_1$	22·2 $2e_1'$	18·29 $2e_2'$				D_{5h}	133
$(CH_2)_6$	25·7 $2a_{1g}$	23·1 $2e_{1u}$	19·49 $2e_{2g}$	18·06 $2b_{1u}$			D_{6h}	133
$(CH_2)_7$	— $2a_1$	23·7 $2e_1$	20·25 $2e_2$	18·07 $2e_3$			D_{7h}	133
$(CH_2)_8$	— $2a_{1g}$	23·9 $2e_{1u}$	21·4 $2e_2$	18·75 $2e_3$	17·66 $2b_2$		D_{8h}	133
$(CH_2)_{10}$	— $2a_{1g}$	≈24·3 $2e_{1u}$	22·5 $2e_{2g}$ / 22·2	20·2 $2e_{3u}$	19·0 $2e_{4g}$ / 18·8	— $2b_{2u}$	D_{10h}	74
$CH_2{=}CH_2$	23·70 $2a_g$	19·1 $2b_{1u}$					D_{2h}	134
$CH_3CH{=}CH_2$	24·16 $4a'$	21·88 $5a'$	18·30 $6a'$				C_s	134
$(CH_3)_2C{=}CH_2$	25·2 $3a_1$	22·3 $3b_2$	21·8 $4a_1$	17·4 $5a_1$			C_{2v}	137
$CH_3CH{=}CHCH_3(t)$	24·9 $3a_g$	23·1 $3b_u$	20·9 $4a_g$	17·9 $4b_u$			C_{2h}	137
$CH_3CH{=}CHCH_3(c)$	24·9 $3a_1$	22·9 $3b_2$	21·2 $4a_1$	17·5 $4b_2$			C_{2v}	137
$(CH_3)_2C{=}CHCH_3$	— $6a'$	22·9 $7a'$	21·9 $8a'$	20·4 $9a'$	16·8 $10a'$		C_s	137
$(CH_3)_2C{=}C(CH_3)_2$	— $3a_g$	23·6 $3b_{1u}$	22·3 $2b_{2u}$	21·7 $2b_{3g}$	20·0 $4a_g$	≈16·5 $1b_{3u}$	D_{2h}	137
$CH_2{=}CH{-}CH{=}CH_2$	24·8 $3a_g$	22·56 $3b_u$	19·20 $4a_g$	18·11 $4b_u$			C_{2h}	134

TABLE IX (cont.)

Molecule							Point group	Ref
$CH_2{=}CCH_3{-}CCH_3{=}CH_2$	— $4a_g$	23·5 $4b_u$	22·0 $5b_u$	21·9 $5a_g$	18·8 $6a_g$	≈16·8 $6b_u$	C_{2h}	137
$HC{\equiv}CH$	23·55 $2\sigma_g$	18·70 $2\sigma_u$					$D_{\infty h}$	134
$CH_3C{\equiv}CH$	24·5 $4a_1$	22·2 $5a_1$	17·4 $6a_1$				C_{3v}	137
$CH_3{-}C{\equiv}C{-}CH_3$	26·0 $3a_{1g}$	23·6 $3a_{2u}$	21·1 $4a_{1g}$	16·3 $4a_{2u}$			D_{3d}	137
$HC{\equiv}C{-}C{\equiv}CH$	25·6 $3\sigma_g$	23·3 $3\sigma_u$	20·0 $4\sigma_g$	17·5 $4\sigma_u$			$D_{\infty h}$	137
△	— $2a_1'$	19·6, 8·3 $2e'$					D_{3h}	141
(cyclopentadiene ring)	26·0 $2a_1'$	22·04 $2e'$	18·75 $2e_2'$				D_{5h}	134
(cyclohexadiene ring)	— $2a_{1g}$	23·49, 22·78 $2e_{1u}$	19·83, 19·29 $2e_{2g}$	16·73 $2b_{1u}$			D_{6h}	134
(benzene ring)	25·9 $2a_{1g}$	22·97 $2e_{1u}$	19·02 $2e_{2g}$	15·45 $2b_{1u}$			D_{6h}	134

The values listed are taken from refs 133 and 134 and work from the authors' laboratory. The $I_{m,j} \approx I_{v,j}$ values given are all in eV. When no $I_{v,j}$ values are given the band was either not detected or identified. The symmetry according to which the orbitals are labelled, with the irreducible representations given, is chosen for convenience for the $C2s$ orbitals rather than to represent the true molecular point group.

that they are independent of R. In (67) the $I_{v,j}(R)$ are the ionization energies corresponding to the bands in the $C2s$-region of the photoelectron spectrum. The observed variation in the α_μ values implies that they incorporate the effect of mixing with the hydrogen $1s$ and with the carbon $2p(\tau)$ atomic orbitals. Thus α_μ is associated with a linear combination η_μ, rather than with a pure $2s$ atomic orbital:

$$\alpha_\mu = \langle \eta_\mu | H | \eta_\mu \rangle$$
$$\eta_\mu = a_\mu 2s + \sum_\rho b_{\mu\rho} 1s_\rho + \sum_\tau c_{\mu\tau} 2p_\tau \qquad (68)$$

Obviously $a_\mu \gg b_{\mu\rho}, c_{\mu\tau}$. The close agreement of the α_μ values from molecules of different symmetry (cf. Table IX) suggests that the separation of the two regions is reasonable as far as parameterization of the data is concerned. The average β value derived from these data was found to be $-1\cdot87$ eV for the C—C single bond. To account for the variation $\beta_{\mu\nu}$ as a function of the CC-distance of $r_{\mu\nu}$, e.g. when unsaturation is present, the relationship $\beta_{\mu\nu} = kS_{\mu\nu}$, where $S_{\mu\nu} = S(r_{\mu\nu})$ is the overlap integral, between two $2s$ atomic orbitals separated by a distance $r_{\mu\nu}$. This approximation also bears out well as is shown in Fig. 37 where the experimentally observed values $\beta_{\mu\nu}$ are plotted as a function of the interatomic distance $r_{\mu\nu}$ and compared to the $\beta_{\mu\nu} = kS_{\mu\nu}$ dependence.[134]

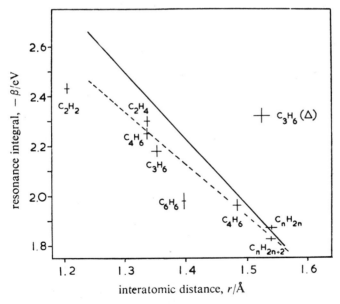

Fig. 37. Dependence of the experimentally derived $\beta_{\mu\nu}$ values on the interatomic distance $r_{\mu\nu}$ (dotted line). The $\beta_{\mu\nu} = kS_{\mu\nu}$ correlation is represented by the solid line.[134]

Overall it would appear that the Hückel approach yields an adequate model which reproduces the $C2s$ band pattern well and yields good quantitative agreement with the experiment. This is in contrast to semi-empirical SCF-procedures such as MINDO, CNDO, and to some extent also *ab initio* results.[134, 139] For the hydrocarbons, SPINDO also yields quite reasonable results for the $C2s$ energies,[139] but not as good as the Hückel model which is of course calibrated *ad hoc* to reproduce the data. It is not surprising that the latter model appears satisfactory although finer details indicate more complexity.

This is evidenced by the vibrational fine structure observed on the band associated with the highest antibonding $C2s$-level. It is found that modes corresponding to C—C vibrations are but weakly excited on the whole whereas the C—H stretching vibrations are strongly excited with appreciable decrease in frequency with respect to ground molecular state values.[133, 134] With the ions of high symmetry, e.g. neopentane, the vibrational analysis of the a_1 band (at ≈ 18 eV)[140] indicates that the situation prevailing is undoubtedly more complicated than represented by the model. The vibrational fine structure of the a_1 band in neopentane[133, 140] suggests a slight C—C bonding character, probably arising from "through-space" interactions of the methyl groups, as well as the excitation of a C—H deformation mode. In some other species C—C—C angle deformations were also evident.[133, 134] Such data reflect some admixture of the $2p$-type orbitals with the $2s$-type as expressed in the linear combination (68).

A good example illustrating the type of differences observed relative to those predicted by the Hückel approach are the ionization energy data for cis- and trans-2-butene,[137] which are given in Table IX. There is a significant difference in the $I_{m,j} \approx I_{v,j}$ values of the $C2s$ bands, which is most pronounced for the band associated with the most antibonding $C2s$ molecular orbitals (~ 0.4 eV). The effect is smaller for the deeper $C2s$ orbitals. In the framework of the Hückel theory, the values of the $C2s$ orbitals should have been the same for both isomers, similar results are observed for other isomeric hydrocarbons.[74] These data suggest that the $2s$- and $2p$-type regions are in general not separable to the extent it would appear at first. Obviously the data available are not yet sufficient to allow an assessment of the range of systems for which the simple Hückel treatment is adequate or to yield answers concerning the importance of long-range "through-space" and "through-bond" interactions for orbitals which are dominantly of $2s$-character.

ACKNOWLEDGEMENTS

We wish to thank Dr. D. W. Turner to whom we owe our introduction to the field of photoelectron spectroscopy and we extend our apologies to him if we have abused the technique in some of the applications to organic chemistry. Our gratitude goes to the Schweizerischer Nationalfonds zur Förderung der wissenschaftlichen Forschung[147] who made this possible.

REFERENCES

1. G. Herzberg, "Molecular Spectra and Molecular Structure", I. Spectra of Diatomic Molecules, III. Electronic Spectra and Electronic Structure of Polyatomic Molecules (D. van Nostrand Co. Inc., New York, 1950, 1966).
2. D. W. Turner, C. Baker, A. D. Baker, and C. R. Brundle, "Molecular Photo-electron Spectroscopy" (Wiley–Interscience, London, 1970).
3. J. N. Murrell, "The Theory of the Electronic Spectra of Organic Molecules" (Methuen & Co. Ltd., London, 1963).
4. E. Heilbronner and J. P. Maier, Organic photoelectron spectroscopy, in: "Electron Spectroscopy: Theory, Techniques and Applications" (Academic Press, London, 1976).
5. M. I. Al-Joboury and D. W. Turner, *J. Chem. Soc.* 5141 (1963).
6. C. R. Brundle and D. W. Turner, *J. Mass Spectrom. Ion Phys.* **2**, 195 (1969).
7. P. Bischof, R. Gleiter, and E. Heilbronner, *Helv. Chim. Acta*, **53**, 1425 (1970).
8. G. Bieri, E. Heilbronner, M. J. Goldstein, R. S. Leight, and M. S. Lipton, *Tetrahedron Letters* 1975, 581; G. Bieri, M. J. Goldstein, and E. Heilbronner, *Helv. Chim. Acta*, **59**, 2657 (1976).
9. D. W. Turner, *Advances Phys. Org. Chem.* **4**, 31 (1966); P. Mollère, H. Bock, G. Becker, and G. Fritz, *J. Organomet. Chem.* **46**, 89 (1972).
10. J. H. D. Eland, *J. Mass Spectrom. Ion Phys.* **2**, 471 (1969); C. R. Brundle and M. B. Robin, *J. Amer. Chem. Soc.* **92**, 5550 (1970); C. R. Brundle, M. B. Robin, N. A. Kuebler, and H. Basch, *ibid.* **94**, 1451 (1972); M. Beez, B. Bieri, H. Bock, and E. Heilbronner, *Helv. Chim. Acta*, **56**, 1028 (1973).
11. C. Baker and D. W. Turner, *Proc. Roy. Soc.* A308, 19 (1968); *Chem. Commun.* 1967, 797; D. C. Frost, F. G. Herring, C. A. McDowell, and I. A. Stenhouse, *Chem. Phys. Letters*, **4**, 533 (1970).
12. S. F. Nelsen, J. M. Buschek, and P. J. Hinz, *J. Amer. Chem. Soc.* **95**, 2013 (1973); S. F. Nelsen and J. M. Buschek, *ibid.* **95**, 2011 (1973); P. Rademacher, *Angew. Chem.* **85**, 447 (1973), *Angew. Chem. Int. Ed.* **12**, 408 (1973).
13. A. Schweig, H. Vermeer, and U. Weidner, *Chem. Phys. Letters*, **26**, 229 (1974).
14. T. Kobayashi, private communication.
15. Ch. Batich, G. Bieri, E. Heilbronner, and E. Vogel, unpublished results.
16. A. D. Baker, C. Baker, C. R. Brundle, and D. W. Turner, *J. Mass Spectrom. Ion Phys.* **1**, 285 (1968); D. W. Turner, *Ann. Rev. Phys. Chem.* **21**, 107 (1970); A. W. Potts and W. C. Price, *Proc. Roy. Soc. Lond.* A326, 181 (1972).
17. Ch. Batich, E. Heilbronner, V. Hornung, A. J. Ashe, III, D. T. Clark, U. T. Cobley, D. Kilkast, and I. Scanlan, *J. Amer. Chem. Soc.* **95**, 928 (1973).
18. H. Oehling, W. Schäfer, and A. Schweig, *Angew. Chem. Int. Ed.* **10**, 656 (1971).
19. G. Distefano, S. Pignataro, G. Innorta, F. Fringuelli, G. Marino, and A. Taticchi, *Chem. Phys. Letters*, **22**, 132 (1973); W. Schäfer, A. Schweig, S. Gronowitz, A. Taticchi, and F. Fringuelli, *J.C.S. Chem. Commun.* **1973**, 541.
20. E. Haselbach, E. Heilbronner, and G. Schröder, *Hevl. Chim. Acta*, **54**, 153 (1971).

21. P. Bischof, J. A. Hashmall, E. Heilbronner, and V. Hornung, *Tetrahedron Letters* **1969**, 4025; E. Heilbronner and K. A. Muszkat, *J. Amer. Chem. Soc.* **92**. 3818 (1970).
22. P. A. Clark, *Theoret. Chim. Acta*, **28**, 75 (1972), and references therein.
23. P. Bruckmann and M. Klessinger, *J. Electr. Spectr.* **2**, 31 (1973); W. Ensslin, H. Bock, and G. Becker, *J. Amer. Chem. Soc.* **96**, 2757 (1974).
24. C. R. Brundle, M. B. Robin, N. A. Kuebler, and H. Basch, *J. Amer. Chem. Soc.* **94**, 1451 (1972); C. R. Brundle, M. B. Robin, and N. A. Kuebler, *ibid.* **94**, 1466 (1972); D. G. Streets and G. P. Ceasar, *Mol. Physics*, **26**, 1037 (1973); D. G. Streets, *Chem. Phys. Letters*, **28**, 555 (1974).
25. The following are particularly useful sources for further references: L. Salem, "Molecular Orbital Theory of Conjugated Systems" (Benjamin, New York, 1966); J. N. Murrell and A. J. Harget, "Semi-empirical Self-consistent Molecular-orbital Theory of Molecules" (Wiley–Interscience, 1972); J. A. Pople and D. L. Beveridge, "Approximate Molecular Orbital Theory" (McGraw-Hill, New York, 1970).
26. R. B. Woodward and R. Hoffmann, *J. Amer. Chem. Soc.* **87**, 395, 2046, 2511 (1965); R. B. Woodward and R. Hoffmann, "Die Erhaltung der Orbital-symmetrie" (Verlag Chemie, Weinheim, 1970).
27. H. tom Dieck and K. Wittel, *Nachrichten Wiss. u. Technik*, **22**, 196 (1974).
28. T. Koopmans, *Physica*, **1**, 104 (1934); W. G. Richards, *J. Mass Spectrom. Ion Phys.* **2**, 419 (1969).
29. C. C. J. Roothaan, *Rev. Mod. Phys.* **23**, 69 (1951).
30. K. H. Johnson, *J. Chem. Phys.* **45**, 3085 (1966); K. H. Johnson and F. C. Smith, *Phys. Rev. Letters*, **22**, 1168 (1969); O. K. Andersen and R. G. Woolley, *Mol. Phys.* **26**, 905 (1973).
31. W. Kutzelnigg, *Topics Current Chem.* **41**, 31 (1973).
32. E. Heilbronner, K. A. Muszkat, and G. Schäublin, *Helv. Chim. Acta*, **54**, 58 (1971); J. M. Hollas and T. A. Sutherley, *Mol. Phys.* **21**, 183 (1971), **22**, 213 (1971).
33. J. A. Pople, D. P. Santry, and G. A. Segal, *J. Chem. Phys.* **43**, S129 (1965); J. A. Pople and G. A. Segal, *ibid.* **43**, S136 (1965).
34. M. J. S. Dewar and E. Haselbach, *J. Amer. Chem. Soc.* **92**, 590 (1970); N. Bodor, M. J. S. Dewar, A. Harget, and E. Haselbach, *ibid.* **92**, 3854 (1970); M. J. S. Dewar, E. Haselbach, and S. D. Worley, *Proc. Roy. Soc. Lond.* **A315**, 431 (1970).
35. B. C. Bingham, M. J. S. Dewar, and K. H. Lo, *J. Amer. Chem. Soc.* **97**, 1285, 1294, 1302, 1307 (1975).
36. C. Fridh, L. Åsbrink, and E. Lindholm, *Chem. Phys. Letters*, **15**, 282 (1972); L. Åsbrink, C. Fridh, and E. Lindholm, *J. Amer. Chem. Soc.* **91**, 5501 (1972).
37. L. Åsbrink, C. Fridh, and E. Lindholm, in: "Chemical Spectroscopy and Photochemistry in the Vacuum-Ultraviolet", C. Sandorfy, P. J. Ausloos, and M. B. Robin, eds. (D. Reidel Publishing Comp., Dordrecht, 1974).
38. C. Edmiston and K. Ruedenberg, *Rev. Mod. Phys.* **35**, 457 (1963); *J. Chem. Phys.* **43**, 597 (1965); K. Ruedenberg, "Modern Quantum Chemistry". (Academic Press, New York, 1965) Vol. I, p. 85; W. England, L. S. Salmon, and K. Ruedenberg, *Fortschr. chem. Forschg.* **23**, 31 (1971).
39. W. England, M. S. Gordon, and K. Ruedenberg, *Theoret. Chim. Acta*, **37**, 177 (1975); A. Schmelzer, R. Pauncz, and E. Heilbronner, unpublished results.

40. G. G. Hall, *Proc. Roy. Soc.* A **205**, 541 (1951); J. E. Lennard-Jones and G. G. Hall, *Trans. Faraday Soc.* 48 (1952).
41. D. F. Brailsford and B. Ford, *Mol. Phys.* **18**, 621 (1970); J. N. Murrell and W. Schmidt, *J.C.S. Faraday II* **1972**, 1709.
42. B. Pullman and A. Pullman, "Les theories électroniques de la chimie organique" (Masson, Paris, 1952); A. Streitwieser, Jr., "Molecular Orbital Theory for Organic Chemists" (John Wiley, New York, 1961); E. Heilbronner and H. Bock, "Das HMO-Modell und seine Anwendung" (Verlag Chemie, Weinheim, 1968).
43. R. Pariser and R. G. Parr, *J. Chem. Phys.* **21**, 466, 767 (1953); J. A. Pople, *Trans. Faraday Soc.* **49**, 1375 (1953).
44. A. Streitwieser, Jr. and P. M. Nair, *Tetrahedron*, **5**, 149 (1959); A. Streitwieser, Jr., *J. Amer. Chem. Soc.* **82**, 4123 (1960).
45. J. H. D. Eland and C. J. Danby, *Zeitschr. Naturforsch.* **23a**, 355 (1968); J. H. D. Eland, *Int. J. Mass Spectrom. Ion Phys.* **9**, 214 (1972).
46. H. C. Longuet-Higgins and L. Salem, *Proc. Roy. Soc. Lond.* **A251**, 172 (1959), **A257**, 445 (1960); L. Salem and H. C. Longuet-Higgins, **A255**, 435 (1960).
47. F. Brogli and E. Heilbronner, *Theoret. Chim. Acta*, **26**, 289 (1972).
48. F. Brogli and E. Heilbronner, *Angew. Chem.* **84**, 551 (1972), *Angew. Chem. Int. Ed.* **6**, 538 (1972).
49. P. A. Clark, F. Brogli, and E. Heilbronner, *Helv. Chim. Acta*, **55**, 1415 (1972).
50. R. Boschi, J. N. Murrell, and W. Schmidt, *Disc. Faraday Soc.* **54**, 116 (1972); R. Boschi, E. Clar, and W. Schmidt, *J. Chem. Phys.* **60**, 4406 (1974).
51. R. H. Martin, S. Obenland, and W. Schmidt, in preparation.
52. C. A. Coulson, *Proc. Roy. Soc. Lond.* **A169**, 413 (1939), **A207**, 91 (1951); *J. Phys. Chem.* **56**, 311 (1952).
53. C. Baker and D. W. Turner, *Chem. Commun.* **1969**, 480; F. Brogli, J. K. Crandall, E. Heilbronner, E. Kloster-Jensen, and S. A. Sojka, *J. Electr. Spectr.* **2**, 455 (1973); P. Bischof, R. Gleiter, H. Hopf, and F. T. Lenich, *J. Amer. Chem. Soc.*, **97**, 5467 (1975).
54. F. Brogli, E. Heilbronner, E. Kloster-Jensen, A. Schmelzer, A. S. Manocha, J. A. Pople, and L. Radom, *Chem. Physics*, **4**, 107 (1974).
55. E. Heilbronner, V. Hornung, J. P. Maier, and E. Kloster-Jensen, *J. Amer. Chem. Soc.* **96**, 4252 (1974).
56. M. J. S. Dewar, "The Molecular Orbital Theory of Organic Chemistry" (McGraw-Hill Book Company, New York, 1969).
57. R. W. Taft, Separation of polar, steric and resonance effects in reactivity, in: "Steric Effects in Organic Chemistry", M. S. Newman, ed. (Chapman and Hall, New York, 1956); P. R. Wells, Linear Free Energy Relationship" (Academic Press, London, 1968); C. D. Johnson, "The Hammett Equation" (Cambridge University Press, Cambridge, 1973); J. Shorter, *Quart. Reviews*, **24**, 433 (1970).
58. K. Watanabe, T. Nakayama, and J. Mottle, *J. Quant. Spectrosc. Radiat. Transfer*, **2**, 369 (1962).
59. J. J. Kaufman and W. S. Koski, *J. Amer. Chem. Soc.* **82**, 3262 (1960).
60. D. W. Turner, *Advances Phys. Org. Chem.* **4**, 31 (1966).
61. F. H. van Cauwelaert, *Bull. Soc. Chim. Belges*, **80**, 181 (1971).
62. F. Brogli, P. A. Clark, E. Heilbronner, and M. Neuenschwander, *Angew. Chem.* **85**, 414 (1973).

63. G. A. Olah and P. von R. Schleyer, "Carbonium Ions" (Interscience, New York, 1968, 1969), Vols I and II.

64. J. D. Baldeschwieler and S. S. Woodgate, *Acc. Chem. Res.* **4**, 114 (1971).

65. B. J. Cocksey, J. H. D. Eland, and C. J. Danby, *J. Chem. Soc.* (B) **1971**, 790.

66. J. H. D. Eland, "Photoelectron Spectroscopy" (Butterworths, London, 1974).

67. J. L. Ragle, I. A. Stenhouse, D. C. Frost, and C. A. McDowell, *J. Chem. Phys.* **53**, 178 (1970); F. Brogli and E. Heilbronner, *Helv. Chim. Acta*, **54**, 1423 (1971); A. W. Potts, H. J. Lempka, D. G. Streets, and W. C. Price, *Phil. Trans. Roy. Soc. Lond.* **A268**, 59 (1970).

68. A. D. Baker, D. Betteridge, N. R. Kemp, and R. E. Kirby, *Analyt. Chem.* **43**, 375 (1971); M. B. Robin and N. A. Kuebler, *J. Electr. Spectr.* **1**, 13 (1972).

69. P. Masclet, O. Grosjean, G. Mouvier, and J. Dubois, *J. Electr. Spectr.* **2**, 225 (1973).

70. P. Carlier, J. E. Dubois, P. Masclet, and G. Mouvier, *J. Electr. Spectr.* **7**, 55 (1975).

71. M. Klessinger, *Angew. Chem. int. Edit.* **11**, 525 (1972); J. P. Maier and D. W. Turner, *Faraday Trans. II*, **69**, 196 (1973); F. Brogli, E. Giovannini, E. Heilbronner, and R. Schurter, *Chem. Ber.* **106**, 961 (1973); R. A. Wielesek and T. Koenig, *Tetrahedron Letters*, **1974**, 2424.

72. H. J. Lempka, T. R. Passmore, and W. C. Price, *Proc. Roy. Soc. Lond.* **A304**, 53 (1968); A. W. Potts and W. C. Price, *Faraday Trans. II*, **67**, 1242 (1971).

73. F. Brogli and E. Heilbronner, *Helv. Chim. Acta*, **54**, 1423 (1971).

74. Unpublished results from this laboratory.

75. R. S. Mulliken, C. A. Rieke, and W. G. Brown, *J. Amer. Chem. Soc.* **63**, 41 (1941).

76. G. W. Wheland and L. Pauling, *J. Amer. Chem. Soc.* **57**, 2086 (1935).

77. R. Hoffmann and R. A. Olofson, *J. Amer. Chem. Soc.* **88**, 943 (1966).

78. R. Hoffmann, C. C. Levin, and R. A. Moss, *J. Amer. Chem. Soc.* **95**, 629 (1973).

79. A. D. Baker, D. Betteridge, N. R. Kemp, and R. E. Kirby, *J. Mol. Structure*, **8**, 75 (1971); A. W. Potts and D. G. Streets, *ibid.* **70**, 875 (1974).

80. R. Hoffmann, A. Imamura, and W. J. Hehre, *J. Amer. Chem. Soc.* **90**, 1499 (1968); W. Adam, A. Grimison, and R. Hoffmann, *ibid.* **91**, 2590 (1969); R. Hoffmann, E. Heilbronner, and R. Gleiter, *ibid.* **92**, 706 (1970); R. Hoffmann, *Accounts Chem. Res.* **4**, 1 (1971).

81. R. Hoffmann and W. N. Lipscomb, *J. Chem. Phys.* **36**, 2179, 3489 (1962); **37**, 2872 (1962); R. Hoffmann, *ibid.* **39**, 1397 (1963).

82. P. Bischof, J. A. Hashmall, E. Heilbronner, and V. Hornung, *Helv. Chim. Acta.* **52**, 1745 (1969); for further references: R. Gleiter, *Angew. Chem.* **86**, 770 (1974); H. Bock and B. G. Ramsey, *Angew. Chem. int. Ed.* **12**, 734 (1973); C. Batich, P. Bischof, and E. Heilbronner, *J. Electr. Spectr.* **1**, 333 (1973); J. P. Maier, *Annual Reports Chem. Soc.* 1974, B.

83. P. Bischof, J. A. Hashmall, E. Heilbronner, and V. Hornung, *Tetrahedron Letters* **1969**, 4025; E. Heilbronner and K. A. Muszkat, *J. Amer. Chem. Soc.* **92**, 3818 (1970); for further references: E. Heilbronner, J. P. Maier, and E. Haselbach, UV photoelectron spectra of heterocyclic compounds, in: "Physical Methods of Heterocyclic Chemistry", A. Katritzky, ed. (Academic Press, New York, 1974), Vol. VI.

84. D. A. Sweigart and D. W. Turner, *J. Amer. Chem. Soc.* **94**, 5599 (1972); H. Bock and G. Wagner, *Angew. Chem.* **84**, 119 (1972); C. Batich and W. Adam, *Tetrahedron Letters* **1974**, 1467.

85. D. O. Cowan, R. Gleiter, J. A. Hashmall, E. Heilbronner, and V. Hornung, *Angew. Chem. Intern. Ed.* **10**, 401 (1971); C. R. Brundle, M. B. Robin, and N. A. Kuebler, *J. Amer. Chem. Soc.* **94**, 1466 (1972).
86. G. Laner, W. Schäfer, and A. Schweig, *Chem. Phys. Letters* **33**, 312 (1975).
87. G. Wagner, H. Bock, R. Budenz, and F. Seel, *Chem. Ber.* **106**, 1285 (1973).
88. P. Bischof, Thesis, University of Basel, 1971; E. Heilbronner, *J. Pure Appl. Chem. XIII Int. Congr.* **7**, 9 (1971).
89. C. Batich, E. Heilbronner, E. Rommel, M. F. Semmelhack, and J. S. Foos, *J. Amer. Chem. Soc.* **96**, 7662 (1974).
90. A. Schweig, U. Weidner, R. K. Hill, and D. A. Cullison, *J. Amer. Chem. Soc.* **95**, 5426 (1973).
91. R. Boschi and W. Schmidt, *Angew. Chem. Intern. Ed.* **12**, 402 (1973).
92. V. Boekelheide and W. Schmidt, *Chem. Phys. Letters,* **17**, 410 (1972).
93. M. Allan, E. Heilbronner, P. M. Keehn, and J. P. Maier, unpublished results.
94. R. Gleiter, *Tetrahedron Letters,* **1969**, 4453.
95. E. Heilbronner and J. P. Maier, *Helv. Chim. Acta,* **57**, 151 (1974).
96. E. Heilbronner, *Israel J. Chem.* **10**, 143 (1972).
97. C. Baker and D. W. Turner, *Proc. Roy. Soc. Lond.* **A308**, 19 (1968); L. S. Cederbaum, W. Domcke, and W. von Niessen, *Chemical Physics,* **10**, 459 (1975).
98. R. Gleiter, E. Heilbronner, and V. Hornung, *Helv. Chim. Acta,* **55**, 255 (1972); C. Fridh, L. Åsbrink, B. O. Jonsson, and E. Lindholm, *J. Mass Spectrom. Ion Phys.* **8**, 101 (1972).
99. E. Heilbronner, V. Hornung, and E. Kloster-Jensen, *Helv. Chim. Acta,* **53**, 331 (1970).
100. P. Bischof, J. A. Hashmall, E. Heilbronner, and V. Hornung, *Helv. Chim. Acta,* **52**, 1745 (1969).
101. M. J. Goldstein, E. Heilbronner, V. Hornung, and S. Natowsky, *Helv. Chim. Acta,* **56**, 294 (1973).
102. P. Bischof, J. A. Hashmall, E. Heilbronner, and V. Hornung, *Tetrahedron Letters* **1970**, 1033.
103. W. Adam, A. Grimson, and R. Hoffmann, *J. Amer. Chem. Soc.* **91**, 2590 (1971); W. Adam, *Jerusalem Symp. Quantum Chem. Biochem. (Israel Acad. Sci. and Humanities),* **2**, 118 (1970).
104. W. R. Wadt and W. A. Goddard, III, *J. Amer. Chem. Soc.* **97**, 2034 (1975).
105. A. Hammett and A. F. Orchard, Photoelectron spectroscopy, in: "Electronic Structure and Magnetism of Inorganic Compounds" A Specialist Periodical Report (Chem. Soc., London, 1972), Vol. 1.
106. L. A. Carreira, R. O. Carter, and J. R. Durig, *J. Chem. Phys.* **59**, 812 (1973); P. W. Rabideau and J. W. Paschal, *J. Amer. Chem. Soc.* **96**, 272 (1974).
107. R. W. Hoffmann, R. Schüttler, W. Schäfer, and A. Schweig, *Angew. Chem. Intern. Ed.* **11**, 512 (1972).
108. E. Heilbronner and A. Schmelzer, *Helv. Chim. Acta,* **58**, 936 (1975).
109. A. Schmelzer, R. Pauncz, and E. Heilbronner, unpublished results.
110. J. Daintith, J. P. Maier, D. A. Sweigart, and D. W. Turner, in: "Electron Spectroscopy", D. Shirley, ed. (North-Holland, Amsterdam, 1972); J. P. Maier and D. W. Turner, *Disc. Faraday Soc.* **54**, 149 (1972).
111. C. R. Brundle and M. B. Robin, *J. Amer. Chem. Soc.* **92**, 5550 (1970).
112. C. H. Chang, A. L. Andreassen, and S. H. Bauer, *J. Org. Chem.* **36**, 920 (1971).

113. C. Batich, O. Ermer, E. Heilbronner, and J. R. Wiseman, *Angew. Chem.* **85**, 302 (1973); *Angew. Chem. Int. Ed.* **12**, 312 (1973).
114. H. Schmidt, A. Schweig, and A. Krebs, *Tetrahedron Letters* **1974**, 1471.
115. J. Haase and A. Krebs, *Z. Naturforsch.* **26a**, 1190 (1971).
116. E. Kloster-Jensen and J. Wirz, *Angew. Chem.* **85**, 723 (1973); *Angew. Chem. Int. Ed.* **12**, 671 (1973); Ch. Römming, private communication, cf. above reference.
117. G. Bieri, E. Heilbronner, E. Kloster-Jensen, A. Schmelzer, and J. Wirz, *Helv. Chim. Acta*, **57**, 1265 (1974).
118. H. Hope, J. Bernstein, and K. N. Trueblood, *Acta Cryst.* **B28**, 1733 (1972).
119. R. Hoffmann, A. Imamura, and W. J. Hehre, *J. Amer. Chem. Soc.* **90**, 1499 (1968).
120. E. Haselbach, *Helv. Chim. Acta*, **54**, 1981 (1971).
121. B. M. Gimarc, *J. Amer. Chem. Soc.* **92**, 266 (1969).
122. P. Millie, L. Praud, and J. Serrè, *Int. J. Quantum Chem.* **4**, 187 (1971).
123. W. L. Mock, *Tetrahedron Letters*, **1972**, 475; L. Radom, J. A. Pople, and W. L. Mock, *ibid.* **1972**, 479.
124. R. Boschi, W. Schmidt, and J.-C. Gfeller, *Tetrahedron Letters*, **1972**, 4107.
125. M. Dobler and J. D. Dunitz, *Helv. Chim. Acta*, **48**, 1429 (1965).
126. C. Batich, E. Heilbronner, and E. Vogel, *Helv. Chim. Acta*, **57**, 2288 (1974).
127. G. Lauer, C. Müller, K. W. Schulte, A. Schweig, and A. Krebs, *Angew. Chem.* **86**, 597 (1974).
128. G. Lauer, C. Müller, K. W. Schulte, A. Schweig, G. Maier, and A. Alzérreca, *Angew. Chem.* **87**, 194 (1975).
129. C. Batich, P. Bischof, and E. Heilbronner, *J. Electr. Spectr.* **1**, 333 (1972–73).
130. E. Heilbronner, II. IUPAC on Non-benzenoid Aromatic Compounds (ISNA), Lindau, 1974, *Pure Appl. Chem.* **44**, 829 (1975).
131. W. C. Price, A. W. Potts, and D. G. Streets, in: "Electron Spectroscopy", D. A. Shirley, ed. (North Holland, Amsterdam, 1972), p. 187.
132. A. W. Potts, T. A. Williams, and W. C. Price, *Faraday Disc. Chem. Soc.* **54**, 104 (1972).
133. A. W. Potts and D. G. Streets, *J. C. S. Faraday II* **70**, 875 (1974).
134. D. G. Streets and A. W. Potts, *J.C.S. Faraday II* **70**, 1505 (1974).
135. U. Gelius, *J. Electr. Spectr.* **5**, 895 (1974).
136. A. W. Potts and T. A. Williams, *J. Electr. Spectr.* **3**, 3 (1974).
137. G. Bieri, Thesis, Universität Basel, 1975.
138. E. Lindholm, *Faraday Disc. Chem. Soc.* **54**, 200 (1972).
139. E. Lindholm, C. Fridh, and L. Åsbrink, *Faraday Dis. Chem. Soc.* **54**, 127 (1972).
140. S. Evans, J. C. Green, P. J. Joachim, A. F. Orchard, D. W. Turner, and J. P. Maier, *J.C.S. Faraday II* **68**, 905 (1972).
141. M. B. Robin, C. R. Brundle, N. A. Kuebler, G. B. Ellison, and K. B. Wiberg, *J. Chem. Phys.* **57**, 1758 (1972).
142. F. Brogli, E. Heilbronner, V. Hornung, and E. Kloster-Jensen, *Helv.Chim. Acta* **56**, 2171 (1973).
143. F. Brogli, E. Heilbronner, E. Kloster-Jensen, J. Wirz, R. G. Bergmann, P. Vollhardt, and A. J. Ashe, III, *Helv. Chim. Acta*, **58**, 2620 (1975).
144. E. Heilbronner, R. Gleiter, H. Hopf, V. Hornung, and A. de Meijere, *Helv. Chim. Acta.* **54**, 783 (1971).
145. F. Brogli, E. Heilbronner, and M. Neuenschwander, *Helv. Chim. Acta*, in press.
146. T. Bally and E. Haselbach, *Helv. Chim. Acta*, **58**, 321 (1975).
147. This is part 90 of project No. 2.159.74. Part 89: J. Bastide and J. P. Maier, *Chem. Physics*, **12**, 177 (1976).

6

Ultraviolet Photoelectron Spectroscopy of Inorganic Molecules

R. L. DEKOCK*

Department of Chemistry,
American University of Beirut, Beirut, Lebanon

* Present address: Department of Chemistry, Calvin College, Grand Rapids, Michigan 49506, USA.

I. INTRODUCTION

The ultraviolet photoelectron spectroscopy (UPS) of gaseous inorganic molecules is a field which has developed even more recently than the field of UPS itself. If one omits the atoms, diatomic, and triatomic molecules from the monograph by Turner and co-workers,[1] which are more properly assigned to "general" chemistry rather than "inorganic" chemistry, one finds that inorganic molecules are treated in the last chapter entitled "Miscellaneous Inorganic Compounds". There are at least two reasons for this lag in the study of inorganic systems: (1) Many of the simple inorganic molecules are reactive and difficult to handle from an experimental viewpoint (e.g. F_2, HF, XeF_2); (2) the low symmetry molecules often contain diverse atoms so that spectral assignments are difficult unless spectral correlations between related molecules can be carried out (e.g. NSF, HOF). Prior to 1970, very few publications appeared dealing with inorganic systems. With the advent of commercial instruments in 1970 the field has blossomed so that in five years it is not an exaggeration to say that 15–20 different research groups are interested in the UPS of inorganic systems.

An excellent monograph covering the basic theory in the field of UPS has been written by Eland.[2] Baker and Betteridge[3] have also written a monograph covering general aspects of both XPS and UPS. Review articles which may be of interest to the reader are: (1) The excellent series published by Orchard and co-workers in "Specialist Periodical Reports".[4-6] This survey is not only complete but critical. (2) The semi-annual report written by Betteridge and co-workers.[7-9] (3) A review article covering exclusively the UPS of inorganic molecules.[10] (4) An informative review written by Bock and Ramsey[11] relating the results of UPS to molecular orbital (MO) theory. This article has also been adapted to illustrate the use of UPS as an experimental approach to teaching MO models.[12] (5) An article by Ballard[13] dealing with the problem of band assignment in UPS.

Looking back over the past five years one can say that the development of the field has been rapid but smooth. Considerable duplication of effort has resulted. However, this has generally been healthy in that alternative assignments of spectra have been resolved, or impurity lines in the spectra have been pinpointed. There has been no central dominant controversy, only skirmishes dealing with select aspects such as the usefulness of calculations in ordering energy states, the application of Koopmans' Theorem in this respect, and the participation of outer d orbitals in the bonding of nontransition elements.

During the past two decades inorganic chemistry is said to have undergone a renaissance which was highlighted in the decade 1955–65 by the application of crystal field and ligand field theory to the understanding of the electronic spectra of transition metal complexes. These theories provided an understanding of only a small part of the total electronic structure of a molecule. In contrast, the combined techniques of UPS and x-ray photoelectron spectroscopy (XPS) provide the chemist with a probe to understand *all* of the occupied energy levels in a molecule. The results of UPS have also been useful in assigning absorption spectra of molecules[14, 15] and cations.[16]

The emphasis in this chapter will not be on the experimental techniques, the strictly spectroscopic information, or small molecules since these aspects are dealt with in other chapters. Rather, the emphasis will be on spectral assignment criteria and on the usefulness of UPS to aid in an understanding of chemical bonding.

Although UPS could refer to any work in the vacuum ultraviolet, for experimental reasons we shall be mainly interested in work utilizing the He I (21·22 eV) and He II (40·81 eV) photon sources.

II. THEORY

The reader is referred to Chapter 4 by Price for fundamental aspects relating to small molecules and to the monograph by Eland[2] for a treatment of the underlying theory of UPS. Important features which will not be discussed in this chapter are determination of the structural changes in the ion compared to the molecule from vibrational fine structure, effects of autoionization on enhancement of band intensities, and the angular distribution of photoelectrons. In this section we delineate only those aspects which will be useful in an understanding of the results on typical inorganic systems.

A. Ionic States

When a photon of energy $h\nu$ ionizes an electron from an isolated atom or molecule, the kinetic energy of the ejected electrons is given by

$$E_n = h\nu - I_n$$

where I_n is the ionization potential (IP) to the nth ionized state of the atom or molecule. So long as the photons are monochromatic and ν is known, measurement of E_n leads directly to I_n. For atoms and molecules containing more than two electrons, there will necessarily be more than one ionic state available. The general situation is shown in Fig. 1 for a molecule with three ionic states accessible to photons with energy $h\nu$. It is necessary to distinguish the different ionizations shown in Fig. 1, I_1, I_2, and I_3, each of which produces a *singly charged ion*, and by convention these are referred to as first, second, and third IP. This is different from the customary usage for atoms where the second IP refers to the minimum energy required to remove a second electron from a singly charged ion.

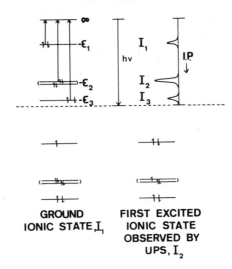

Fig. 1. Energy levels and ionization potentials.

For atoms, sharp IP are observed as shown in Fig. 1. However, for molecules we must include the rotational and vibrational structure which results in

$$I_n = I_0 + I_{\text{rot}} + \Delta I_{\text{vib}}$$

where I_0 is referred to as the adiabatic IP. We will neglect the effect of rotational fine structure which has been observed only for small molecules. The effects of vibrational structure are illustrated in Fig. 2 for a diatomic molecule.[17] In the $\tilde{X}$ ionic state the adiabatic and vertical (most intense transition) IP are identical. This corresponds to electron ejection from a nonbonding orbital so that the bond length in the $\tilde{X}$ ionic state is little changed from that in the molecular ground state, and most of the Franck–Condon intensity is concentrated in the 0–0 adiabatic transition. The $\tilde{A}$ ionic state shows a lengthy vibrational progression resulting from ionization out of a bonding orbital with a subsequent increase in bond length. The ionic states $\tilde{B}$ and $\tilde{C}$ illustrate the expected structure for transitions reaching unbound portions of the potential curve. Finally, the $\tilde{D}$ state illustrates the effect of curve crossing resulting in pre-dissociation.

Vibrational fine structure is often not observed for large molecules. This may result from reaching a repulsive portion of the potential energy curve as shown in Fig. 2 for ionic state $\tilde{C}$. In polyatomic molecules where a large number of vibrational modes are possible, unresolved bands could also be the result of many closely spaced vibrational lines. The loss of fine structure on the tail of state $\tilde{D}$ in Fig. 2 is due to lifetime shortening in the molecular ion, if the process corresponds to an allowed pre-dissociation. The fact that transitions

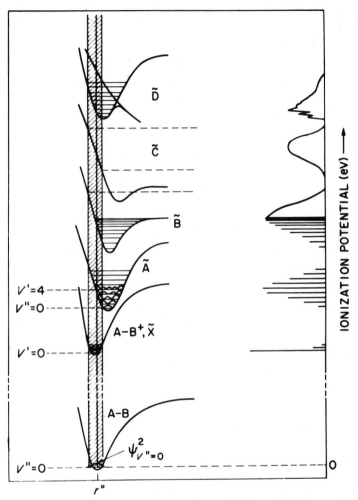

Fig. 2. Correlation between the Franck–Condon principle and the shapes of the bands observed in the photoelectron spectrum for the removal of electrons in different molecular orbitals. Reproduced from Brundle and Robin[17] with permission and by courtesy of Academic Press, New York.

occur to repulsive portions of potential energy curves such as in ionic states $\tilde{B}$, $\tilde{C}$, and $\tilde{D}$ (Fig. 2) should not be taken to mean that we are observing the UPS of the fragmented molecule. The time scale of the experiment must be at least as fast as that of ultraviolet spectroscopy (10^{-15} s) and may even be 10^{-17} s [18] whereas molecular dissociation times are generally 10^{-13} s. It has been suggested[19] that in molecules where no parent ion is observed in the mass spectrum the UPS is that of the fragment ion rather than the molecule, but this is clearly false since ion transit times in a mass spectrometer are at least 10^{-7} s.

B. Selection Rules and Photoionization Cross-sections

The photoelectric effect is not restricted by any symmetry selection rule since the ejected electron can carry any necessary angular momentum to make the process electric dipole allowed. Generally, only one-electron transitions are observed but this results from negligible configuration interaction rather than from a symmetry effect. The He I and He II sources do not have sufficient energy to cause two-electron ejection for most molecules. However, it is possible to observe "forbidden" transitions due to simultaneous ionization and excitation, particularly with the He II source.[20] Usually, these combined transitions result in weak bands but one notable exception is the spectrum of CS which exhibits a strong "forbidden" band with a vertical IP of 16·05 eV.[21] The corresponding ionic state ($^2\Sigma$) consists predominantly of the configuration $(1\sigma)^2 (2\sigma)^2 (3\sigma)^2 (4\sigma)^2 (1\pi)^3 (5\sigma)^1 (2\pi)^1$ and since the 1π orbital is bonding and the 2π antibonding, the vibrational progression shows a spacing of ~ 840 cm^{-1} compared to the ground-state molecular frequency of 1272 cm^{-1} (configuration: $(1\sigma)^2 (2\sigma)^2 (3\sigma)^2 (4\sigma)^2 (1\pi)^4 (5\sigma)^2$).

Although all one-electron ionizations are allowed, this does not mean that all bands will be equally intense. Figure 1 illustrates the intuitive result that band intensity should be directly proportional to electron occupancy.

Ballard[13] has shown that the photocurrent i for quantum yield Q is approximately

$$i \cong QI_0 n\sigma l$$

where I_0 is the incident photon intensity, l the path length of a substance containing n molecules per unit volume, and σ the photoionization cross-section. We are interested only in the variation of σ here. For ionization from a closed-shell molecule and assuming the orbitals of the ion are unchanged from those of the molecular ground state (frozen orbitals), Cox and Orchard[22] have shown that σ is proportional to $\omega_a |\langle \phi_a | \mu | \chi_p \rangle|^2$. In this expression ω_a is the orbital degeneracy, ϕ_a the orbital from which ionization takes place, μ the electric dipole moment operator, and χ_p the wave function of the ejected electron. The intuitive result that intensity is proportional to degeneracy is borne out if $\langle \phi_a | \mu | \chi_p \rangle$ is relatively constant for different orbitals; that is, if the different orbitals have similar localization properties.

Aside from the simple criterion of orbital degeneracy ω_a, one needs to consider variations in the integral $\langle \phi_a | \mu | \chi_p \rangle$. Price[23, 24] has stated that this integral will be roughly maximized when the half-width of the orbital ϕ_a is of the same dimensions as one-fourth the wavelength of the ejected photoelectron.

This allows one to understand the change in relative intensity of He I spectra compared to He II. For a given ionization, the ejected photoelectron will have a shorter wavelength with a He II source than with He I. Hence, orbitals with smaller size should be favored in He II spectra and vice versa.

The above discussion has dealt with relative intensity in closed-shell molecules. The situation for open-shell molecules has been discussed by Cox *et al.*[25] For molecules having a single open-shell they summarize their rules as follows:

"(i) If a closed shell is ionised, all states arising from the coupling of the positive hole with the open-shell ground state will be realized, the relative cross sections for production of these states being in proportion to their spin-orbital degeneracies.

(ii) If the open-shell is ionised, the relative probabilities of producing different ion states will reflect the squares of the fractional parentage coefficients, which may, but will not in general, be proportional to the spin-orbital degeneracies.

(iii) If orbitals belonging to different subshells are assumed to have the same one-electron cross sections, the integrated ionisation cross section . . . of a particular subshell is simply proportional to the occupancy of that subshell in the molecule."

As an example of the application of these rules consider first a closed-shell octahedral molecule (e.g. $Cr(CO)_6$) with configuration . . . $t_{1u}^6 t_{2g}^6$. We expect only two states for ionization from these orbitals with equal relative cross sections

$$t_{1u}^6 t_{2g}^5 \qquad {}^2T_{2g}$$

$$t_{1u}^5 t_{2g}^6 \qquad {}^2T_{1u}$$

For the open-shell molecule $V(CO)_6$ with configuration . . . $t_{1u}^6 t_{2g}^5$ we obtain

$$t_{1u}^6 t_{2g}^4 \qquad {}^1A_{1g}, \ {}^1E_g, \ {}^1T_{2g}, \ {}^3T_{1g}$$
$$\phantom{t_{1u}^6 t_{2g}^4 \qquad} 1 \qquad 2 \qquad 3 \qquad 9$$

and

$$t_{1u}^5 t_{2g}^5 \qquad {}^1T_{1u}, \ {}^3T_{1u}, \ {}^1T_{2u}, \ {}^3T_{2u},$$
$$\phantom{t_{1u}^5 t_{2g}^5 \qquad} 3 \qquad 9 \qquad 3 \qquad 9$$

$$\phantom{t_{1u}^5 t_{2g}^5 \qquad} {}^1E_u, \ {}^3E_u, \ {}^1A_{2u}, \ {}^3A_{2u}$$
$$\phantom{t_{1u}^5 t_{2g}^5 \qquad} 2 \qquad 6 \qquad 1 \qquad 3$$

where the relative intensities are indicated beneath each state. Since the ground state configuration of $V(CO)_6$ generates only one term, ${}^2T_{2g}$, the coefficients of fractional parentage are equal to the spin–orbital degeneracies.

III. SOME EXPERIMENTAL POINTS

In principle the equipment required is quite simple, consisting of a photon source, an ionization chamber, an electron energy analyzer, and an electron multiplier detector with appropriate recording system. Because of the lack of window materials at wavelengths below 100 nm, the various sections of the apparatus are separated by differential pumping. In studying the UPS of gases, the vacuum requirements are not stringent (10^{-5}–10^{-6} Torr) and consequently oil diffusion pumps are satisfactory.

A. Photon Sources

The requirements of the photon source in UPS are (1) sufficient energy to ionize valence electrons and (2) copious output of monochromatic photons. These requirements are met by the He I emission at 58·4 nm with a photon energy of 21·22 eV involving the transition $^1S(1s^2) \leftarrow {}^1P(1s\,2p)$. The helium discharge also emits a few per cent of other lines; the most intense of these lesser lines is He I β (23·09 eV) to be distinguished from the He I α source mentioned above. The source can also contain impurity lines due to H Lyman α (10·2 eV) and N I (10·9 eV) but careful attention to experimental detail can remove these.[26] By increasing the voltage on the discharge tube and lowering the helium pressure, it is possible to produce several per cent of He II emission at 30·4 nm (40·81 eV). Polystyrene[27] and aluminum[28] windows have been used to increase the per cent of He II photons relative to He I. The He II source allows access to all valence levels of a molecule except those derived predominantly from fluorine $2s$.[29] Lower-energy photons, particularly those from discharges in neon and argon, are used occasionally for the study of autoionization[30] or to increase resolution.[31]

B. Sample Introduction

In order to record spectra in a reasonable period of time, sample pressures in the region of 10^{-1}–10^{-3} Torr are required. Samples that have pressures lower than this at ambient temperature can be heated in the inlet chamber. In the Perkin Elmer PS–18 spectrometer the photon source is used to heat the sample, temperatures of 250 °C being readily attained. Several transient species have also been studied (e.g. HBS,[32, 33] SiF$_2$,[34, 35] S$_2$O[36, 37]) by preparing them in the inlet system just prior to their entrance into the ionization chamber.

Thermal decomposition of unstable or reactive species can occur in the inlet system, and this may be responsible for the different spectra obtained for Pt(PF$_3$)$_4$.[38, 39] The intense peak observed in the spectra of Green et al.[38] at 12·3 eV probably corresponds to contamination of the Pt(PF$_3$)$_4$ spectrum by PF$_3$. Reaction of hydrolytically unstable compounds with adsorbed water in the instrument is common, but spectra of the most likely volatile products are well known and easily recognized. "Memory" effects[40] wherein the sample is adsorbed on the walls of the ionization chamber occur often for polar inorganic molecules. The sample may be subsequently desorbed when passing a different sample through the instrument resulting in spectral contamination. While moisture and sample adsorption are problems, decomposition of the sample by the photon beam is unlikely to alter spectra since the output of the lamp is 10^{10}–10^{11} photons s^{-1} (ref. 41) and the flow of molecules through the ionization region is at least 10^6 times greater than this.

C. Electron Analyzers and the Presentation of Spectra

Most UPS work is carried out using the 127° electric deflection analyzer which was first developed by Turner[1] for applications to UPS and later produced commercially by the Perkin Elmer Corporation. Other analyzers of note are the 180° hemispherical[42] and the cylindrical mirror[41] types, both operating on the electric deflection principle.

The kinetic energies of the electrons can be differentiated in either of two ways: (1) the voltage between the cylindrical or hemispherical plates can be varied, or (2) the voltage across the plates can be maintained at a constant and the potential difference between the target chamber and analyzer entrance slit varied. Theory indicates that for deflection analyzers $E/\Delta E$ is a constant so that the effective resolution deteriorates (ΔE increases) with increasing kinetic energy, E, of the electrons. Consequently, at high IP, there is a fall-off in intensity which is due to this instrumental factor. The advantage of maintaining constant analyzer voltage is that all electrons passing through the analyzer have the same energy and the $E/\Delta E$ effect is constant. However, scanning the retarding potential usually results in a broad intense band of rising background at high IP due to scattered electrons. Fehlner and Turner[34] found that they could improve the resolution and reduce the scattered electron intensity by shortening the length of the analyzer entrance slit and placing a grounded slit between the source slit and analyzer entrance slit. Their spectrum of SiF_2, taken on an instrument with constant analyzer potential, shows no effect of scattered electrons.[34]

The adverse effect of scattered electrons is shown in Fig. 3 where the UPS of ozone is compared for a 180° hemispherical analyzer[43] scanning the retarding voltage and a 127° analyzer[44] scanning the analyzer voltage. In the former, the authors observed a band at 20.3 ± 0.1 eV although there is some hint of a band between 17 eV and 18 eV in the rising background. In the spectrum obtained on the 127° analyzer there is definite evidence for a band between 16.5 and 18.5 eV in addition to the band ranging from 18.7 eV to 21.5 eV centered at 20.1 eV. It is clear from Fig. 3 that band intensities in published spectra can only be compared if the method of scan is known. To compare band intensities within a given spectrum wherein the analyzer voltage is scanned, one must compare A/E where A is the area of the band.[45] This gives a qualitative[46] correction to the fact that bands at high IP appear less intense because of increasing resolution.

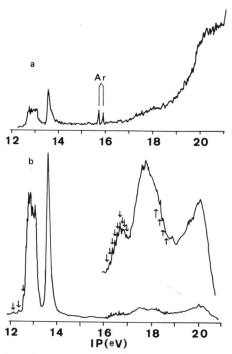

Fig. 3. The He I photoelectron spectrum of ozone. (a) Recorded on a hemispherical analyzer, scanning the retarding voltage. Reproduced from Frost *et al.*[43] with permission. (b) Recorded on a 127° analyzer, scanning the analyzer potential. Arrows identify features in the spectrum attributable to O_2. Reproduced from Brundle[44] with permission and by courtesy of North-Holland Publishing Company, Amsterdam.

According to the constancy of $E/\Delta E$ it is possible to improve resolution ΔE by reducing E, and some work has been carried out using lower-energy photons,[31, 47] which reduces E and hence ΔE. Electron retardation before the analyzer has also been used for this purpose,[42, 48] but in either case it is difficult to obtain a routine practical resolution of much better than 20 meV because of contamination of the slit surfaces by the compounds being studied. "High resolution" in inorganic UPS usually means a resolution between 25 meV and 75 meV for He I ionization of argon, i.e. for electrons of 5 eV kinetic energy. With such a resolution the best reproducibility one can expect is 10 meV but since bands not exhibiting fine structure have a considerable bandwidth the reproducibility may differ by 50 meV and even 100 meV for very broad bands. Lloyd[26] has discussed the problem of calibration with standard compounds which give known sharp peaks in the He I ionization region. In particular, one should choose a calibrant close to the peak of interest since the analyzer field may not be linear with IP.

IV. ASSIGNMENT OF BANDS

A. Optical Spectroscopy

The most definitive assignment criterion is available from a study of the optical absorption and emission of the molecular ions themselves. These studies have generally been carried out only on atoms, diatomic, and triatomic molecular ions. Thus the separations of and symmetry classification of the states of N_2^+, CO_2^+, N_2O^+, and similar small molecules are in no doubt.[49]

For larger molecules, where the optical spectrum of the ions has been recorded as a salt or in solution, the UPS may provide additional information on the problem of assignment of the optical spectrum. This approach has been taken by Herring and McLean[16] in assigning the spectra of I_2^+, Br_2^+, $Fe(C_5H_5)_2^+$, and ClO_2^+. One cannot expect to be able to assign all bands in the optical spectrum from the UPS. Any excited state in the optical spectrum derived from one-electron excitation to a virtual orbital of the cation will not be observed in the UPS. However, by mutual exclusion this can be very useful. While in the above cases, the UPS was used to shed light on the optical spectrum, the reverse can also be true. Such an example has been amply illustrated by Warren[50] for chromocene and manganocene where ligand field calculations were able to aid in assigning the proper ground state and the expected order of excited states for the corresponding cations.

B. Calculation of Ionization Energies

Implicit in Fig. 1 is Koopmans' Theorem[51] (KT) which states that the negative of the one-electron orbital energy obtained in a Hartree–Fock self-consistent field (SCF) calculation on a closed-shell molecule corresponds to the vertical ionization potential (VIP) in the zeroth-order approximation: $IP^0(i) = -\varepsilon(i)$. It is well known[52] that this result neglects relaxation (R) and correlation (C) so that the true VIP is given by $IP(i) = IP^0(i) - R + \Delta C$, where ΔC refers to the change in correlation energy between molecule and ion. Fortunately, R and ΔC tend to cancel and for many molecules KT is a good approximation if one wishes to compute IP to no better than a few electron volts. Unfortunately, even for simple molecules such as N_2[53] and F_2[54, 55] KT predicts the wrong order of electronic states for the ion. For many molecules, if calculations are to be of use in predicting the experimental order of states, the computed IP must be able to reproduce the experimental IP to within 0·1 eV.

In general, the application of KT is more dangerous in transition-metal chemistry than in main-group chemistry. Very large relaxation effects are exhibited in some transition-metal complexes upon ionization of electrons which are mainly metal in character. Examples include ferrocene,[56] bis-(π-allyl) nickel,[57] $Fe(CO)_2(NO)_2$,[58] $Co(CO)_3NO$,[58] and $(C_5H_5)NiNO$.[59] The reasons for this large relaxation are not clear, although it seems invariably to arise when the ground-state calculation places the predominantly metal MO's more stable than some of the ligand MO's.

Evans et al.[59] have attributed large relaxation effects to localized orbitals; the larger the localization the larger the relaxation upon ionization. This rationale is intuitively appealing and has also been employed by Heilbronner and co-workers[60] in interpreting the UPS of alkylfulvenes. Based on the common occurrence of large relaxation effects for metal localized orbitals, the ground ionic states of $Mn(CO)_5H$, $Mn(CO)_5CH_3$, and $Mn(CO)_5Cl$ are probably ^2E and not $^2A_1$ as predicted[61] by application of KT (see Section V-F). It seems that transition metal complexes containing "non-innocent"[62] or "soft" ligands exhibit the largest relaxation effects upon ionization.

Since the advent of photoelectron spectroscopy there has been renewed interest in improvements on KT. Initial work involved a calculation on the molecular ground state and a separate calculation on each of the observed ionic states. Subtraction of the total energy for the ground state and the ion then yields the desired IP. Such a procedure, ΔSCF, takes account of relaxation but makes no correction for the correlation energy. This fact, the inherent danger involved in subtracting two large energies, and the large amount of computer time required to do separate calculations on each state of the ion make the ΔSCF procedure unsatisfactory for large molecules.

Numerous workers have employed the ΔSCF procedure to obtain better prediction of IP. For N_2,[53] the calculated order of electronic states still does not agree with the experimental order, unless the basis functions are re-optimized for the ion.[63] For the transition metal complexes discussed above,[56–59] the ΔSCF method places the ionizations of metal and ligand electrons in the proper order, which is not achieved by KT.

Once relaxation has been taken into account, the only remaining effect is the correlation energy. This can be corrected for by including configuration interaction (CI). Inclusion of ΔSCF and CI is enormously time consuming for ab initio SCF calculations on large molecules. However, it has been applied to relatively large molecules using the semi-empirical INDO method including first-order CI(ΔINDO+FOCI).[64] As Chong et al. have stated,[64] this method gives no better results than KT for most closed-shell molecules, but may possibly be useful for open-shell molecules such as NF_2 and NO_2, where KT does not apply.[65] In cases where KT gave incorrect assignments (e.g. F_2O and HOF) the assignment was also incorrect by ΔINDO + FOCI, as shown by the same authors in later work[66, 67] employing Rayleigh–Schrödinger perturbation theory.

Recently, several alternatives to direct application of KT and the ΔSCF method have been put forward. One method involves an entirely different type of calculation: the Hartree–Fock–Slater (HFS) method.[68, 69] This method has several variations, but in all of them it is possible to carry out calculations on both the ground state of a molecule and on a "transition state" which corrects for relaxation effects.

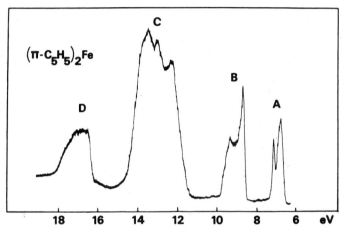

Fig. 4. The He I photoelectron spectrum of ferrocene. Reproduced from Evans et al.[48] with permission and by courtesy of Elsevier Publishing Company, Amsterdam.

The advantage of the transition state is that one needs to do only a single calculation on each ionic state without the problem of subtracting two large energies as in the ΔSCF method. The HFS eigenvalues themselves cannot be directly related to IP as in Hartree–Fock SCF since KT is not applicable to these eigenvalues.[68] Nonetheless, it has been empirically observed that the ground-state eigenvalues normally do not change in order compared to the transition state so that they are often compared directly with IP.[70, 71]

A comparison of HFS, KT, and ΔSCF methods is provided by calculations on ferrocene. The experimental spectrum is shown in Fig. 4 and the calculated results in Fig. 5. Treating ferrocene as Fe^{2+} and $C_5H_5^-$, we consider first the formally d^6 electrons for Fe which reside in the $e_{2g}(3d_{xz, yz})$ and $a_{1g}(3d_{z^2})$ orbitals of the molecule. As shown in the first column of Fig. 5, the eigenvalues of an SCF calculation[56] place these orbitals more stable than the ligand orbitals $e_{1u}(\pi)$ and $e_{1g}(\pi)$. This situation is rectified by carrying out calculations on the individual states of the ion (ΔSCF) in which case ionization from the mainly metal orbitals requires less energy than from the ligand orbitals, as one might expect intuitively (column 2, Fig. 5). The calculated ordering of states is $IP(e_{2g}) < IP(a_{1g}) < IP(e_{1u}) < IP(e_{1g})$. This computed ordering of the mainly metal e_{2g} and a_{1g} orbitals is in agreement with the original[72] and subsequent[73, 74] interpretations of the UPS of ferrocene based upon the observed[74] intensity ratio of 5 : 2 for the first and second peaks of band A, electron occupancy alone predicting a ratio of 2 : 1. Whereas the ΔSCF calculation appears to produce the correct *order* of the top two states, the calculated *splitting* of 1·8 eV is far more than the experimental splitting of 0·3 eV (column 3, Fig. 5).

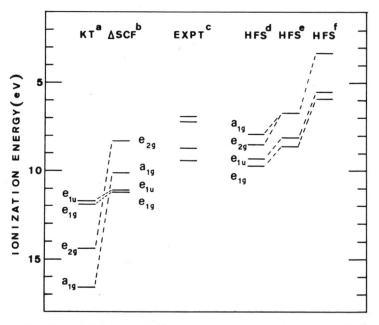

Fig. 5. Correlation of the low-lying e_{1u}, e_{1g}, e_{2g}, and a_{1g} energy levels and ionic states for ferrocene. Results obtained by: (a) Koopmans' Theorem,[56] (b) Δ SCF procedure,[56] (c) Experiment,[73, 74] (d) Hartree–Fock–Slater calculation on the transition state according to Rösch and Johnson,[75] (e) Hartree–Fock–Slater calculation on the transition state according to Baerends and Ros,[76] (f) Hartree–Fock–Slater calculation on the ground state according to Baerends and Ros.[76]

Two groups of workers have carried out HFS calculations on ferrocene. The results of Rösch and Johnson[75] employing the "transition state" show good agreement between calculated and observed bands but with the ordering of e_{2g} and a_{1g} just inverted from the ΔSCF calculations. Rösch and Johnson claim that $IP(a_{1g}) < IP(e_{2g})$ and that intensities should not be relied upon since the $3d_z^2(a_{1g})$ orbital is less involved in bonding than $3d_{xz,yz}(e_{2g})$, and $3d_{z^2}$ is also mixed with Fe $4s$ character. However, as noted by Evans *et al.*[74] the relative intensity of the first two states (e_{2g} and a_{1g}) to the second two states (e_{1u} and e_{1g}) is 1 : 2, whereas the occupation ratio is only 6 : 8 or 1 : 1·3. One would then expect that stronger mixing of $3d_{xz, yz}$ with ligand orbitals would tend to *increase* the e_{2g} intensity even more, not decrease it as argued by Rösch and Johnson.[75] In fact, an increase in the expected e_{2g} intensity was mentioned above in that the experimental ratio of the first two states is 5 : 2, compared to the expected 2 : 1 if one adopts the assignment $IP(e_{2g}) < IP(a_{1g})$. Finally, the HFS results of Baerends and Ros[76] are presented for both the transition-state and ground-state calculation in the last two columns of Fig. 5. Their results predict the a_{1g} and e_{2g} orbitals to be degenerate in both calculations.

Other experimental results which pertain to the a_{1g}, e_{2g} assignment controversy are the *d–d* electronic spectrum, the electron spin resonance (ESR) spectrum, and the electronic Raman spectrum of the ferricenium cation. The *d–d* electronic spectrum requires the ordering ... $(e_{2g})^4 (a_{1g})^2$. This is not in disagreement with the opposite ordering obtained from UPS since the one-electron energy is defined *differently* in the SCF and ligand field theories.[74, 77] The ligand field theory does not include electron repulsion terms in the definition of orbital energy whereas these are included in the SCF definition. For the d^6 ferrocene system the relationship is

$$\varepsilon^{SCF}(a_{1g} - e_{2g}) = \varepsilon^{LF}(a_{1g} - e_{2g}) - 20B$$

where ε^{LF} derived from the ligand field treatment is 7100 cm^{-1} and $B = $ 390 cm^{-1}. Consequently, $\varepsilon^{SCF}(a_{1g} - e_{2g}) = -700$ cm^{-1}, which order is in agreement with the experimental assignment and the ΔSCF calculations.

Both the ESR[78] and the electronic Raman[79] spectrum of the ferricenium cation can be interpreted only in terms of a $^2E_{2g}$ ground state, in disagreement with the contention of Rösch and Johnson.[75]

To sum up the controversy surrounding the ferrocene assignment, the experimental results and the ΔSCF calculation all favor the assignment $IP(e_{2g}) < IP(a_{1g})$. The author feels that, at present, the available evidence leaves the original assignment unchanged in spite of the calculations of Rösch and Johnson. The calculations lack absolute authority since the two HFS calculations[75, 76] do not agree on the assignment of Rösch and Johnson and a thorough study of calculational sensitivity to variation in the exchange parameter α has not been underatken.

Aside from the direct calculational techniques of Hartree–Fock SCF and HFS discussed above, several other procedures have been employed to improve on KT. These methods require an SCF calculation to be carried out with subsequent adjustment of the orbital energy. The types of adjustment include utilization of Green's functions,[80] Rayleigh–Schrödinger perturbation theory,[66] and the Heisenberg equation of motion (EOM) theory.[81] So far, these methods have only been applied to relatively small molecules. It remains to be seen whether these techniques will be useful in resolving disputes such as the case of ferrocene discussed above.

The group of Cederbaum, Hohlneicher and von Niessen[80] has been active in developing the method of Green's functions and Cederbaum[82] has shown how simple symmetry considerations can be used to rationalize the breakdown of KT for F_2 and N_2. The results obtained by the use of Green's functions for F_2 [80] and N_2 [83] and the EOM theory for N_2 [81] are compared with experiment and KT in Table I. In both of these molecules KT predicts the

TABLE I
Calculated and Experimental Ionization Potentials[a]

	F_2			N_2			
	Expt.[b]	KT[c]	Green's[c] function	Expt[d]	KT[e]	Green's[f] function	EOM[g]
$1\pi_g$	15·83	18·52	15·87	—	—	—	—
$1\pi_u$	18·80	22·31	18·93	16·98	17·10	16·83	17·03
$3\sigma_g$	~21·0	20·21	20·64	15·60	17·36	15·50	15·69
$2\sigma_u$	—	—	—	18·78	20·92	18·59	18·63

[a] In eV.   [b] Ref. 55.   [c] Ref. 80, Koopmans' Theorem.   [d] Ref. 1.
[e] Ref. 53, Koopmans' Theorem.   [f] Ref. 83.
[g] Ref. 81, Equations of motion.

wrong ordering of ionic states with absolute errors as large as 3 eV. The new methods, on the other hand, not only predict the correct ordering of states but are able to calculate the absolute value to within a few tenths of an electron volt. For F_2 the computed result is more accurate than the experiment for ionization from the $3\sigma_g$ orbital.

Further application of the Green's function has been carried out by utilizing CNDO/2 SCF wavefunctions rather than the more expensive and elaborate *ab initio* wavefunctions.[84] Results have been obtained for F_2, CO_2, BF_3, NF_3, cyclopropane, and ethylene oxide. For BF_3 and NF_3 the assignment is identical to that arrived at by Bassett and Lloyd[85] in their experimental studies, except for the inconsequential ordering of $4e$ and $1a_2$ in NF_3 which are nearly degenerate and lie under the same band in the spectrum.

C. Fine Structure

Three types of fine structure are exhibited in UPS: vibrational, spin–orbit, and Jahn–Teller. Theoretically, vibrational fine structure could be observed in any ionic state. For a nondegenerate state, single quanta vibrational excitation involves the symmetric modes only, whereas for a degenerate ionic state nonsymmetric modes are also allowed leading to Jahn–Teller distortion of the molecular ion. Spin–orbit coupling can also occur in degenerate ionic states, since coupling of electron angular momentum with a nonzero orbital angular momentum is required. The presence of spin–orbit or Jahn–Teller effects indicates degeneracy of the ionic state. The vibrational fine structure is not such a definitive assignment criterion; however, a short vibrational progression is indicative of ionization from a nonbonding orbital, a long progression of ionization from a bonding or antibonding orbital (Fig. 2). If the vibrational spacing in the ion is reduced from that in the molecular ground state, a bonding orbital is deduced and vice versa. Hollas and Sutherley[86] have pointed out that precautions should be observed when making deductions from intensity patterns in Franck–Condon envelopes about the bond lengths in the ion and about bonding in the molecules.

Spectra which illustrate the combined effects of vibrational excitation and spin–orbit coupling are the ground ionic states of the cyanogen halides, Fig. 6.[86] It is immediately obvious from the spin–orbit coupling that ionization is taking place from a degenerate π orbital mainly localized in the halogen. As expected, the spin–orbit coupling increases in the order $Cl < Br < I$ and is less than the corresponding value in the hydrogen halides, indicating some delocalization of the halogen p_π electrons.

A good example of the operation of the Jahn–Teller effect is provided by the first UPS band of SnH_4,[47] Fig. 7. In tetrahedral symmetry, the band consists of a single state, $^2T_2$. However, it is obvious from the spectrum that three distinct states are present as a result of Jahn–Teller distortion. Other examples of the Jahn–Teller effect are provided by the first ionic states of P_4,[87–88] OsO_4,[89–91] and BH_3CO.[92]

A detailed analysis of spin–orbit splitting is essential if several heavy atoms surround a central atom, since the combined spin–orbit splittings can yield a plethora of bands. Examples include the spectra of BX_3,[93] PX_3,[40, 94, 95] AsX_3,[96] and SbX_3[96] where X is a halogen atom. Analysis of the spin–orbit splitting can only be carried out within the extended double group of the molecular point group. Spin–orbit coupling effects in molecules can be divided into first-order effects, which cause splitting of degenerate orbitals, and into second-order smaller contributions.

Consider BI_3 where the $^2E'$ and $^2E''$ states transform as $E_{\frac{3}{2}} + E_{\frac{3}{2}}$ and $E_{\frac{3}{2}} + E_{\frac{1}{2}}$, respectively, in the extended point group. Yet, only the $^2E'$ state exhibits a first-order splitting; the $^2E''$ state is split only by the interaction of the two $E_{\frac{3}{2}}$ states (second-order effect). The presence of a first-order splitting in the $^2E'$ state but lack of it in $^2E''$ can be intuitively understood by reference to Fig. 8. The degenerate pair of e' molecular orbitals contains a contribution from a *different* atomic orbital on each iodine atom. This situation is analogous to the

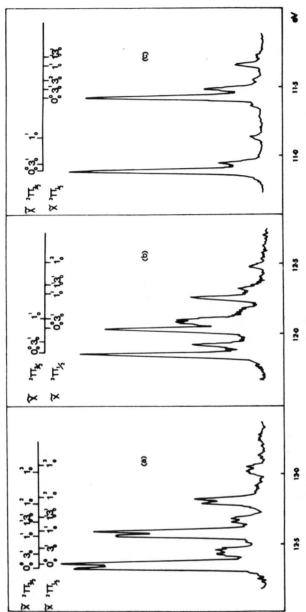

Fig. 6. Photoelectron spectra showing the first two band systems associated with the ionization producing the $\tilde{X}\,^2\Pi_{\frac{3}{2}}$ and $\tilde{X}\,^2\Pi_{\frac{1}{2}}$ components of the ground states of (a) ClCN+, (b) BrCN+, and (c) ICN+. The notation n_β^α refers to the nth vibrational mode arising in the β electronic state with α vibrational quanta. Reproduced from Hollas and Sutherley[86] with permission and by courtesy of Taylor and Francis Ltd., London.

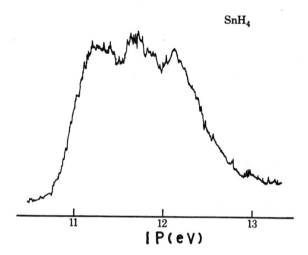

Fig. 7. The first band in the photoelectron spectrum of stannane, showing the three intensity maxima produced by the Jahn–Teller effect. Reproduced from Potts and Price[47] with permission of the Royal Society.

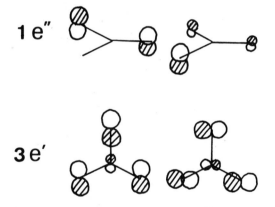

Fig. 8. Orbital symmetry character of the $1e''$ and $3e'$ orbitals in trigonal planar BX_3. The e'' orbitals are perpendicular to the plane of the molecule. Adapted from Bassett and Lloyd with permission[104] and by courtesy of the Chemical Society, London.

degenerate pair p_x and p_y in the $^2\Pi$ state of HI where spin–orbit interaction produces $^2\Pi_{\frac{3}{2}}$ and $^2\Pi_{\frac{1}{2}}$ states. However, the degenerate pair of e'' molecular orbitals contains contributions from the *same* atomic orbital on each iodine atom. The net result of the first- and second-order splitting is shown in Fig. 9 for BI_3.

A somewhat different example of spin–orbit splitting is provided by XeF_2 where the $^2\Pi_g$ state exhibits no splitting while the $^2\Pi_u$ state is split by 0·47 eV.[97]

The heavy atom xenon has no first-order contribution to the π_g orbital while the degenerate pair Xe $5p_x$ and $5p_y$ contribute to the π_u orbital. The magnitude of the splitting can be related to the localization of the appropriate atomic orbitals, and this approach has been investigated by Grimm using CNDO/2 wavefunctions.[98]

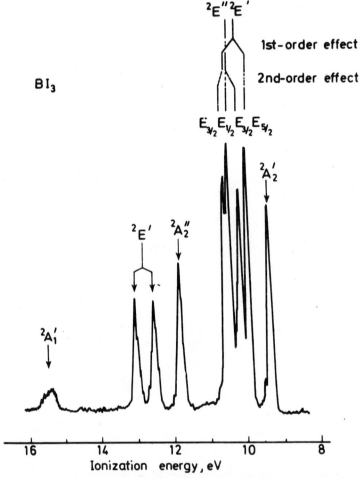

Fig. 9. Photoelectron spectrum of boron triiodide, showing a second-order spin-orbit splitting. Reproduced from Eland, "Photoelectron Spectroscopy", with permission and by courtesy of Butterworths, London.

D. Bandwidth

In the absence of fine structure, bandwidths (measured as full width at half maximum, fwhm) can vary over a range of a few tenths to several eV. A simple rule of thumb states that the narrower the band, the more nonbonding is the orbital from which the electron was ejected. This rule holds qualitatively for molecules which are predominantly covalent, but is just inverted for predominantly ionic molecules as discussed by Berkowitz.[41] Narrow, structureless bands are often observed for halogen lone pairs. One of the narrowest reported is that of the F p_π lone pair[67] at 16·0 eV in the HOF spectrum[99] with a fwhm of 0·2 eV. Other than this example and HF,[100] narrow bands have not been observed for fluorine lone pairs, indicating considerable mixing of these orbitals with other orbitals in the molecule. This is not unexpected since the F atomic IP is 17·42 eV and a typical value in molecules is 15·8 eV.[101] Hence, in many molecules the fluorine lone pairs are energetically close to the other orbitals, in contrast to the other halogens whose lone pair IP are usually the first IP in a molecule.

E. Symmetry Considerations

With the application of group theoretical principles[102] one can readily derive the number and degeneracy of the occupied molecular orbitals. Consideration of the averaged valence state ionization energies for atoms[29] allows one to estimate the number of orbitals expected in a given energy region, e.g. 0–20 eV or 0–40 eV. This approach is especially useful for molecules containing a threefold or higher axis of symmetry, since the resultant degeneracies simplify the spectrum considerably.

As an example consider the 0–30 eV range for BF_3. We can neglect the fluorine $2s$ electrons since they are expected to ionize near 40 eV.[29] The remaining eighteen valence electrons should occupy the orbitals $a_2' + 2e' + e''$ $+ a_2'' + a_1'$ according to the application of group theory (D_{3h} symmetry) and nodal arguments. Indeed, the UPS of BF_3 does exhibit six bands between 16 eV and 22 eV, one for each orbital concerned.[103, 104] The actual ordering of these orbitals must then be decided on the basis of spectral features (e.g. relative intensity, fine structure), optical spectroscopy of the cation (if known), and/or molecular orbital calculations.

The ideal situation of one band per orbital as illustrated in Fig. 1 and mentioned for BF_3 does not occur often for large molecules or for small molecules of low symmetry. For XeF_6 the 38 valence electrons (excluding fluorine $2s$) are expected to occupy the orbitals $2a_{1g} + 2t_{1u} + t_{2g} + t_{2u} + t_{1g} + e_g$ (assuming O_h symmetry). One might then hope to see eight bands in the UPS while the observed spectrum exhibits only three bands (albeit one of them has three peaks superimposed on it).[97] There must be overlapping of states in XeF_6, not altogether surprising when one considers that many of these molecular orbitals consist mainly of fluorine $2p$ character and consequently are expected to have energies which are nearly degenerate.

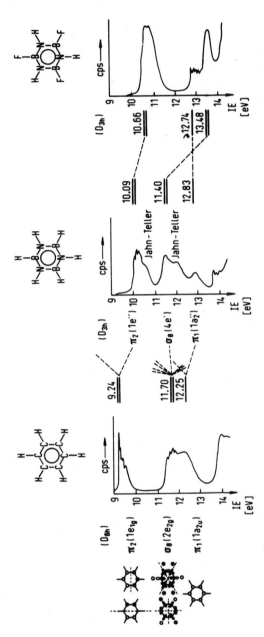

Fig. 10. Photoelectron spectra in the region 9–14 eV of benzene, borazine, and B-trifluoroborazine with band assignments. Reproduced from Bock and Ramsey[11] with permission and by courtesy of Verlag Chemie, GmbH, W. Germany.

Thiazyl chloride, NSCl, is an example of a small molecule of low symmetry (C_s). The 18 valence electrons are expected to occur in nine molecular orbitals ($7a' + 2a''$), whereas only four bands are observed before 20 eV in the He I UPS. Exclusion of the s valence electrons on each atom leaves six molecular orbitals. It is easy to imagine two of the chlorine lone pairs being nearly degenerate in the bent molecule since they would be strictly degenerate if the molecule were linear. One electronic state still remains unaccounted for in the assigned spectrum.[105]

Of course, symmetry effects also can be utilized to ascertain the correct order of ionic states, within the confines of KT. This approach has been taken in correlating the UPS of benzene, borazine, and B-trifluoroborazine (Fig. 10). Bock and co-workers[11, 106] reason that the $1e''$ MO of borazine should be more stable than that of benzene since the symmetry lowering now allows an interaction of $1e''$ in borazine with an empty orbital of the same symmetry, likewise for $1a_2''$. On the other hand, $4e'$ will have interactions with both lower and higher e' orbitals to leave it unshifted. The order of orbital energies arrived at for borazine is then $\pi_2(1e'') > \sigma_8(4e') > \pi_1(1a_2'')$.

On fluorine substitution in B-trifluoroborazine, the σ level is stabilized by 2·08 eV so that the order now becomes $\pi > \pi > \sigma$. This order is further substantiated by the relative intensities (Fig. 10).

A final application of symmetry effects is provided by the spectrum of ferrocene (Fig. 4). In section IV-B we discussed the first band system (A) extensively. We wish now to show how one can assign the second band system (B), which is obviously split into two states. The metal d^6 electrons occurred in the first band system; application of simple Hückel theory predicts that the next orbitals should consist of the ligand π nonbonding set. Two of these combinations are shown in Fig. 11. Consideration of ligand–ligand interaction would predict $IP(e_{1g}) < IP(e_{1u})$, the assigned[74] order in $Mg(C_5H_5)_2$, where e_{1u} is further stabilized by interaction with $3p_x$ and $3p_y$ orbitals on magnesium. However, in ferrocene it is the e_{1g} orbital which is stabilized with the Fe $3d_{xz}$ and $3d_{yz}$ orbitals so that the order becomes $IP(e_{1g}) > IP(e_{1u})$. This assignment finds experimental justification in the greater breadth of the e_{1g} component, suggesting that this orbital has substantial metal–ligand bonding character.[74] The calculations on ferrocene also agree on this assignment.[56, 75, 76]

F. Substituent Effects

The most widely used substituent effect is the so-called *perfluoro effect*[107, 108] defined as "the substitution of fluorine for hydrogen in a planar molecular (sic) has a much larger stabilizing effect on the σ MO's than on the π MO's" The magnitude of the shifts are 0–1·5 eV for the π electrons and 1·5–4 eV for the σ electrons. Its application to inorganic chemistry includes a comparison

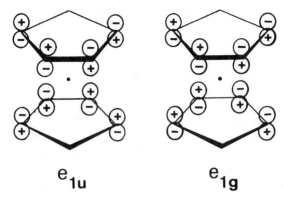

$$e_{1u} \qquad e_{1g}$$

Fig. 11. Orbital symmetry character for one component of the e_{1u} and e_{1g} orbitals in a staggered metal sandwich compound (e.g. ferrocene) showing contributions from the ring orbitals only.

of borazine with B-trifluoroborazine (Fig. 10) and water with oxygen difluoride (Fig. 12). In each of these four molecules, the highest occupied MO is shown to be of π symmetry since substitution by fluorine results in a shift of the first UPS band by only a few tenths of an eV, whereas the σ levels were shifted by 2 eV or more.[107] The perfluoro effect is better applied to the small shift in π MO's than to large shifts in σ MO's. This is because the fluorine π level occurs at about 16 eV and is not too intermixed with the highest occupied π levels of the molecular framework, since the latter generally occur at 13 eV or less. The σ levels, however, become intermixed with those of the molecular framework making the correlation less unique than in the case of the π levels. This is illustrated in Fig. 12 for the pair $H_2O–F_2O$, where the σ levels in H_2O ($1b_1, 3a_1$) are correlated with $3b_1$ and $5a_1$ in F_2O rather than $4b_1$ and $6a_1$. This correlation was based on a population analysis of the orbitals.[107]

If one applies the perfluoro effect to the small shift in the π levels only, it need not be called the "perfluoro" effect but can be simply dubbed the "fluoro" effect. For example, the first IP of H_2O is 12·61 eV,[1] that of HOF is 13·00 eV,[99] and of F_2O is 13·25 eV,[107] indicating that in all three molecules the highest occupied MO is of π symmetry.

From a theoretical viewpoint the perfluoro effect is not well represented by KT. For the above three molecules H_2O, HOF, and F_2O, the computed first IP is 13·77 eV,[66] 15·73 eV,[67] and 16·51 eV,[67] respectively. Whereas the computed results predict shifts of 1 eV or more on substitution of each fluorine, the experimental IP shifts are only 0·3–0·4 eV. Indeed, to interpret the entire UPS spectra of HOF and F_2O, it has been necessary to abandon KT.[66, 67] In an attempt to understand this discrepancy, Chong et al.[64] carried

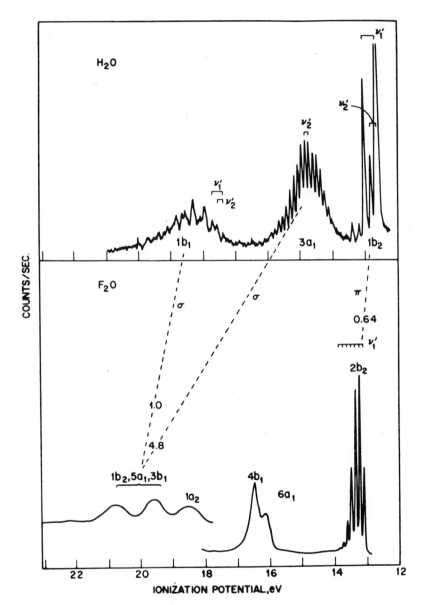

Fig. 12. The photoelectron spectra of water and oxygen difluoride. In the latter, the spectrum above 18 eV was determined using He II excitation. Reproduced from Brundle *et al. J. Amer. Chem. Soc.* **94**, 1451. Copyright 1972 by the American Chemical Society. Reprinted with permission of the copyright owner.

out ΔINDO + FOCI calculations on H_2O and F_2O but achieved no better results than KT as regards the trend in first IP. They conclude, "there are substantial reorganization and correlation effects and a fuller study should be made." Intuitively, one can understand the lack of shift in π levels as being due to the stabilizing effect of charge withdrawal and the destabilizing effect of the antibonding mixing by the fluorine π orbitals with the π orbitals in the molecular framework: the familiar inductive and mesomeric effects.[101] Further examples of molecules illustrating the fluoro effect are HF : F_2 (16·05[100] : 15·8[55] eV), CH_4 : CH_3F (14·0[109] : 13·1[109] eV), and HCl : ClF : ClF_3 (12·78[100] : 12·91[31] : 12·88[31] eV), where the first vertical IP is given in parentheses with spin–orbit states averaged for HCl and ClF. The importance of the mesomeric effect is immediately obvious; in some cases substitution of fluorine for hydrogen results in a *lowering* of IP, the most dramatic effect occurring in the first IP of CH_3F compared to CH_4. Although CH_3F is not a planar molecule, the fluoro effect is operating here in the highest occupied MO (2e) which has the same symmetry properties as π_g in F_2.

The importance of the mesomeric effect of fluorine is observed in more general situations than replacement of H by F. Consider the spin–orbit averaged first IP of the series Xe : XeF_2 : XeF_4 : XeF_6 (12·56[1] : 12·65[110] : 13·06[97] : 12·51[97] eV). The shift in first IP with fluorination is negligible whereas the shift in Xe $3d_{\frac{5}{2}}$ core electrons relative to free Xe varies from 2·87 eV for XeF_2 to 7·07 eV for XeF_6.[111]

Betteridge and Thompson[112] have written an article entitled "Interpretation of ultraviolet photoelectron spectra by simplified methods" in which they present five rules for interpreting spectra. These rules involve the "fingerprint" idea along with substituent effects, symmetry considerations, and orbital degeneracy. If fingerprinting proves useful, it would be a great help in the analytical applications of UPS.[7, 40, 113] The fingerprint idea is presented in Fig. 13 and illustrates the limited usefulness of this idea in UPS for inorganic systems. In fact, the IP span larger regions than indicated in Fig. 13. Consider the O_{2p} orbital: its IP can occur from a low of 8·25 eV in $(CH_3)_3NO$[114] to a high of 14·11 eV in F_3NO.[85] Although the fingerprint idea may be of use in organic chemistry it is unlikely to be useful for inorganic, where charge effects can be very important. Thiazyl chloride, NSCl, presents a case in point; four bands are observed between 10·96 eV and 14·46 eV.[105] Figure 13 shows that the fingerprint idea is of no help in interpreting the spectrum. This is because only the chlorine lone pairs are reasonably localized, contributing to the second band in the UPS spectrum at 11·80 eV.

A useful and intuitively appealing approach to IP trends has been given by Baker et al.[115] who observed a straight-line plot of IP(HX) versus the electronegativity of X where X = F, Cl, Br, and I. The second IP of HF was predicted to occur at about 20 eV before it was actually observed with confidence.[116, 117] Such linear correlations are not unexpected since according to the Mulliken definition,[118] electronegativity is nothing but a suitable average of IP and electron affinity. The electron affinity of the halogen atoms varies only slightly,[119] so that one is really plotting IP(HX) versus IP(X). Lempka et al.[100] observed a straight-line correlation between IP(HX) and IP of the isoelectronic rare gas and predicted the $^2\Sigma$ state of HF$^+$ to occur at 20 eV.

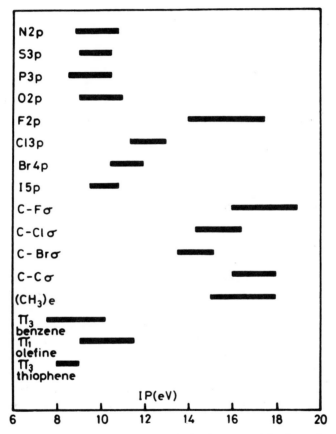

Fig. 13. Some typical ranges of orbital ionization potentials. Reproduced from Betteridge and Thompson[112] with permission and by courtesy of Elsevier Publishing Company, Amsterdam.

Aside from correlations of IP with electronegativity, Baybutt et al.[120] have shown that the first IP of the hydrogen halides exhibits a linear correlation with the corresponding dipole moments. Finally, substituent effects in the first IP of the copper acetylacetonates have been investigated and found to be related to the Taft σ constants.[121]

G. Relative Intensity

The discussion in Section II-B illustrated that for orbitals with similar localization properties the relative intensity should be proportional to the degeneracy or electron occupancy of the orbitals. The first two ionic states of each of the molecules $Ni(CO)_4$, $Fe(CO)_5$, and $Fe(C_5H_5)_2$ should satisfy the orbital localization criterion since in each case ionization occurs from a predominantly metal d orbital. The order of states and expected intensities are:

$$Ni(CO)_4^+ \qquad {}^2T_2 < {}^2E \qquad 3:2$$

$$Fe(CO)_5^+ \qquad {}^2E' < {}^2E'' \qquad 1:1$$

$$Fe(C_5H_5)_2^+ \qquad {}^2E_{2g} < {}^2A_{1g} \qquad 2:1$$

The predicted intensities are borne out experimentally to within 10–20%.[58, 74, 122] One should not be too exhuberant, however, since in $Fe(PF_3)_5$ where one expects the first two ionic states to be similar to $Fe(CO)_5$, it is stated that "the observed intensities deviate considerably from the expected 1 : 1 pattern . . . ".[123]

Another example where the expected degeneracy is not observed is the He I spectrum of BF_3.[93, 104] The ionic states occur in the order[84, 104]

$${}^2A_2' < {}^2E'' < {}^2E' < {}^2A_2'' < {}^2E' < {}^2A_1'$$

and one predicts the first three bands to fall in the ratio 1 : 2 : 2. The observed intensity ratio, however, indicates that the second band is the weakest. Use of the He II photon source results in some improvement of the observed intensities relative to the expected.[23] This points out the necessity of using more than one photon source before making assignments based on intensity alone, since band intensities can be inordinately enhanced in a spectrum if their energies lie near Rydberg bands of the molecule.[23]

Even more dangerous than making assignments based on relative intensities of bands within a given molecule is the comparison of relative intensities between related molecules. However, in some cases it can still prove useful as shown in Fig. 14, where the UPS of the isoelectronic molecules $XeOF_4$[124, 125] and IF_5[126] are displayed. The spectra appear very similar in number and position of bands except that the first band in $XeOF_4$ is considerably more intense than that in IF_5 relative to the remainder of the spectrum. This is expected if ionization of the oxygen p_π lone pairs occurs in the first band along with ionization of the lone pair on xenon. If one normalizes all of the bands between 14·5 eV and 20 eV to the same intensity for the two molecules, one would then predict a relative intensity of 1 : 3 for the first band; the experimental ratio is 1 : 3·5 for IF_5 : $XeOF_4$.

The variation of relative intensity with incident photon wavelength was also discussed in Section II-B. It was concluded that He I photons should favor diffuse orbitals and He II photons contracted orbitals. These qualitative ideas are borne out in the He I and He II UPS of the xenon fluorides;

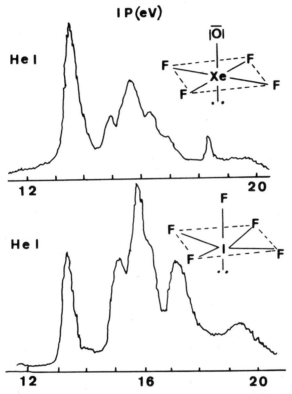

Fig. 14. The He I photoelectron spectra of xenon oxide tetrafluoride and iodine pentafluoride. Adapted from Brundle and Jones[124] and DeKock et al.[126] with permission and by courtesy of Elsevier Publishing Company, Amsterdam.

representative spectra for XeF_4 are shown in Fig. 15. It is clear that the first two bands must correspond to ionization from orbitals containing large contributions of the diffuse Xe $5s$ and $5p$ orbitals whereas the remainder of the bands consist mainly of the relatively contracted fluorine $2p$ orbitals.

Orbital size effects are also exhibited by the He I UPS of BrF_5 and IF_5, Fig. 16. In each case the first band is assigned to the nondegenerate lone pair on the central atom. Normalization of the bands between 14·5 eV and 20 eV results in an experimental ratio of nearly 1 : 2 for BrF_5 : IF_5. The observed intensities support the assignment and show that orbital size effects can be as important as the degeneracy effect in affecting relative intensities.

A common problem in UPS is the low cross-section of ionization from orbitals containing s atomic orbital character. For most molecules the s-type orbital is too tightly bound to be ionized by He I photons. In SiH_4, GeH_4, and SnH_4, however, both the $^2T_2$ and $^2A_1$ states are observable by He I. For

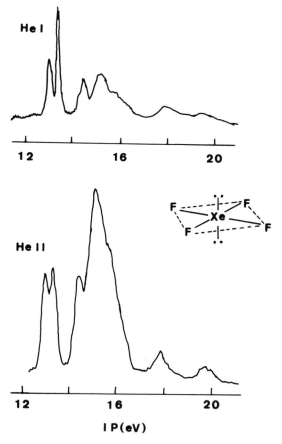

Fig. 15. The He I and He II photoelectron spectra of xenon tetrafluoride showing the decrease in relative intensity of the first two bands which are assigned to the predominantly xenon lone-pair orbitals. Adapted from Brundle *et al. J. Chem. Phys.* **55**, 1098 (1971) with permission of the authors and the American Institute of Physics.

SiH_4, the $^2\bar{A}_1$ band has only 0·6% of the intensity of the $^2T_2$ band.[127] Simple degeneracy arguments predict 33% intensity for the band assigned to $^2A_1$. Potts and Price[47] have attributed this weak intensity to the small size of s orbitals resulting in a low photoionization cross-section. If this is the case, the band should be enhanced considerably under He II radiation. Unfortunately, the intensity ratio using the He II photon source has not been published.

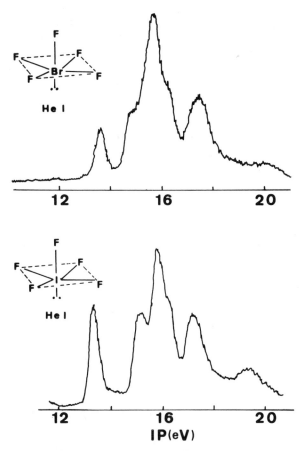

Fig. 16. The He I photoelectron spectra of bromine pentafluoride and iodine penta-fluoride. Adapted from DeKock, Higginson, and Lloyd[126] with permission and by courtesy of the Chemical Society, London.

Further understanding of the low cross-section of these s type bands is provided by the theoretical and experimental work of Schweig and co-workers.[128–130] Their results for PH_3 are presented in Table II. For this molecule the P $3p_z$ orbital can mix with the P $3s$ in the a_1 MO's. However, both a_1 MO's and particularly the one at higher binding energy (19·0 eV) contain large $3s$ contributions. Experimentally, this a_1 band has less than 15% of the intensity of the e band. Schweig and co-workers[128–130] have calculated the relative cross-sections by partitioning the total cross-section into one-center and two-center terms. They find that for lone pairs and other localized MO's, the one-center terms are by far the most important. Figure 17 exhibits typical one-center contributions for P $3s$ and $3p$ orbitals. It is clear that the *s*-type orbital has a small intensity due to the rapid fall-off in photoionization cross section of the 3*s* orbital, compared to the 3*p*, with increasing kinetic energy of the free photoelectron.

TABLE II

Vertical Ionization Potentials and Relative Band
Intensities for Phosphine[a]

| | | Relative intensity[b] | |
MO	IP (eV)	Calc.	Expt.
a_1	10·60	23	39 ± 1
e	13·60	100	100
a_1	19·0	10	< 15

[a] Ref. 128.   [b] He II photon source.

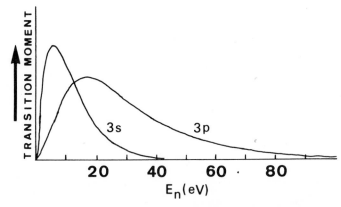

Fig. 17. One-center transition moment of the phosphorus 3*s* and 3*p* basis AO's as a function of the kinetic energy E_n of the free photoelectron ($E_n = E_{ph} - IP_n$). Adapted from Dechant *et al.*[128] with permission and by courtesy of Verlag Chemie, GmbH, W. Germany.

H. Sum Rule

Potts *et al.*[27] were the first to apply the sum rule which can be stated as follows: the total sum of eigenvalues of a secular equation is mathematically equal to its diagonal sum and therefore, assuming Koopmans' Theorem, the total orbital energies should be unchanged before and after taking orbital interaction into account. One can arbitrarily divide the MO's into two subsets, those containing predominantly *s* character on the one hand and *p* character on the other. The sum rule is then best applied to the *s*-type MO's since they form a completely occupied set whereas the p-type are incomplete because not all of the antibonding MO's are occupied in a minimal basis set. The sum rule as employed by Potts et $al.$[27] was not used as an assignment criterion but unexpected deviations in the resultant sum were ascribed to multiple bonding and d_π bonding.

Kimura and co-workers[131] have applied the sum rule to the incomplete p-type MO's observed in He I spectra of hydrogen peroxide and hydrazine.[132, 133] The UPS of H_2O_2 exhibits four peaks at 11·69 eV, 12·69 eV,

TABLE III

Sum Rule applied to Hydrogen Peroxide[a]

Localized orbitals		Sum rule
$I_n(O) = 12\cdot61$ eV[b]	Species b	Expt. $I_1 + I_4 = 29\cdot0_9$ eV
$I_\sigma(O—O) = 16\cdot10$ eV[c]		Calcd. $I_n + I_\sigma(O—H) = 29\cdot21$ eV
$I_\sigma(O—H) = 16\cdot60$ eV[d]	Species a	Expt. $I_2 + I_3 + (I_5) = 45\cdot42$ eV
		Calcd. $I_n + I_\sigma(O—O) + I_\sigma(O—H)$ $= 45\cdot31$ eV

[a] Ref. 132.
[b] Taken from $^2B_2$ state of H_2O^+.
[c] Chosen so as to reproduce the experimental total sum $I_1 + I_2 + I_3 + I_4 + I_5 = 74\cdot5$ eV.
[d] Taken as the average of $^2B_1$ and $^2A_1$ states of H_2O^+.

15·33 eV, and 17·4 eV. Application of the sum rule, Table III, is made not only to the total sum of IP but also to the sum within each irreducible representation. The calculated and experimental sums agree best if the first IP is taken to be of b symmetry in the C_2 point group. This assignment agrees with the calculated CNDO/2[132] and ab $initio$[134] eigenvalues.

V. DISCUSSION OF SELECTED RESULTS

One is faced with a monumental task in any attempt at choosing which results to discuss and there can be no doubt that personal bias is a factor. The discussion is oriented around the volatile inorganic molecules which are found in Groups III–VIII with boron compounds the main representative of Group III. The less volatile molecules found in Groups I–III are discussed in Chapters 4 and 7 by Price and Berkowitz, respectively. The interhalogen and rare gas compounds of Groups VII and VIII have already been given extensive coverage in Section IV. In this section we classify the molecules according to the group of the central atom and discuss Groups III, IV, and V, VI as two units. This is convenient since groups III and IV do not contain "lone pairs" on the central atom whereas lone pairs are usually present on the central atoms of Groups V and VI, depending on the coordination number.

There is one point that deserves comment concerning the general nature of reported results. Many publications tend to place undue attention on the assignment and interpretation of the spectra with little regard to specifying the complete experimental details. A particularly annoying aspect is that many authors do not report adiabatic IP. In the absence of vibrational fine structure the adiabatic IP is not well defined; in that case the onset of photoelectrons should be reported. Although this onset may not be important for the assignment or interpretation of spectra, its value is very important in other fields of spectroscopy. Two such fields are optical spectroscopy (Section IV-A) and ion cyclotron resonance spectroscopy.[135] One aspect of the latter field is the measurement of proton affinity which requires a knowledge of the *adiabatic* IP.

A. *The Concept of Localized Electron Pairs and the Ionization Process*

Implicit in the discussion on calculation of IP (Section IV-B) is that the interpretation via Koopmans' Theorem must be based on canonical Hartree–Fock orbitals.[109] It is these delocalized orthonormal molecular orbitals which are eigenfunctions of the Fock operator. Whereas localized orbitals may prove extremely useful in understanding ground-state properties of molecules, they are not the most useful starting point for a discussion of the ionization process[136–138] because one cannot derive an analogy of Koopmans' Theorem for the localized orbitals.

This can be understood intuitively by a consideration of the familiar example of methane. In a simplified valence bond description, the four bonds of methane are described as localized sp^3 hybrids. In the ionization process, however, there is no reason why one particular sp^3 hybrid should ionize and not the others, since the photon is interacting with the entire molecule. One can describe the ionization process by use of these sp^3 hybrid orbitals only if their interaction is considered, so that the resulting orbitals are orthogonal. In a localized bond orbital description one obtains $\alpha_{CH} = -16\cdot4$ eV as the energy of an isolated C—H localized orbital and $\beta_{CH} = -2\cdot2$ eV as the interaction parameter between two localized orbitals.[11] The eigenvalues of the 4×4 matrix in a Hückel model are

$$\varepsilon_1 = \alpha + 3\beta = -23\cdot0 \text{ eV}$$

$$\varepsilon_{2,3,4} = \alpha - \beta = -14\cdot2 \text{ eV}$$

where ε_1 corresponds to the orbital with no nodes (a_1 symmetry) and $\varepsilon_{2,3,4}$ have one node (t_2 symmetry). The same result could have been achieved by starting with the atomic orbital basis set on the carbon and hydrogen atoms and forming the LCAO's which again have the symmetry $a_1 + t_2$. The choice of starting basis set, atomic orbitals or localized bond orbitals, is mainly a matter of preference depending upon the system at hand. For methane the interpretation of the UPS is readily achieved by either method. For neopentane, $C(CH_3)_4$, the interpretation is facilitated by considering localized CH_3 symmetry orbitals which are then allowed to interact with the other CH_3 groups and with the central carbon atom.[139]

Further insight into the concept of hybridization and the ionization process is provided by lone-pair ionizations. Sweigart has discussed the Group VI and VII hydrides.[140] Consider H_2O which in the simplified valence bond approach would have two equivalent bond orbitals and two equivalent lone-pair orbitals. The UPS of H_2O exhibits four bands at 12·61 eV, 14·73 eV, 18·55 eV, and 32·2 eV.[107] The assignment of these is

$$IP(1b_1) < IP(3a_1) < IP(1b_2) < IP(2a_1)$$

where the predominant character is $1b_1(O\,p_\pi)$, $3a_1$ (symmetric O—H), $1b_2$(antisymmetric O—H), and $2a_1(O_{2s})$. The interaction between the two localized bonding orbitals has caused them to split by about 4 eV and the lone-pair orbitals $1b_1$ and $2a_1$ are split by about 20 eV. These qualitative conclusions are borne out by the *ab initio* calculations on H_2O.[107] The situation for the isovalent molecule H_2S is not so straightforward as that for H_2O. The S $3s$ orbital enters into the bonding much more than the O $2s$ orbital. This is not surprising since the energy difference between the $2s$ and $2p$ valence orbital ionization potentials of oxygen is 16·5 eV whereas for S the corresponding $3s$ and $3p$ difference is only 9·05 eV.[29] The extent of s–p hybridization in the corresponding $4a_1$ and $5a_1$ valence orbitals is not obvious from the UPS, but calculations[141, 142] on H_2S indicate that these two orbitals cannot be divided into lone-pair (S $3s$) and bonding orbitals as they were for H_2O. The valence electron a_1 matrix is 3×3, containing the basis functions S $3s$, S $3p_x$, and the symmetric combination of H $1s$ functions. The lower two eigenfunctions will be occupied; S $3s$ involvement will certainly stabilize the lowest $(4a_1)$ but one cannot predict *a priori* whether $5a_1$ will be stabilized or destabilized. Intuitive arguments would predict destabilization, since S $3s$ mixes into the highest occupied a_1 orbital $(5a_1)$ in an antibonding fashion.[143]

The effect of s–p hybridization on orbital energies is not immediately obvious in molecules as was just mentioned for H_2S. Considering the central atom alone, s–p mixing will not cause the original s and p orbitals to move energetically closer together. This is readily illustrated by the following simple example. Define α_s and α_p to be the valence orbital ionization potentials of an atom. One can then form the sp hybrids

$$\phi_1 = 1/\sqrt{(2)}\,(s+p)$$
$$\phi_2 = 1/\sqrt{(2)}\,(s-p)$$

and assign each hybrid the energy $H_{11} = H_{22} = (\alpha_s + \alpha_p)/2$. If we now calculate the interaction $H_{12} = \langle \phi_1 | H | \phi_2 \rangle$ we find it to be $\frac{1}{2}(\alpha_s - \alpha_p)$.[137] Solution of the 2×2 matrix to obtain the eigenvalues gives $E_1 = H_{11} + H_{12} = \alpha_s$ and $E_2 = H_{22} - H_{12} = \alpha_p$. That is, the original eigenvalues are returned.

To return to the question of lone-pair ionization, one cannot predict from the Lewis electron dot structure the number of low-energy IP to be expected from lone pairs. If a molecule contains more than one lone pair on the central atom (e.g. H_2O), one of them will invariably belong to the completely symmetric representation (a_1 or σ) and the other to an unsymmetric representation which has one node passing through the central atom (e.g. b_1 or π). The unsymmetric state will have lower IP. Examples of this include (order of binding energy)

$$HF: \quad 1\pi(F\ p) < 3\sigma(H{-}F) < 2\sigma(F_{2s})$$

$$H_2O: \quad 1b_1(O\ p) < 3a_1(O{-}H) < 1b_2(O{-}H) < 2a_1(O_{2s})$$

These simple conclusions do not readily apply in the electron-rich inter-halogen and rare-gas halides. In these molecules the "lone pairs" are really antibonding orbitals and the completely symmetric combination can occur at lower IP (less stable) than the unsymmetric. For example, the three "lone pairs" in XeF_2,[97] can be assigned to the first two ionic states $IP(\pi_u) < IP(\sigma_g)$ and indeed the unsymmetric combination appears first. On the other hand, the two lone pairs for XeF_4 [97] are assigned $IP(a_{1g}) < IP(a_{2u})$ although there is no experimental proof for this assignment. Based on the large error in Koopmans' Theorem for XeF_2 [97] in the σ_g orbital, however, it is possible that the ground ionic state of XeF_4^+ is $^2A_{2u}$ and not $^2A_{1g}$.

B. Lone-pair Interactions in Lewis Acid–Base Adducts

Molecules which may be considered as Lewis acid–base adducts are ideal to study the perturbation of the "lone-pair" orbital on the base

$$A + :B \longrightarrow A:B$$

This approach is most suited when the acid fragment has few valence electrons and the spectrum of the individual fragments is understood. Ideal acids include transition-metal atoms, the oxygen atom, and the isoelectronic borane molecule. We shall be mainly interested in the stabilization of the lone pair upon bond formation and will show that this depends critically upon three factors: (1) the value of the lone-pair IP in the free base; high IP are stabilized less than low IP for the same acid, (2) the localization of the lone pair on the donor atom; the greater the localization the greater the stabilization upon bond formation, and (3) the electron affinity of the acid fragment; a larger electron affinity will correspond to a larger lone-pair stabilization.

First consider the transition metal compounds $Ni(PF_3)_4$, $Pd(PF_3)_4$, and $Pt(PF_3)_4$ where the M—P σ bonds transform as $t_2 + a_1$ in T_d symmetry. Relative to the PF_3 lone pair IP of $12 \cdot 3$ eV, the t_2 orbital is stabilized by $0 \cdot 9$ eV, $1 \cdot 4$ eV, and $2 \cdot 2$ eV for the Ni, Pd, and Pt complexes, respectively.[144] Since the PF_3 lone-pair IP and lone-pair localization are constant, these shifts must be due to increasing electron affinity of the metal atom, which is indeed the case as shown by Nyholm.[145] These shifts can be related to the differing amounts of σ bonding in the complexes although this is not the only factor involved in the overall stability, since it is known that $Pd(PF_3)_4$ is the least thermally stable of the three.

In contrast to the metal trifluorophosphine complexes the metal carbonyls (e.g. $Ni(CO)_4$, $Cr(CO)_6$) show a destabilization or negligible stabilization of the M—C σ bonding combinations (e.g. t_2 for $Ni(CO)_4$ or t_{1u} for $Cr(CO)_6$) relative to the free CO lone pair. The IP of CO is $14 \cdot 01$ eV; t_2 (M—C) for $Ni(CO)_4$ is not definitely assigned but possibly corresponds to the band at $14 \cdot 12$ eV.[58] The corresponding t_{1u} orbital in $Cr(CO)_6$ occurs at $13 \cdot 38$ eV.[146] The lack of σ M—C stabilization in metal carbonyls may be due to the higher IP of CO compared to PF_3 (point (1) above) or differing amounts of charge relaxation upon ionization for free CO compared to the metal carbonyls.

Next we turn to the main group Lewis acid–base adducts which are exemplified by the following compounds (lone-pair stabilization compared to free base in parentheses): $O-PF_3$ ($3 \cdot 4$ eV),[85] H_3B-PF_3 ($0 \cdot 3$ eV),[92] $O-CO$ ($4 \cdot 0$ eV),[1] H_3B-CO ($0 \cdot 1$ eV),[92] H_3B-NH_3 ($3 \cdot 1$ eV),[92] $H_3B-N(CH_3)_3$ ($3 \cdot 1$ eV),[92] and $F_3B-N(CH_3)_3$ ($3 \cdot 7$ eV).[147] Comparing OPF_3 with BH_3PF_3 we see the effect of changing the acid. Although the oxygen atom and BH_3 are isoelectronic, the electron affinity of the oxygen atom must be considerably larger than that of BH_3. (Compare also the molecules OCO and BH_3CO.) Indeed, there is no need to propose significant backbonding from the BH_3 e orbitals to the $CO\pi^*$ orbital or to the phosphorus d orbitals to understand the slight stabilization in BH_3CO and BH_3PF_3. The negligible stabilization is a result of the weak bonds in these adducts. Whereas the dissociation energies of BH_3PF_3 and BH_3CO are about 20 kcal/mole, those for $O-PF_3$ and $O-CO$ (to oxygen ^1S) are over 150 kcal/mole. Certainly the back bonding from the oxygen atom to CO on OCO is greater than the corresponding amount in BH_3CO, yet OCO exhibits a stabilization of $4 \cdot 0$ eV. Also, the fact that H_3B-NH_3 has a much larger lone-pair stabilization than H_3B-CO is not an indication of backbonding in the latter molecule. One would expect a much larger stabilization in H_3B-NH_3 (even though the resulting bond is weak) for two reasons: (1) the lone pair on NH_3 is more localized than the lone pair on CO, and (2) the vertical IP of NH_3 ($10 \cdot 8$ eV) is much less than that of CO ($14 \cdot 0$ eV).

listed above. The lone-pair vertical IP of $N(CH_3)_3$ is 8·45 eV,[92] considerably less than the value of 10·8 eV for NH_3. Both bases, however, exhibit the same lone-pair stabilization (3·1 eV) when bonded to BH_3. The fact that $N(CH_3)_3$ did not exhibit a larger stabilization is probably because its lone pair is more delocalized than the corresponding lone pair of NH_3. The stabilization of 3·7 eV for $F_3B—N(CH_3)_3$ illustrates that BF_3 is a stronger Lewis acid than BH_3.

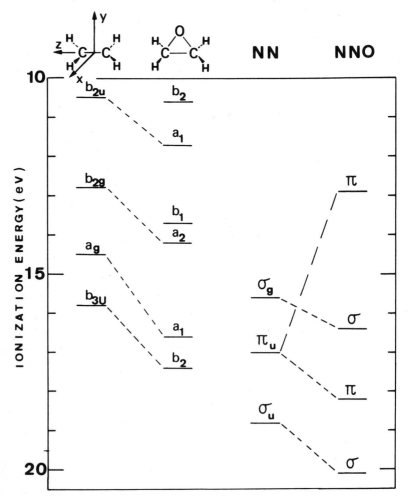

Fig. 18. Correlation diagram for the ionic states of ethylene : ethylene oxide and dinitrogen : nitrous oxide. The Y axis is the two-fold axis for ethylene oxide. The uncorrelated b_1 and b_2 orbitals are predominantly oxygen lone pairs.

In Fig. 18 correlations of ionic states are presented for the acid–base pairs dinitrogen : nitrous oxide[1] and ethylene : ethylene oxide.[148, 149] These correlations are in sharp contrast to the published correlations for numerous acid–base pairs including those discussed earlier. The usual behavior, typified by $PF_3 : OPF_3$, is that the lone pair is stabilized by a much larger amount than the other orbitals in the molecule. For OPF_3 the lone pair undergoes a stabilization of 3·4 eV, whereas the remaining orbitals undergo a stabilization of about 1 eV.[85] The results presented in Fig. 18 indicate a nearly uniform stabilization of all orbitals. For ethylene oxide the constancy of shift is not surprising since the acid approaches the base "side-one" and evidently charge is withdrawn from the entire molecule. The nearly equal stabilization of $3\sigma_g$ and $2\sigma_u$ for $N_2 : N_2O$ can be rationalized as due to their symmetry-required delocalization in N_2.

C. Groups III and IV

As examples of Group III we shall discuss selected boron compounds. The UPS of diborane itself is not particularly interesting but when compared with that of the isoelectronic ethylene molecule the changes in electronic structure become immediately apparent (Fig. 19).[148] As pointed out originally by Pitzer,[150] diborane can be considered as being formed from diprotonation of the π orbital in hypothetical $B_2H_4{}^{2-}$, the latter being isoelectronic and isostructural with C_2H_4. The net effect is that in the conversion $C_2H_4 \rightarrow B_2H_6$ one expects a destabilization of all orbitals derived from ethylene σ orbitals, but a stabilization of the C_2H_4 $\pi(1b_{2u})$ orbital to form the bridge bonds in B_2H_6. These expectations are fully borne out by the UPS and the *ab initio* calculations on the two molecules.[148]

The boron halides have been studied extensively; the spectrum of BI_3 is displayed in Fig. 9, and the assignment of the BF_3 spectrum was discussed in Section IV-G. One interesting aspect of the spectra is the energy difference between the E'' and A_2'' states. Assuming KT, the former corresponds to degenerate nonbonding halogen p_π orbitals (Fig. 8) whereas the latter are p_π bonding with the boron p_π orbital. The stabilization of the A_2'' state relative to the E'' can be qualitatively related to the amount of π stabilization,[104] $\ddot{X}$—B. For the series BI_3, BBr_3, BCl_3, and BF_3 this difference is 1·4 eV, 1·6 eV, 1·8 eV, and 2·05 eV.[11] This order is paralleled by the order of decreasing Lewis acidity $BI_3 \approx BBr_3 > BCl_3 > BF_3$.

The UPS of the boron subhalides B_2F_4,[151] B_2Cl_4,[151] and B_4Cl_4[152] have also been reported. The spectra of the B_2X_4 species are difficult to assign in their entirety due to their low symmetry (D_{2d}) and the large number of ionic states which are accessible. One feature which is obvious, however, is the assignment of the first IP at 13·26 eV in B_2F_4 and 10·97 eV in B_2Cl_4. This band

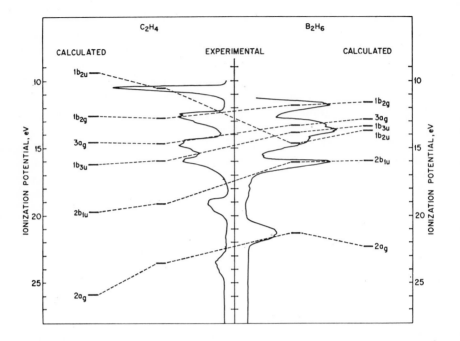

Fig. 19. Comparison of the He II photoelectron spectra of ethylene and diborane with those calculated using Koopmans' Theorem. Reproduced from Brundle *et al. J. Amer. Chem. Soc.* **92**, 3863. Copyright 1970 by the American Chemical Society. Reprinted with permission of the copyright owner.

exhibits a shift of only 2·3 eV between the two molecules whereas the others shift by about 4 eV. The first band must correspond largely to the B—B σ bond in contrast to the other orbitals which are mainly halogen in character.[151]

We turn next to the UPS of the higher symmetry (T_d) B_4Cl_4 molecule (Fig. 20).[152] The first band is assigned to a t_2 orbital which by its symmetry contains one node in each of the degenerate orbitals and is expected to be relatively B—B nonbonding. What is perhaps surprising is that this orbital has an IP of 10·60 eV, only slightly less than the a_1 B—B bonding orbital of B_2Cl_4 at 10·97 eV. The next three intense, sharp peaks are assigned to the $t_1 + e + t_2$ set of mainly Cl p_π orbitals.

One interesting aspect is the large splitting between the t_1 and e orbitals of 0·5 eV which is comparable with the 0·6 eV splitting in $SnCl_4$, although the Cl—Cl distance in B_4Cl_4 is greater than that in $SnCl_4$. Lloyd and Lynaugh pointed out that the most likely explanation involves π back donation of charge from Cl to B.[152] The following two bands between 15–17 eV are assigned to $t_2 + a_1$ orbitals, the a_1 is nonbonding whereas the t_2 contains a sizeable B—Cl bond character.[153] The last band at 19·5 eV is assigned to an a_1 orbital containing mainly boron orbital character; it is this orbital which holds the B_4 tetrahedron together.

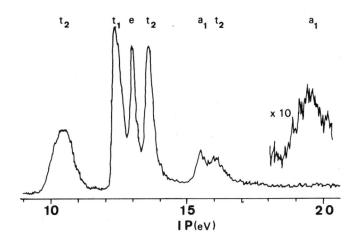

Fig. 20. The He I photoelectron spectrum of B_4Cl_4. Adapted from Lloyd and Lynaugh[152] with permission and by courtesy of the Chemical Society, London.

Most of the UPS work on Group IV compounds involves stable tetra-coordinate molecules although spectra of the high-energy species CH_3[154] and SiF_2[34, 35] have been obtained. The latter clearly exhibits a band due to the "lone pair" on Si at about 11 eV, considerably lower than the onset of ionization from orbitals which are mainly F lone pair at 15 eV. The first vertical IP of the saturated molecule SiF_4 is 16·45 eV.[155, 156] The UPS of $Sn[CH(SiMe_3)_2]_2$ shows two bands at low ionization energy (7·42 eV and 8·33 eV).[157] By way of contrast the first vertical IP of $Sn(CH_3)_4$ is 9·70 eV,[139] of $Sn[CH_2C(CH_3)_3]_4$ is $8·5_8$ eV,[158] and of $Sn[CH_2CSi(CH_3)_3]_4$ is $8·7_1$ eV[158, 159] which in each case is assigned to the Sn—C t_2 bond. The band at 7·42 eV in $Sn[CH(SiMe_3)_2]_2$ must correspond to ionization of the "lone pair" on the Sn(II) atom.

In the case of the tetracoordinate molecules, generally only the valence s and p have been observed by UPS, but Cradock has observed the "core" $3d$ electrons of GeH_4 at 26·9 eV and 27·4 eV by using the He II source.[160] This illustrates the sharp increase in $3d$ orbital energy across the periodic table since the $3d$ ionization of ZnX_2 occurs at ≈ 19 eV.[161]

Extensive work has been done on the high-symmetry species of Group IV, namely MF_4[103, 155, 156, 160, 162, 163] and $M(CH_3)_4$,[139, 162, 164]. For the fluorides the generally agreed orbital energy sequence for the 24 valence electrons access-ible to He I photons is $t_1 > t_2 > e > t_2 > a_1$ which also appears to hold for the isoleptic transition metal halides.[165] The transition metal oxides RuO_4 and OsO_4[166] apparently retain the same sequence except that the high energy t_2 and a_1 orbitals are reversed. This result can be rationalized by noting that the major σ bonding in the main group elements involves central atom s and p orbitals with $s(a_1)$ being more stable than $p(t_2)$. This effect and the fact that the t_2 orbital has one node whereas the a_1 has none always places $t_2 > a_1$. In RuO_4 and OsO_4, however, it is the $s(a_1)$ and $d(e + t_2)$ orbitals which are mainly responsible for the σ bonding, but now the d orbital is more stable than the s so the order $a_1 > t_2$ is not surprising.[166]

The highest occupied molecular orbital (HOMO) for all of the tetrahalides and tetroxides is of t_1 symmetry whereas for $M(CH_3)_4$ the HOMO is assigned to t_2, derived mainly from the M—C σ bond orbitals, rather than t_1 which is derived from the CH_3 e localized orbitals. This is a common occurrence in molecules containing CH_3 and probably occurs in all compounds except CH_3F,[109] where the perfluoro effect of M ($=$ F) has produced a large stabiliza-tion of the C—F σ bond.

Ramsey et al.[14] have studied the UPS of the acyl-silanes and -germanes, $(CH_3)_3MC(O)CH_3$ (M $=$ Si, Ge) and compared the spectra to acetone. The first vertical IP of acetone is a very narrow band at 9·7 eV assigned to the lone pair on oxygen. By way of contrast the first IP of $(CH_3)_3SiC(O)CH_3$ is 8·6 eV and that of $(CH_3)_3GeC(O)CH_3$ is 8·5 eV, lower by about 1 eV relative to acetone. Furthermore, the bands are considerably broadened with a half-width of 0·5 eV. The shift in energy of the first UPS band and its breadth could be due to (1) d_π–p_π bonding, (2) $R_3M \rightarrow$ electron releasing inductive effects, or (3) mixing of the oxygen lone-pair σ orbital with the M—C σ orbital. Point (1) could explain the width of the band but not its shift to lower IP. Inductive effects could rationalize a shift to lower IP but not the increased width of the band. Mixing of σ orbitals within the molecule remains the only viable alternative.[14] Since there is considerable mixing of the highest occupied molecular orbital in the molecule, the first UV transition is more properly assigned as $\sigma \rightarrow \pi^*$ than $n_0 \rightarrow \pi^*$.[14]

Related to the acyl-silanes and -germanes are the carbonyl halides, X_2CO,[167, 168] and the thiocarbonyl halides, X_2CS,[168–172] for which an extensive amount of work has been reported. Wittel et al.[172] have presented a qualitative MO diagram for thiophosgene which seems to typify the results for all of these compounds. The HOMO is b_2, mainly sulfur lone pair in the plane of the molecule and the penultimate orbital is b_1, mainly C—S π bonding but antibonding with the halogen p_π orbitals. It is this antibonding effect which is responsible for the negligible shift of π levels in the "perfluoro" effect (Section IV-F). Indeed, in comparing H_2CS and F_2CS,[169] the b_2 level is stabilized by 1·5 eV compared to a destabilization of 0·1 eV for the $b_1 \pi$ level.

One of the most encouraging aspects of UPS is the ability to relate the IP data to other molecular properties. One remarkable instance has been the work of Arbelot et al.[173] in relating the rates of the following reaction to the IP of some heterocyclic thiocarbonyl compounds.

$$\begin{array}{c} X \\ \diagdown \\ \diagup \\ Y \end{array} C{=}S + CH_3I \xrightarrow{\text{acetone}} \begin{array}{c} X \\ \diagdown \\ \diagup \\ Y \end{array} C^+{-}CH_3 + I^-$$

They found that the rate of reaction was directly related to the first $IP(b_2)$ via Klopman's frontier orbital equation[174]

$$\Delta E = \Delta_{\text{solv}} + \sum_{m \,(\text{occ})} \sum_{n \,(\text{unocc})} \frac{2(Cr^m)^2 (Cs^n)^2 \beta_{rs}}{E_m{}^* - E_n{}^*}$$

where Cr and Cs are the coefficients of atomic orbitals r and s in molecular orbitals m and n. The parameter β_{rs} represents the amount of interaction between the nucleophile (CH_3^+) and donor (sulfur lone pair). The energy of the sulfur lone-pair orbital is $E_m{}^*$ and is measured in UPS. Arbelot et al.[173] chose their system so as to maintain Cr^m constant; that is, the localization of the sulfur lone pair is relatively unchanged. Also, differences in solvation and β_{rs} can be considered as negligible so that Klopman's equation simplifies to

$$E = a + [b/(E_m{}^* - E_n{}^*)]$$

An excellent relationship was found for $a = 21·34$, $b = 52·68$, and $E_n{}^* = -5$ eV. The rate of reaction is shown to decrease by almost three orders of magnitude due to an increase in sulfur lone-pair IP from 7·8 eV to 8·3 eV.[173]

D. Groups V and VI

Elbel et al.[175] have studied the UPS of Me_3X where $Me = CH_3$ and $X = N$, P, As, or Sb and compared the results to H_3X, H_2Y, or Me_2Y where $Y = O$, S, Se, and Te. The main interest involves trends in the first IP's which are presented in Fig. 21. The expected trend that IP should decrease with descent in the periodic table is shown clearly for all series except Me_3X for which the first IP is constant to within 0·2 eV throughout the series. This constancy can be rationalized on the basis of two factors: (1) differing amounts of central atom s orbital character in the highest occupied MO, and (2) mixing of CH_3 group orbitals of a_1 symmetry with the central atom s and p_z orbitals. No s orbital mixing can occur in the series Me_2Y where the first

IP is assigned to a b_1 orbital. Such mixing can occur, however, in Me_3X where the first IP and the s orbital on the central atom both have a_1 symmetry.

Another interesting aspect of Group V and VI molecules is the study of lone-pair interactions on adjacent atoms. Perhaps the simplest example of a "through-bond" interaction[176] is provided by N_2 where the lone pairs on the adjacent atoms are split by 3·2 eV[1] and belong to the representations $3\sigma_g$ and $2\sigma_u$. Such lone-pair interactions have been studied to a large extent in organic

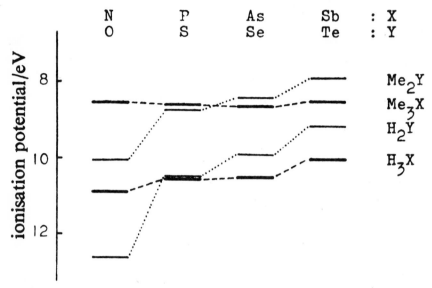

Fig. 21. Trends in the first ionization potential of Group V and VI compounds. Reproduced from Elbel *et al.*[175] with permission and by courtesy of the Chemical Society, London.

chemistry (e.g. pyrazine) where the lone-pair splitting is caused by a combination of through-bond and through-space effects.[177, 178] We shall be interested in examples provided by lone-pair interactions on adjacent atoms in the molecules N_2R_4 and S_2R_2. These species have been studied extensively and the ability to predict the conformation of the molecule based on the splitting of the first two bands in the UPS is very encouraging. Consider first the disulfides which have been studied by several groups.[179–182] The approach of Wagner and Bock[180] was to carry out semi-empirical calculations (extended Hückel and CNDO/2) on $(CH_3)_2S_2$ for the dihedral angle θ varying between $0°$ (s-*cis*-conformation, symmetry C_{2v}) and $180°$ (s-*trans*-conformation, symmetry C_{2h}). Typical results for variation of orbital energy with dihedral angle are shown in Fig. 22. The sulfur lone pairs are the two highest occupied molecular orbitals and their interaction causes a splitting which is maximized in C_{2v} and

C_{2h} symmetry but minimized for the skewed C_2 configuration at 90°. The experimental splitting of the first two bands in R_2S_2 for R = H, CH_3, CH_2CH_3, $CH_2CH_2CH_3$, or $CH(CH_3)_2$ was relatively constant at $\Delta E = 0.25 \pm 0.03$ eV.

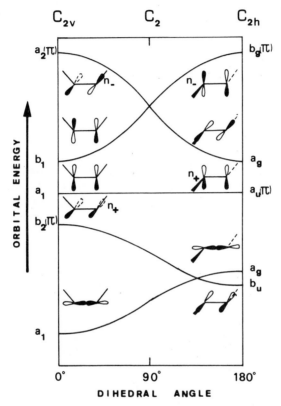

Fig. 22. Qualitative orbital correlation diagram for H_2O_2 and H_2S_2 as a function of the dihedral angle. The two-fold axis is held constant in both C_{2h} and C_{2v}. The correlation diagram is best understood by starting with C_{2v} symmetry and moving the two H atoms in opposite directions simultaneously as one might open a book starting with it in an upright position. According to group theory, more correlation lines can be drawn than those shown in the figure. Also, the diagram does not exhibit possible mixing of orbital character for a_1 symmetry in C_{2v} or a_g symmetry in C_{2h}. Adapted from Wagner and Bock[180] with permission and by courtesy of Verlag Chemie, GmbH, W. Germany, and Batich and Adam, *Tetrahedron Lett.* 1467 (1974) with permission and by courtesy of Pergamon Press Ltd., Oxford.

For R = tert-butyl, however, the observed splitting was 0.65 eV which by interpolation of the calculated splitting for $(CH_3)_2S_2$ predicted $\theta = 110°$, considerably greater than the average value of about 85° for the other disulfides. The largest disulfide splitting (0.95 eV) was observed for 1,2-dithiane, in which the cyclic six-membered ring constrained $\theta \approx 60°$. Based on the calculations for $(CH_3)_2S_2$ the authors would have predicted a splitting of 0·85 eV. Calculations on 1,2-dithiane itself predicted a smaller splitting of 0·42 eV. Consequently, Wagner and Bock[180] conclude that part of this splitting must be due to interaction of the n_+ lone-pair combination with a ring σ orbital.

Baker et al.[181] take an entirely different approach in employing UPS to determine the dihedral angle. They use an equation developed by Bergson

$$\Delta E = \frac{2\gamma |\cos \theta|}{1 - (0·129 |\cos \theta|)^2}$$

where γ is empirically determined from ultraviolet absorption spectra. Their calculated values of θ are in good agreement with those obtained by Wagner and Bock.

A considerable amount of work has also gone into the hydrazines.[183–189] Rademacher[183, 184] has calibrated the calculated MINDO/2 curve of lone pair splitting, $\Delta E = E(n_-) - E(n_+)$ (analogous to Fig. 22) by using the known splittings ΔE and the known conformations of three compounds. He obtained the formula

$$\Delta E = A \cos \theta + B$$

where $A = 2·17$ eV and $B = -0·35$ eV. For the acyclic compounds the results showed little dependence of the dihedral angle on substituent.[187] For example, nine tetralkylhydrazines with Me, Et, iPr, nPr, and nBu substituents showed ΔE values of $-0·51$ eV to $-0·61$ eV for which θ is calculated to be near 90°. For cyclic and bicyclic hydrazines, however, the values of ΔE vary between $+1·57$ eV and $-2·32$ eV for which the derived values of θ are 30° and 164°.[188] For some of the cyclic compounds, two conformations were detected in the UPS.

We should also mention the work of Cowley et al.[190] on the UPS of diphosphines and diarsines. For $(CH_3)_4P_2$, $(CH_3)_4As_2$, and $(CF_3)_4P_2$ they have detected trans and gauche rotamers but only the trans rotamer of $(CF_3)_4As_2$. The trans rotamers invariably exhibit two IP due to lone pairs, split by about 1·5 eV. The gauche rotamers, on the other hand, show only a single band due to both n_+ and n_- which IP corresponds roughly to the average IP of the trans rotamer. By measuring the relative intensities of the bands due to trans and gauche rotamers, they were able to estimate the relative percentage of the two isomers.[190] Although not rigorous, the use of relative intensity in this case is probably justified since the localization of the orbitals is similar in both rotamers.

In summary, UPS is a useful tool for qualitative conformational analysis. Its usefulness in detecting different conformers of the same molecule is directly related to the short time scale of the measurement (Section IV-A).

It is perhaps surprising that the analysis of ΔE is so straightforward, depending as it does on the assumption of Koopmans' Theorem and that the major interaction of the orbitals is through space and not through bond. No doubt the latter effect is partly responsible for the fact that real ΔE curves pass through zero at values other than $\theta = 90°$.

E. Evidence for Outer d Orbital Involvement

One of the aspects of chemical bonding which inorganic chemists find interesting is the possible involvement of outer d orbitals in compounds containing Si, Ge, P, S, Cl, Kr, Xe, or any other element with empty low-lying d orbitals. Unfortunately, outer d orbital involvement in the occupied molecular orbitals is not a direct physical observable and can only be inferred by making certain logical deductions, all of which are based on an assumed validity of Koopmans' Theorem. As with the conclusions drawn from other physical methods, a controversy has developed concerning the importance of d orbitals.

The author feels that the following two statements will suffice to set the tone for the subsequent discussion.

"The question of the d-function participation is to a certain degree a fictitious problem."[191]

"The main concern for the chemist might be not so much whether $3d_{Si}$ participation is real and provable or not but rather which model will be most useful and easily applicable to rationalize experimental data."[192a]

Possible d orbital involvement does not appear to be critical for the interpretation of the UPS of ClF,[31] ClF_3,[31] ClO_2,[193] Cl_2O,[194] KrF_2,[195] XeF_2,[110] XeF_4,[97] XeF_6,[97] $(CH_3)_nSiCl_{4-n}$ ($n = 0-4$),[196] $M(CH_3)_4$ (M = C, Si, Ge, Sn, or Pb),[139, 164] $SiH_3C\equiv CH$,[192] and $SiH_3C\equiv CSiH_3$.[192] Interpretations invoking d orbitals, however, have been employed for vinyl silanes,[197, 198] H_nSiCl_{4-n} ($n = 0-4$),[199, 200] phenylsilane,[201] polyphosphines,[202] silyl and germyl pseudo-halides,[203] silyl and germyl amines, phosphines, and arsines,[204] silyl and germyl derivatives of Group VI elements,[205] $(NPF_2)_n$ ($n = 3-8$),[206] PF_3,[85, 207] PF_3O,[85, 208] PF_3BH_3,[92,] SiF_4,[155, 156] and SF_6.[126] The above lists are not meant to be complete since virtually every UPS publication dealing with an element in Groups IV-VIII mentions the aspect of possible d orbital involvement. As can be seen, no clear trend emerges as to which compounds utilize d orbitals; this points to the pertinence of the above two quotations.

To choose some specific examples, we consider SF_6 first. At least twenty theoretical and experimental treatments have appeared in the past decade on the electronic structure of this molecule. The experimental UPS spectrum cannot be assigned definitively because only one of the bands exhibits fine structure and several of the bands are overlapping.[1, 103] Comparison of the spectrum with that of SF_5Cl is considerably more straightforward if the topmost occupied MO is t_{1g}.[126] The experimental ordering which was deduced is

$$t_{1g} > 3t_{1u} > 2e_g > 1t_{2u} > 1t_{2g} > 2t_{1u} > 2a_{1g} \quad {}^{126}$$

This ordering is in excellent agreement with an HFS calculation[209] employing the "transition state" to correct for relaxation effects. The only difference is that the calculation reverses the order of $2e_g$ and $1t_{2u}$; there was no definitive criteria in the experimental assignment of these two orbitals. Concerning d orbital involvement, the HOMO is consistently calculated to be t_{1g} when S $3d$ orbitals are included, whether using HFS,[209] *ab initio* Hartree–Fock–Roothaan,[210] or semi-empirical calculations.[211] Removal of the d orbitals places $2e_g$ as HOMO. Rösch *et al.*[209] have concluded that inclusion of d orbitals is important to reproduce the experimental level ordering but its effect on the charge distribution and possibly also on the bonding is minor.

Another group of molecules which provide evidence that is consistent with d orbital involvement is the series $E(XH_3)_n$ where E is an element of Groups V to VII and X = C, Si, or Ge. The majority of this work has been carried out by Cradock, Ebsworth and co-workers[203–205] in which they study the UPS of 28 such compounds in addition to many related systems (e.g. XH_3SH) and RQ (R = H, CH_3, SiH_3, GeH_3, or Me_3Si; Q = NCO, NCS, or N_3).[203] The following evidence can be interpreted as being "consistent" with d orbital participation:[192a] (1) the first IP of the compounds $E(XH_3)_n$ always reach their maximum value in the silyl derivative $IP_{CH_3} < IP_{SiH_3} > IP_{GeH_3}$, (2) spin–orbit coupling of heavy atoms is reduced in the silyl relative to the methyl derivative, and (3) UPS bands assigned to lone-pair ionizations are usually broadened in the spectra of silyl compounds. Typical results are presented in Fig. 23 for the series CH_3OCH_3, CH_3OSiH_3, SiH_3OSiH_3, CH_3SCH_3, CH_3SSiH_3, and SiH_3SSiH_3 where the assignments of Mollère *et al.*[192c] are presented for the first three bands. The trend in $2b_1$ shows stabilization upon substitution of CH_3 by SiH_3 as mentioned in point (1) above. On the other hand, the $3b_2$ orbital exhibits an inductive destabilization since the Si $3d$ orbitals cannot participate to stabilize this orbital. In the sulfur compounds the $4a_1$ orbital is relatively unshifted due to a competition of poor Si d orbital overlap and the opposing inductive effects. The peculiar crossing of the $4a_1$ and $3b_2$ in passing from CH_3OSiH_3 to SiH_3OSiH_3 has been interpreted by Bock and co-workers[192b] as being due to the "doubly eclipsed" conformation of disiloxane whereas methoxysilane is "singly eclipsed".

Ensslin *et al.*[192a] have pointed out that each of the three points mentioned earlier as being "consistent" with d orbital involvement can also be rationalized without d orbital involvement. They use the linear combination of bond orbital model (LCBO) and reason that each of the three points could be due to: (1) hyper-conjugation, (2) second-order spin–orbit coupling effects, and

(3) increased mixing of symmetry equivalent orbitals due to a decrease in their energy separation. In short, although a wide variety of UPS data is "consistent" with d orbital involvement, it by no means "proves" its existence.

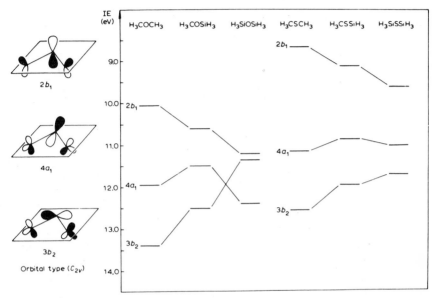

Fig. 23. Correlation of the three highest occupied orbitals in ROR' and RSR'. Reproduced from Mollère *et al.*[192a] with permission and by courtesy of Elsevier Sequoia S.A., Lausanne.

Cowley *et al.*[202] studied the UPS of $(CF_3P)_5$, $(CF_3P)_4$, $(CF_3)_4P_2(trans)$, and $(CF_3)_3P$ and observed mean phosphorus lone pair IP of 10·70 eV, 10·93 eV, 11·44 eV, and 11·70 eV, respectively. They reasoned that, if the phosphorus lone pairs were interacting with d orbitals on adjacent phosphorus atoms, the mean lone pair IP of the first three should be stabilized relative to $(CF_3)_3P$. Since no such stabilization was observed, they concluded that p_π–d_π bonding must be relatively unimportant. This reasoning is incorrect since in $(CF_3)_3P$ there are three electron-withdrawing CF_3 groups attached to each phosphorus, whereas the other compounds contain only one or two CF_3 groups per phosphorus. Obviously, the mean lone-pair IP will be affected more by the nature of the attached ligands than by any amount of p_π–d_π bonding.

F. Transition-metal Compounds

Relatively little work has been done on compounds of the transition elements, mainly because of their low volatility. The compounds of Groups

IVB–VIIIB with formulation MX_4, MOX_3, MO_2X_2, and MO_4 have been studied by Orchard and co-workers.[166] The interpretation of the UPS is analogous to the isoleptic Group IVA molecules with some inversion of orbital ordering due to the strong involvement of metal d orbitals (Section V-C). In this section we shall be interested only in compounds having "non-bonding" d electrons localized mainly on the transition-metal atom.

1. *Pentacarbonyl Halide Compounds*, $M(CO)_5X$. In the binary metal carbonyls (e.g. $Cr(CO)_6$,[146] $Fe(CO)_5$,[122] $Ni(CO)_4$[58]) the first bands in the UPS at 8–10 eV are readily assigned to the "metal" d electrons. The onset of the carbonyl region does not begin until 13–14 eV. By way of contrast, the first UPS band of $Mn(CO)_5X$ (X = Cl, Br, or I) was initially assigned to a predominantly halogen p_π orbital.[212, 213] The assignment of the low energy UPS of these compounds and the analogous $Mn(CO)_5H$, $Mn(CO)_5CH_3$, and $Mn(CO)_5CF_3$ has been under discussion for five years as regards the proper sequence of $e(X\,p_{x,y})$, $a_1(Mn{-}X)$, $e(Mn\,d_{xz,yz})$, and $b_2(Mn\,d_{xy})$.[61, 214–220] We will abbreviate these orbitals as $e(X)$, $a_1(X)$, $e(M)$, and $b_2(M)$. Recent work by Ceasar et al.[219] and Higginson et al.[220] has shown that a very useful approach is to compare the manganese pentacarbonyl halides with their rhenium analogues. Ceasar et al.[219] conclude that the correct orbital energy sequence is $e(X) > b_2(M) > e(M) > a_1(X)$. The supporting evidence is: (1) the band assigned to b_2 varies least with X for both the Mn and Re series. This is expected since the b_2 level is exclusively metal carbonyl bonding and should be shifted only by inductive effects. (2) In the case of X = Br and I, spin–orbit splitting due to the halogen atoms is observable in the first band, hence it must be $e(X)$. (3) The energy difference between the $e(X)$ and $a_1(X)$ levels averages to 2·22 eV for the chlorides, 1·97 eV for the bromides, and 1·91 eV for the iodides. These splittings are reasonable when compared to those observed in CH_3X, HX, and X_2.[219] (4) In the $M(CO)_5$ fragment, symmetry considerations would predict $e(M) > b_2(M)$. The interaction between the $e(X)$ and $e(M)$ levels is sufficient to reverse the order in $M(CO)_5X$. (5) The relative band intensities indicate that the first and third bands are more intense than the second and fourth.

Higginson et al.[220] argue that the sequence of $e(X)$ and $e(M)$ is reversed in $Mn(CO)_5Cl$ and $Mn(CO)_5Br$. They point out that with the proposed sequence of Ceasar et al.,[219] the $e(X)$ IP occurs at 8·94 eV, 8·86 eV, and 8·59 eV for $Mn(CO)_5Cl$, $Mn(CO)_5Br$, and $Mn(CO)_5I$, respectively. This is a very small shift for mainly halogen-based orbitals. Their assignment,[220] however, places a very small splitting between $e(X)$ and $a_1(X)$. The conflict in different assignments may be semantic since $e(X)$ contains M character and $e(M)$ contains X character. Furthermore, Koopmans' Theorem is assumed to hold throughout. If there is a breakdown in Koopmans' Theorem, it is possible that the ionic state sequence is $^2E(M) < {}^2B_2(M) < {}^2E(X) < {}^2A_1(X)$, whereas the orbital energy level sequence in the molecular ground state may be $e(X)$

$> b_2(M) > e(M) > a_1(X)$. Indeed, Spiess and Sheline[221] have concluded from the small anistropy in the ^{55}Mn chemical shift obtained in broadline NMR studies that the highest occupied orbital in the molecular ground states of $Mn(CO)_5X$ must be primarily halogen in character.

The assignment of $Mn(CO)_5H$ is undoubtedly[217, 220] $e(M) > b_2(M) > a_1$ (M—H) where the metal e and b_2 levels are split by only 0·29 eV and not 1·60 eV as originally assigned.[212, 213] The spectrum of $Mn(CO)_5CH_3$ is most plausibly assigned as $e(M) > b_2(M) > a_1(M—CH_3) > e(CH_3)$.[220] The $e(CH_3)$ orbitals are more stable than $a_1(M—CH_3)$ which is the usual feature in all CH_3-containing compounds except CH_3F where the a_1 orbital is stabilized by the "perfluoro effect".

2. *Metal Sandwich Compounds.* Of the transition-metal organometallic compounds, the vast majority which have been studied are of the metal sandwich type.[25, 50, 59, 73, 74, 222-225] The ligands involved include cyclopentadienyl, benzene, cycloheptatrienyl, and cyclo-octatetraene. The prototype molecule ferrocene has been discussed extensively (Fig. 4 and Section IV-B). The sandwich compounds have the dubious distinction of having the lowest reported IP for a diamagnetic molecule: 5·01 eV for *bis*-π-mesitylene chromium.[224]

Evans et al.[222] have compared the low-energy IP of the 18 electron systems $(\pi\text{-}C_6H_6)_2Cr$, $(\pi\text{-}C_6H_6)(\pi\text{-}C_5H_5)Mn$, and $(\pi\text{-}C_5H_5)_2Fe$, having the formal d^6 configuration. In all cases the low-energy IP bands are assigned to the "metallic" d electrons. For the chromium and manganese compounds the intensity ratio of the first two bands is 1 : 2 (increasing IP), compared to the opposite sequence for ferrocene. It is concluded that the order of ionic states is $^2E_{2g} < ^2A_{1g}$ for ferrocene but $^2A_{1g} < ^2E_{2g}$ for the manganese and chromium analogues. The trend followed by the lowest two energy states is presented in Fig. 24. The $^2A_{1g}$ state follows the same trend as the valence state IP since the a_{1g} orbital is relatively nonbonding. The $^2E_{2g}$ state, however, shows less shift with increasing nuclear charge on the metal since the e_{2g} orbital contains more ligand character than a_{1g}, as pointed out in Section IV-B. The effect of methyl substitution in the π-arene chromium complexes further substantiates these ideas concerning orbital localization.[224] The effect is greatest for ionizations from the mainly ligand e_{1g} and e_{1u} orbitals where decreases of about 0·4 eV are found for monomethylation of each ring. The effect of adding one methyl group in changing from benzene to toluene appears comparable with adding two from toluene to mesitylene. The effect on ionization from the mainly metal orbitals is about 0·2 eV for a_{1g} and 0·3 eV for e_{2g}.[224]

The problem of band assignment for ionic states arising from ionization of the predominantly metal d electrons is particularly difficult for molecules which contain more than one open shell. Consider chromocene for which the most probable ground state is $^3E_{2g}(a_{1g}^{1} e_{2g}^{3})$. Evans et al.[223] have attempted to

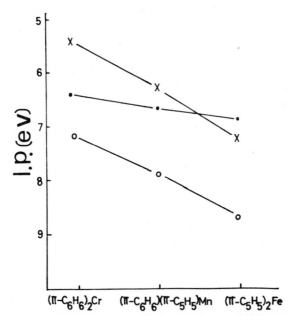

Fig. 24. Changes in ionization energy of the a_1 ($\times$) and e_2 ($\bullet$) levels in the series
bis-π-benzenechromium, π-benzene-π-cyclopentadienylmanganese and *bis*-π-cyclo-
pentadienyl-iron. Also shown, for comparison, are the average *d*-orbital energies for
the neutral atoms ($\bigcirc$). Reproduced from Evans *et al.*[222] with permission and by courtesy
of the Chemical Society, London.

TABLE IV

Ionic States of Chromocene[a]

Ionic state	Calcd.[b] energy	Obsd.[c] energy	Predicted[d] intensity
$^4A_{2g}(a_{1g}{}^1 e_{2g}{}^2)$	0·00	0·00	$\frac{4}{3}$
$^2E_{1g}(a_{1g}{}^1 e_{2g}{}^2)$	1·01	1·33	1
$^2E_{2g}(e_{2g}{}^3)$	1·36	1·59	$k_{a_1}{}^e$
$^2A_{1g}(a_{1g}{}^1 e_{2g}{}^2)$	1·41		$\frac{1}{2}$
		1·87	
$^2A_{2g}(a_{1g}{}^1 e_{2g}{}^2)$	1·48		$\frac{1}{6}$

[a] Energy is expressed in eV relative to the energy of the lowest photoelectron band.
[b] Ref. 50.   [c] Ref. 223.   [d] Refs. 25 and 223.
[e] k_{a_1} is the one-electron cross-section of the $a_{1g}(d)$ orbital relative to the $e_{2g}(d)$ cross-
section.

assign the four bands in its low energy UPS (5–8 eV) on the basis of relative band intensities and a partial ligand field treatment, and Warren[50] has carried out a complete ligand field analysis. The calculated and observed energy differences and expected relative intensities are presented in Table IV. The

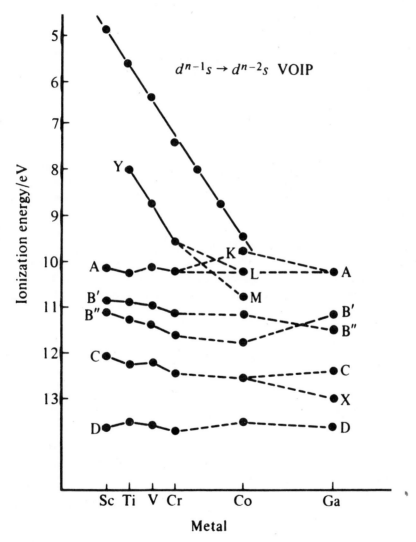

Fig. 25. Plot of ionization energies for the *tris*-hexafluoroacetylacetonato complexes, $M(hfa)_3$, having outer electron configurations $(3d)^n$. The ionization process Y concerns these d electrons, whereas processes $A–D$ relate to electrons mainly localized on the ligands. Reproduced from Evans *et al.*[228] with permission and by courtesy of the Chemical Society, London.

calculated ligand field results[50] gain credence since the order of energy states does not change with variation in the ligand field parameters. This assignment, however, does not agree with the expected intensities, and consequently the assignment presented in Table IV is not certain.

A particularly interesting result in the UPS of open-shell sandwich molecules is provided by 1,1'-dimethylmanganocene. Evans *et al.*[223] noticed that its spectrum contained at least two additional bands compared to manganocene itself and concluded that under the experimental conditions it exists in a high-spin/low-spin equilibrium. This conclusion was subsequently confirmed by magnetic susceptibility and electron spin resonance experiments.[226, 227]

3. *Tris-hexafluoacetylacetonato(hfa) Complexes.* Aside from the metal carbonyls and organometallic compounds, most transition metals compounds containing nonbonding *d* electrons are relatively involatile. The complexes $M(hfa)_3$ (M = Ga, Al, Sc, Ti, V, Cr, Mn, Fe, or Co), however, are sufficiently volatile that their UPS has been obtained.[228] Comparison of the spectra of the *d* complexes (Al, Ga, and Sc) with the other species allows assignment of the bands corresponding to the ionization of the *d* electrons. A schematic presentation of the shift in bands across the series is presented in Fig. 25. Assuming Koopmans' Theorem, there is a rapid stabilization of the metal *d* electrons so that by the time one reaches cobalt, the *d* orbitals are more stable than the highest occupied ligand orbitals. The trend in ionization of the *d* electrons is roughly paralleled by the valence orbital ionization potential.

REFERENCES

1. D. W. Turner, C. Baker, A. D. Baker, and C. R. Brundle, "Molecular Photoelectron Spectroscopy" (Wiley–Interscience, New York, 1970).
2. J. H. D. Eland, "Photoelectron Spectroscopy" (Butterworths, London, 1974).
3. A. D. Baker and D. Betteridge, "Photoelectron Spectroscopy: Chemical and Analytical Aspects" (Pergamon Press, Oxford, 1972).
4. A. Hamnett and A. F. Orchard, in: "Electronic Structure and Magnetism of Inorganic Compounds", Vol. 1, P. Day, ed. (Specialist Periodical Reports, The Chemical Society, London, 1972), p. 1.
5. S. Evans and A. F. Orchard, in: "Electronic Structure and Magnetism of Inorganic Compounds", Vol. 2, P. Day, ed. (Specialist Periodical Reports, The Chemical Society, London, 1973), p. 1.
6. A. Hamnett and A. F. Orchard, in: "Electronic Structure and Magnetism of Inorganic Compounds", Vol. 3, P. Day, ed. (Specialist Periodical Reports, The Chemical Society, London, 1974), p. 218.
7. D. Betteridge and A. D. Baker, *Analyt. Chem.* **42**, 43A (1970).
8. D. Betteridge, *Analyt. Chem.* **44**, 100R (1972).
9. D. Betteridge and M. A. Williams, *Analyt. Chem.* **46**, 125R (1974).

10. R. L. DeKock and D. R. Lloyd, in: "Advances in Inorganic Chemistry and Radiochemistry", Vol. 16, H. J. Emeléus and A. G. Sharpe, eds. (Academic Press, New York and London, 1974), p. 65.
11. H. Bock and B. G. Ramsey, *Angew. Chem. Int. Ed.* **12**, 734 (1973).
12. H. Bock and P. D. Mollère, *J. Chem. Educ.* **51**, 506 (1974).
13. R. E. Ballard, *Appl. Spectrosc. Rev.* **7**, 183 (1973).
14. B. G. Ramsey, A. Brook, A. R. Bassindale, and H. Bock, *J. Organometal. Chem.* **74**, C41 (1974).
15. W. Fuss and H. Bock, *J. Chem. Phys.* **61**, 1613 (1974).
16. F. G. Herring and R. A. N. McLean, *Inorg. Chem.* **11**, 1667 (1972).
17. C. R. Brundle and M. B. Robin, in: "Determination of Organic Structures by Physical Methods", Vol. 3, F. Nachod and G. Zuckerman, eds. (Academic Press, New York, 1971), p. 1.
18. C. K. Jørgenson, in: "Advances in Quantum Chemistry", Vol. 8, P. O. Löwdin, ed. (Academic Press, New York, 1974), p. 137.
19. J. J. Kaufman, E. Kerman, and W. S. Koski, *Int. J. Quantum Chem.* Symposium No. 4, 391 (1971).
20. A. W. Potts and T. A. Williams, *J. Electron Spectrosc.* **3**, 3 (1974).
21. M. Okuda and N. Jonathan, *J. Electron Spectrosc.* **3**, 19 (1974).
22. P. A. Cox and A. F. Orchard, *Chem. Phys. Lett.* **7**, 273 (1970).
23. W. C. Price, in: "Advances in Atomic and Molecular Physics", Vol. 10, D. R. Bates, ed. (Academic Press, New York, 1974), p. 131.
24. W. C. Price, A. W. Potts, and D. G. Streets, in: "Electron Spectroscopy", D. A. Shirley, ed. (North Holland, Amsterdam, 1972), p. 187.
25. P. A. Cox, S. Evans, and A. F. Orchard, *Chem. Phys. Lett.* **13**, 386 (1972).
26. D. R. Lloyd, *J. Phys. E.* **3**, 629 (1970).
27. A. W. Potts, T. A. Williams, and W. C. Price, *Discuss. Faraday Soc.* No. 54, 104 (1972).
28. J. A. Kinsinger, W. L. Stebbings, R. A. Valenzi, and J. W. Taylor, *Analyt. Chem.* **44**, 773 (1972).
29. H. Basch, A. Viste, and H. B. Gray, *Theor. Chim. Acta*, **3**, 458 (1965).
30. P. Natalis, J. Delwiche, and J. E. Collin. *Discuss. Faraday Soc.* No. 54, 98 (1972).
31. R. L. DeKock, B. R. Higginson, D. R. Lloyd, A. Breeze, D. W. J. Cruickshank, and D. R. Armstrong, *Mol. Phys.* **24**, 1059 (1972).
32. H. W. Kroto, R. J. Suffolk, and N. P. C. Westwood, *Chem. Phys. Lett.* **22**, 495 (1973).
33. T. P. Fehlner and D. W. Turner, *J. Amer. Chem. Soc.* **95**, 7175 (1973).
34. T. P. Fehlner and D. W. Turner, *Inorg. Chem.* **13**, 754 (1974).
35. N. P. C. Westwood, *Chem. Phys. Lett.* **25**, 558 (1974).
36. D. C. Frost, S. T. Lee, and C. A. McDowell, *Chem. Phys. Lett.* **22**, 243 (1973).
37. P. Rosmus, P. D. Dacre, B. Solouki, and H. Bock, *Theor. Chim. Acta*, **35**, 129 (1974).
38. J. C. Green, D. I. King, and J. H. D. Eland, *J. Chem. Soc.* D, 1121 (1970).
39. I. H. Hillier, V. R. Saunders, M. J. Ware, P. J. Bassett, D. R. Lloyd, and N. Lynaugh, *J. Chem. Soc.* D, 1316 (1970).
40. D. Betteridge, M. Thompson, A. D. Baker, and N. R. Kemp, *Analyt. Chem.* **44**, 2005 (1972).
41. J. Berkowitz, *J. Chem. Phys.* **56**, 2766 (1972).

42. G. R. Branton, D. C. Frost, T. Makita, C. A. McDowell, and I. A. Stenhouse, *J. Chem. Phys.* **52**, 802 (1970).
43. D. C. Frost, S. T. Lee, and C. A. McDowell, *Chem. Phys. Lett.* **24**, 149 (1974).
44. C. R. Brundle, *Chem. Phys. Lett.* **26**, 25 (1974).
45. J. W. Rabalais, T. P. Debies, J. L. Berkosky, J. J. Huang, and F. O. Ellison, *J. Chem. Phys.* **61**, 516 (1974).
46. J. Berkowitz and P. M. Guyon, *Int. J. Mass Spectrom. Ion Phys.* **6**, 301 (1971).
47. A. W. Potts and W. C. Price, *Proc. Roy. Soc.* **A326**, 165 (1972).
48. S. Evans, A. F. Orchard, and D. W. Turner, *Int. J. Mass Spectrom. Ion Phys.* **7**, 261 (1971).
49. G. Herzberg, *Quart. Rev.* **25**, 201 (1971).
50. K. D. Warren, *Inorg. Chem.* **13**, 1243 (1974).
51. T. Koopmans, *Physica*, **1**, 104 (1934).
52. W. G. Richards, *Int. J. Mass Spectrom. Ion Phys.* **2**, 419 (1969).
53. P. E. Cade, K. D. Sales, and A. C. Wahl, *J. Chem. Phys.* **44**, 1973 (1966).
54. A. C. Wahl, *J. Chem. Phys.* **41**, 2600 (1964).
55. A. B. Cornford, D. C. Frost, C. A. McDowell, J. L. Ragle, and I. A. Stenhouse, *J. Chem. Phys.* **54**, 2651 (1971).
56. M.-M. Coutière, J. Demuynck, and A. Veillard, *Theor. Chim. Acta*, **27**, 281 (1972).
57. M.-M. Rohmer and A. Veillard, *J.C.S. Chem. Commun.* 250 (1973).
58. I. H. Hillier, M. F. Guest, B. R. Higginson, and D. R. Lloyd, *Mol. Phys.* **27**, 215 (1974).
59. S. Evans, M. F. Guest, I. H. Hillier, and A. F. Orchard, *J.C.S. Faraday II*, **70**, 417 (1974).
60. F. Brogli, P. A. Clark, and E. Heilbronner, *Angew. Chem. Int. Ed.* **12**, 422 (1973).
61. M. F. Guest, M. B. Hall, and I. H. Hillier, *Mol. Phys.* **25**, 629 (1973).
62. C. K. Jørgenson, *Structure and Bonding*, **1**, 234 (1966).
63. R. C. Sahni and B. C. Sawhney, *Int. J. Quantum Chem.* **1**, 251 (1967).
64. D. P. Chong, F. G. Herring, and D. McWilliams, *J.C.S. Faraday II*, **70**, 193 (1974).
65. J. L. Dodds and R. McWeeny, *Chem. Phys. Lett.* **13**, 9 (1972).
66. D. P. Chong, F. G. Herring, and D. McWilliams, *J. Chem. Phys.* **61**, 78 (1974).
67. D. P. Chong, F. G. Herring, and D. McWilliams, *Chem. Phys. Lett.* **25**, 568 (1974).
68. J. C. Slater, in: "Advances in Quantum Chemistry", Vol. 6, P.-O. Löwdin, ed. (Academic Press, New York and London, 1972), p. 1.
69. K. H. Johnson, in: "Advances in Quantum Chemistry", Vol. 7, P.-O. Löwdin, ed. (Academic Press, New York and London, 1973), p. 143.
70. T. Parameswaran and D. E. Ellis, *J. Chem. Phys.* **58**, 2088 (1973).
71. A. Rauk, T. Ziegler, and D. E. Ellis, *Theor. Chim. Acta*, **34**, 49 (1974).
72. D. W. Turner, in: "Physical Methods in Advanced Inorganic Chemistry", H. A. O. Hill and P. Day, eds. (Wiley–Interscience, London, 1968), p. 74.
73. J. W. Rabalais, L. O. Werme, T. Bergmark, L. Karlsson, M. Hussain, and K. Siegbahn, *J. Chem. Phys.* **57**, 1185 (1972).
74. S. Evans, M. L. H. Green, B. Jewitt, A. F. Orchard, and C. F. Pygall, *J.C.S. Faraday II*, **68**, 1847 (1972).
75. N. Rösch and K. H. Johnson, *Chem. Phys. Lett.* **24**, 179 (1974).
76. E. J. Baerends and P. Ros, *Chem. Phys. Lett.* **23**, 391 (1973).

77. D. N. Hendrickson, *Inorg. Chem.* **11**, 1161 (1972).
78. R. Prins, *Mol. Phys.* **19**, 603 (1970).
79. B. Gächter, G. Jakubinck, B. E. Schneider-Poppe, and J. A. Koningstein, *Chem. Phys. Lett.* **28**, 160 (1974).
80. L. S. Cederbaum, G. Hohlneicher, and W. von Niessen, *Mol. Phys.* **26**, 1405 (1973).
81. T.-T. Chen, W. D. Smith, and J. Simons, *Chem. Phys. Lett.* **26**, 296 (1974).
82. L. S. Cederbaum, *Chem. Phys. Lett.* **25**, 562 (1974).
83. L. S. Cederbaum, G. Hohlneicher, and W. von Niessen, *Chem. Phys. Lett.* **18**, 503 (1973).
84. B. Kellerer, L. S. Cederbaum, and G. Hohlneicher, *J. Electron Spectrosc.* **3**, 107 (1974).
85. P. J. Bassett and D. R. Lloyd, *J.C.S. Dalton*, 248 (1972).
86. J. M. Hollas and T. A. Sutherley, *Mol. Phys.* **22**, 213 (1971).
87. S. Evans, P. J. Joachim, and A. F. Orchard, *Int. J. Mass Spectrom. Ion Phys.* **9**, 41 (1972).
88. C. R. Brundle, N. A. Kuebler, M. B. Robin, and H. Basch, *Inorg. Chem.* **11**, 20 (1972).
89. S. Evans, A. Hamnett, and A. F. Orchard, *J. Amer. Chem. Soc.* **96**, 6221 (1974).
90. E. Diemann and A. Müller, *Chem. Phys. Lett.* **19**, 538 (1973).
91. S. Foster, S. Felps, L. C. Cusachs, and S. P. McGlynn, *J. Amer. Chem. Soc.* **95**, 5521 (1973).
92. D. R. Lloyd and N. Lynaugh, *J.C.S. Faraday II*, **68**, 947 (1972).
93. G. H. King, S. S. Krishnamurthy, M. F. Lappert, and J. B. Pedley, *Discuss. Faraday Soc.* No. 54, 70 (1972).
94. P. A. Cox, S. Evans, A. F. Orchard, N. V. Richardson, and P. J. Roberts, *Discuss. Faraday Soc.* No. 54, 26 (1972).
95. J. L. Berkosky, F. O. Ellison, T. H. Lee, and J. W. Rabalais, *J. Chem. Phys.* **59**, 5342 (1973).
96. T. H. Lee and J. W. Rabalais, *J. Chem. Phys.* **60**, 1172 (1974).
97. C. R. Brundle, G. R. Jones, and H. Basch, *J. Chem. Phys.* **55**, 1098 (1971).
98. F. A. Grimm, *J. Electron Spectrosc.* **2**, 475 (1973).
99. J. Berkowitz, J. L. Dehmer, and E. H. Appleman, *Chem. Phys. Lett.* **19**, 334 (1973).
100. H. J. Lempka, T. R. Passmore, and W. C. Price, *Proc. Roy. Soc. Ser.* **A304**, 53 (1968).
101. D. G. Streets, *Chem. Phys. Lett.* **20**, 555 (1974).
102. F. A. Cotton, "Chemical Applications of Group Theory", 2nd ed. (Wiley–Interscience, New York, 1970).
103. A. W. Potts, H. J. Lempka, D. G. Streets, and W. C. Price, *Phil. Trans. Roy. Soc. Lond.* **A268**, 59 (1970).
104. P. J. Bassett and D. R. Lloyd, *J. Chem. Soc. A*, 1551 (1971).
105. D. O. Cowan, R. Gleiter, O. Glemser, and E. Heilbronner, *Helv. Chim. Acta*, **55**, 2418 (1972).
106. J. Kroner, D. Proch, W. Fuss, and H. Bock, *Tetrahedron*, **28**, 1585 (1972).
107. C. R. Brundle, M. B. Robin, N. A. Kuebler, and H. Basch, *J. Amer. Chem. Soc.* **94**, 1451 (1972).
108. C. R. Brundle, M. B. Robin, and N. A. Kuebler, *J. Amer. Chem. Soc.* **94**, 1466 (1972).
109. C. R. Brundle, M. B. Robin, and H. Basch, *J. Chem. Phys.* **53**, 2196 (1970).

110. C. R. Brundle, M. B. Robin, and G. R. Jones, *J. Chem. Phys.* **52**, 3383 (1970).
111. T. X. Carroll, R. W. Shaw, Jr., T. D. Thomas, C. Kindle, and N. Bartlett, *J. Amer. Chem. Soc.* **96**, 198 (1974).
112. D. Betteridge and M. Thompson, *J. Mol. Structure*, **21**, 341 (1974).
113. D. Betteridge, A. D. Baker, P. Bye, S. K. Hasanuddin, N. R. Kemp, D. I. Rees, M. A. Stevens, M. Thompson, and B. J. Wright, *Z. Anal. Chem.* **263**, 286 (1973).
114. J. P. Maier and J.-F. Muller, *Tetrahedron Lett.* **35**, 2987 (1974).
115. A. D. Baker, D. Betteridge, N. R. Kemp, and R. E. Kirby, *Int. J. Mass Spectrom. Ion Phys.* **4**, 90, (1970).
116. J. Berkowitz, *Chem. Phys. Lett.* **11**, 21 (1971).
117. C. R. Brundle, *Chem. Phys. Lett.* **7**, 317 (1970).
118. R. S. Mulliken, *J. Chem. Phys.* **2**, 782 (1934), **3**, 573 (1935).
119. R. S. Berry, *Chem. Rev.* **69**, 533 (1969).
120. P. Baybutt, M. F. Guest, and I. H. Hillier, *Mol. Phys.* **25**, 1025 (1973).
121. B. W. Levitt and L. S. Levitt, *J. Coord. Chem.* **3**, 187 (1973).
122. D. R. Lloyd and E. W. Schlag, *Inorg. Chem.* **8**, 2544 (1969).
123. J. F. Nixon, *J.C.S. Dalton*, 2226 (1973).
124. C. R. Brundle and G. R. Jones, *J. Electron Spectrosc.* **1**, 403 (1973).
125. R. L. DeKock, *J. Electron Spectrosc.* **4**, 155 (1974).
126. R. L. DeKock, B. R. Higginson, and D. R. Lloyd, *Discuss. Faraday Soc.* No. 54, 84 (1972).
127. S. Cradock, *J. Chem. Phys.* **55**, 980 (1971).
128. P. Dechant, A. Schweig, and W. Thiel, *Angew. Chem. Int. Ed.* **12**, 308 (1973).
129. A. Schweig and W. Thiel, *J. Chem. Phys.* **60**, 951 (1974).
130. A. Schweig and W. Thiel, *J. Electron Spectrosc.* **3**, 27 (1974).
131. K. Kimura, S. Katsumata, Y. Achiba, H. Matsumoto, and S. Nagakura, *Bull. Chem. Soc. Japan*, **46**, 373 (1973).
132. K. Osafune and K. Kimura, *Chem. Phys. Lett.* **25**, 47 (1974).
133. K. Osafune, S. Katsumata, and K. Kimura, *Chem. Phys. Lett.* **19**, 369 (1973).
134. D. W. Davies, *Chem. Phys. Lett.* **28**, 520 (1974).
135. J. L. Beauchamp, *Annual Rev. Phys. Chem.* **22**, 552 (1971).
136. J. A. Pople, *Quart. Rev. Chem. Soc.* **11**, 273 (1957).
137. D. S. Urch, "Orbitals and Symmetry" (Penguin Books Ltd., Middlesex, England, 1970), p. 108.
138. W. Kutzelnigg, *Angew. Chem. Int. Ed.* **12**, 546 (1973).
139. S. Evans, J. C. Green, P. J. Joachim, A. F. Orchard, D. W. Turner, and J. P. Maier, *J.C.S. Faraday II*, **68**, 905 (1972).
140. D. A. Sweigart, *J. Chem. Educ.* **50**, 322 (1973).
141. A. Rauk and I. G. Csizmadia, *Canad. J. Chem.* **46**, 1205 (1968).
142. F. P. Boer and W. N. Lipscomb, *J. Chem. Phys.* **50**, 989 (1969).
143. R. L. DeKock, *Chem. Phys. Lett.* **27**, 297 (1974).
144. P. J. Bassett, B. R. Higginson, D. R. Lloyd, N. Lynaugh, and P. J. Roberts, *J. Chem. Soc. Dalton Trans.* 2316 (1974).
145. R. S. Nyholm, *Proc. Chem. Soc. (London)*, 273 (1961).
146. B. R. Higginson, D. R. Lloyd, P. Burroughs, D. M. Gibson, and A. F. Orchard, *J. Chem. Soc. Faraday Trans. II*, **69**, 1659 (1973).
147. R. F. Lake, *Spectrochim. Acta*, **27A**, 1220 (1971).
148. C. R. Brundle, M. B. Robin, H. Basch, M. Pinsky, and A. Bond, *J. Amer. Chem. Soc.* **92**, 3863 (1970).

149. H. Basch, M. B. Robin, N. A. Kuebler, C. Baker, and D. W. Turner, *J. Chem. Phys.* **51**, 52 (1969).
150. K. S. Pitzer, *J. Amer. Chem. Soc.* **67**, 1126 (1945).
151. N. Lynaugh, D. R. Lloyd, M. F. Guest, M. B. Hall, and I. H. Hillier, *J. Chem. Soc. Faraday Trans. II*, **68**, 2192 (1972).
152. D. R. Lloyd and N. Lynaugh, *J. Chem. Soc.* D, 627 (1971).
153. M. F. Guest and I. H. Hillier, *J. Chem. Soc. Faraday Trans. II*, **70**, 398 (1974).
154. L. Golob, N. Jonathan, A. Morris, M. Okuda, and K. J. Ross, *J. Electron Spectrosc.* **1**, 506 (1972).
155. W. E. Bull, B. P. Pullen, F. A. Grimm, W. E. Moddeman, G. K. Schweitzer, and T. A. Carlson, *Inorg. Chem.* **9**, 2474 (1970).
156. P. J. Bassett and D. R. Lloyd, *J. Chem. Soc.* A, 641 (1971).
157. P. J. Davidson and M. F. Lappert, *J. Chem. Soc. Chem. Commun.* 317 (1973).
158. S. Evans, J. C. Green, and S. E. Jackson, *J. Chem. Soc. Faraday Trans. II*, **69**, 191 (1973).
159. M. F. Lappert, J. B. Pedley, and G. Sharp, *J. Organometal. Chem.* **66**, 271 (1974).
160. S. Cradock, *Chem. Phys. Lett.* **10**, 291 (1971).
161. J. Berkowitz, *J. Chem. Phys.* **61**, 407 (1974).
162. A. E. Jonas, G. K. Schweitzer, F. A. Grimm, and T. A. Carlson, *J. Electron Spectrosc.* **1**, 29 (1973).
163. M. B. Hall, M. F. Guest, I. H. Hillier, D. R. Lloyd, A. F. Orchard, and A. W. Potts, *J. Electron Spectrosc.* **1**, 497 (1973).
164. R. Boschi, M. F. Lappert, J. B. Pedley, W. Schmidt, and B. T. Wilkins, *J. Organometal. Chem.* **50**, 69 (1973).
165. J. C. Green, M. L. H. Green, P. J. Joachim, A. F. Orchard, and D. W. Turner, *Phil. Trans. Roy. Soc. Lond.* **A268**, 111 (1970).
166. P. Burroughs, S. Evans, A. Hamnett, A. F. Orchard, and N. V. Richardson, *J. Chem. Soc. Faraday Trans. II*, **70**, 1895 (1974).
167. R. K. Thomas, H. Thompson, *Proc. Roy. Soc. Lond.* **A327**, 13 (1972).
168. D. Chadwick, *Canad. J. Chem.* **50**, 737 (1972).
169. H. W. Kroto and R. J. Suffolk, *Chem. Phys. Lett.* **17**, 213 (1972).
170. G. W. Mines, R. K. Thomas, and H. Thompson, *Proc. Roy. Soc. Lond.* **A333**, 171 (1973).
171. H. Bock, K. Wittel, and A. Haas, *Zeit. Anorg. Allg. Chem.* **408**, 107 (1974).
172. K. Wittel, A. Haas, and H. Bock, *Chem. Ber.* **105**, 3865 (1972).
173. M. Arbelot, J. Metzer, M. Chanon, C. Guimon, G. Pfister-Guillouzo, *J. Amer. Chem. Soc.* **96**, 6217 (1974).
174. G. Klopman, *J. Amer. Chem. Soc.* **90**, 223 (1968).
175. S. Elbel, H. Bergmann, and W. Ensslin, *J. Chem. Soc. Faraday Trans. II*, **70**, 555 (1974).
176. R. Hoffman, *Accts. Chem. Res.* **4**, 1 (1971).
177. R. Gleiter, E. Heilbronner, and V. Hornung, *Helv. Chim. Acta*, **55**, 255 (1972).
178. R. L. Ellis, H. H. Jaffé, and C. A. Masmanidis, *J. Amer. Chem. Soc.* **96**, 2623 (1974).
179. G. Wagner, H. Bock, R. Budenz, and F. Seel, *Chem. Ber.* **106**, 1285 (1974).
180. G. Wagner and H. Bock, *Chem. Ber.* **107**, 68 (1974).
181. A. D. Baker, M. Brisk, and M. Gellender, *J. Electron Spectrosc.* **3**, 227 (1974).
182. R. J. Colton and J. W. Rabalais, *J. Electron Spectrosc.* **3**, 345 (1974).
183. P. Rademacher, *Angew. Chem. Int. Ed.* **12**, 408 (1973).

184. P. Rademacher, *Tetrahedron Lett.* 83 (1974).
185. S. F. Nelsen and J. M. Buschek, *J. Amer. Chem. Soc.* **95**, 2011 (1973).
186. S. F. Nelsen, J. M. Buschek, and P. J. Hintz, *J. Amer. Chem. Soc.* **95**, 2013 (1973).
187. S. F. Nelsen and J. M. Buschek, *J. Amer. Chem. Soc.* **96**, 2392 (1974).
188. S. F. Nelsen and J. M. Buschek, *J. Amer. Chem. Soc.* **96**, 6982 (1974).
189. S. F. Nelsen and J. M. Buschek, *J. Amer. Chem. Soc.* **96**, 6987 (1974).
190. A. H. Cowley, M. J. S. Dewar, D. W. Goodman, and M. C. Padolina, *J. Amer. Chem. Soc.* **96**, 2648 (1974).
191. E. Zeeck, *Theor. Chim. Acta*, **35**, 301 (1974).
192. (a) W. Ensslin, H. Bock, and G. Becker, *J. Amer. Chem. Soc.* **96**, 2757 (1974), (b) H. Bock, P. Mollère, G. Becker, and G. Fritz, *J. Organometal. Chem.* **61**, 113 (1973), (c) P. Mollère, H. Bock, G. Becker, and G. Fritz, *J. Organometal. Chem.* **61**, 127 (1973).
193. A. B. Cornford, D. C. Frost, F. G. Herring, and C. A. McDowell, *Chem. Phys. Lett.* **10**, 345 (1971).
194. A. B. Cornford, D. C. Frost, F. G. Herring, and C. A. McDowell, *J. Chem. Phys.* **55**, 2820 (1971).
195. C. R. Brundle and G. R. Jones, *J. Chem. Soc. Faraday Trans. II*, **68**, 959 (1972).
196. M. C. Green, M. F. Lappert, J. B. Pedley, W. Schmidt, and B. T. Wilkins, *J. Organometal. Chem.* **31**, C55 (1971).
197. U. Weidner and A. Schweig, *J. Organometal. Chem.* **37**, C29 (1972).
198. U. Weidner and A. Schweig, *J. Organometal. Chem.* **39**, 261 (1972).
199. S. Cradock and R. A. Whiteford, *Trans. Faraday Soc.* **67**, 3425 (1971).
200. D. C. Frost, F. G. Herring, A. Katrib, R. A. N. McLean, J. E. Drake, and N. P. C. Westwood, *Canad. J. Chem.* **49**, 4033 (1971).
201. R. A. N. McLean, *Canad. J. Chem.* **51**, 2089 (1973).
202. A. H. Cowley, M. J. S. Dewar, D. W. Goodman, and M. C. Padolina, *J. Amer. Chem. Soc.* **96**, 3666 (1974).
203. S. Cradock, E. A. V. Ebsworth, and J. D. Murdoch, *J. Chem. Soc. Faraday Trans. II*, **68**, 86 (1972).
204. S. Cradock, E. A. V. Ebsworth, W. J. Savage, and R. A. Whiteford, *J. Chem. Soc. Faraday Trans. II*, **68**, 934 (1972).
205. S. Cradock and R. A. Whiteford, *J. Chem. Soc. Faraday Trans. II*, **68**, 281 (1972).
206. G. R. Branton, C. E. Brion, D. C. Frost, K. A. R. Mitchell, and N. L. Paddock, *J. Chem. Soc. A*, 151 (1970).
207. J. P. Maier and D. W. Turner, *J. Chem. Soc. Faraday Trans. II*, **68**, 711 (1972).
208. D. C. Frost, F. G. Herring, K. A. R. Mitchell, and I. A. Stenhouse, *J. Amer. Chem. Soc.* **93**, 1596 (1971).
209. N. Rösch, V. H. Smith, Jr., and H. M. Whangbo, *J. Amer. Chem. Soc.* **96**, 5984 (1974).
210. F. A. Gianturco, C. Guidotti, U. Lamanna, and R. Moccia, *Chem. Phys. Lett.* **10**, 269 (1971).
211. D. P. Santry and G. A. Segal, *J. Chem. Phys.* **47**, 158 (1967).
212. S. Evans, J. C. Green, M. L. H. Green, A. F. Orchard, and D. W. Turner, *Discuss. Faraday Soc.* No. 47, 112 (1969).
213. R. F. Fenske and R. L. DeKock, *Inorg. Chem.* **9**, 1053 (1970).
214. M. B. Hall and R. F. Fenske, *Inorg. Chem.* **11**, 768 (1972).
215. M. B. Hall, M. F. Guest, and I. H. Hillier, *Chem. Phys. Lett.* **15**, 592 (1972).

216. D. L. Lichtenberger, A. C. Sarapu, and R. F. Fenske, *Inorg. Chem.* **12**, 702 (1973).
217. S. Cradock, E. A. V. Ebsworth, and A. Robertson, *J. Chem. Soc. Dalton Trans.* 22 (1973).
218. D. L. Lichtenberger and R. F. Fenske, *Inorg. Chem.* **13**, 486 (1974).
219. G. P. Ceasar, P. Milazzo, J. L. Cihonski, and R. A. Levenson, *Inorg. Chem.* **13**, 3035 (1974).
220. B. R. Higginson, D. R. Lloyd, S. Evans, and A. F. Orchard, *J. Chem. Soc. Faraday Trans. II*, **71**, 1913 (1975).
221. H. W. Spiess and R. K. Sheline, *J. Chem. Phys.* **54**, 1099 (1971).
222. S. Evans, J. C. Green, and S. E. Jackson, *J. Chem. Soc. Faraday II*, **68**, 249 (1972).
223. S. Evans, M. L. H. Green, B. Jewitt, G. H. King, and A. F. Orchard, *J. Chem. Soc. Faraday II*, **70**, 356 (1974).
224. S. Evans, J. C. Green, S. E. Jackson, and B. Higginson, *J. Chem. Soc. Dalton Trans.* 304 (1974).
225. D. W. Clack, *Theor. Chim. Acta*, **35**, 157 (1974).
226. M. E. Switzer, R. Wang, M. F. Rettig, and A. H. Maki, *J. Amer. Chem. Soc.* **96**, 7669 (1974).
227. J. H. Ammeter, R. Bucher, and N. Oswald, *J. Amer. Chem. Soc.* **96**, 7833 (1974).
228. S. Evans, A. Hamnett, A. F. Orchard, and D. R. Lloyd, *Discuss. Faraday Soc.* No. 54, 227 (1972).

7

High Temperature UPS Studies and Other Variations

J. BERKOWITZ

Argonne National Laboratory, Argonne, Illinois, U.S.A.

I. INTRODUCTION

Prior to 1970, UPS was largely confined to the study of samples whose vapor pressures were such that measurements could be made at or near room temperature. To be sure, there were some isolated studies on mercury and the mercuric halides in which the inlet line of an apparatus might have to

355

be heated to $\sim 100\,°C$, but the great preponderance of studies were performed on materials that may be characterized in the following manner:

(a) Compounds containing C, O, H, N, occasional volatile sulfides and halides.
(b) Compounds, therefore, which were for the most part in the domain of organic chemistry.
(c) Compounds which tended to bond covalently.
(d) Compounds confined largely to the first two rows of the Periodic Chart.

A perusal of the compendium of Turner et al.,[1] published in 1970, will verify the above observations. A number of generalizations had to be drawn on the basis of limited evidence with a small and unrepresentative sample of the elements in the Periodic Chart. It has been the function of high-temperature UPS work since that time to overcome these limitations.†

The obvious problem in high temperature UPS is to generate a vapor of the material to be studied in the photoionization region, without impairing the performance of the electron analyzer. Attempts to raise the temperature of the entire apparatus to the sample temperature have to be abandoned when one gets beyond $\sim 100\,°C$. Hence, one resorts to molecular beams of the sample. The in-beam molecular number density will generally be much lower than the molecular density conventionally used with an ambient permanent gas. Signal intensity will therefore be more of a problem, and one must aim for a high collection efficiency.

Obviously, one way to enhance the signal is to design an analyzer that accepts a large fraction of the photoelectrons. Retarding field analyzers generally accept a larger solid angle of photoelectrons, but one must then contend with a differentiated spectrum, which often has more limited resolution, and may introduce spurious structure. A few recent investigations[2-4] in high temperature UPS have gone this route, but the published results of two such groups are subjected to question in this monograph (see Timoshenko and Akopyan[3] under alkali halides, Cocksey et al. under Group IIb dihalides).

Potts et al.[5] and Orchard and Richardson[6] have apparently succeeded in achieving adequate signals with a conventional selector analyzer. Our approach[7] has been to construct an analyzer having cylindrical symmetry, so that the solid angle of acceptance for photoelectrons is $2\pi\Delta\theta$, instead of $\Delta\phi\Delta\theta$. It was felt at the time that we might experience a gain in transmission of ~ 10—from $\sim 0.1\%$ to $\sim 1\%$ of the generated photoelectrons.

A concentric spherical system such as described initially by Purcell[8] and commercially originally marketed by Varian Associates is one solution that satisfies our desired criterion. However, we have opted for a concentric

† Most of the literature survey in this chapter was completed in March 1975.

cylindrical geometry which appeared simpler to construct and had some intriguing properties, according to theory.[9, 10] Two types of solutions have been considered for this analyzer system, axial focussing and slit-to-slit focussing. The axial focussing analysis leads to perfect second-order focussing in angle (as well as first-order focussing) at an angle of $\sim 42°$ to the cylinder axis. However, the dispersion of the apparatus increases with this angle rather markedly beyond 55°. We have chosen the slit-to-slit solution at a launching angle of $\sim 60°$, which takes advantage of the increased dispersion, and is rather close to the "magic angle" which makes relative intensities more meaningful (*see* Section A on Atoms). This solution also is not as sensitive to the size of the ionizing region, whereas the 42° second-order focus presupposes a point source. Figure 1 is a schematic diagram of our apparatus, called a cylindrical mirror analyzer. Subsequently Allen *et al.*[11] and Samson *et al.*[12] have built similar devices.

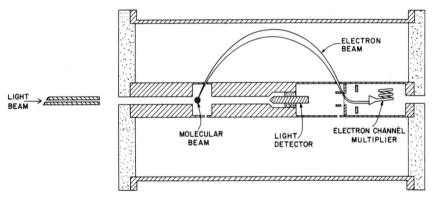

Fig. 1. Schematic diagram of the cylindrical mirror analyzer used in high temperature UPS.

The production of a vapor beam also presents problems not only because of electrical and magnetic interference if one uses resistive, inductive or electron impact heating, but also because vapor invariably deposits on sensitive surfaces in the analyzer and begins to degrade the energy resolution.

Thus far we and most subsequent groups of investigators have employed resistive heating, the heater wire being wound non-inductively. Allen *et al.* have used a CO_2 laser to heat the base of a crucible which contains the sample. There appears to be a limitation in temperature of *ca.* $1000 \pm 200°$ K in these systems.

We have recently embarked upon a program to use laser irradiation directly on the sample surface, which is located near the ionization region. Preliminary experiments indicate that the method should be capable of studying at least some materials which require temperatures in excess of 2000° K for generation of the necessary vapor pressure.

In the ensuing pages, we present data and analyses for representatives of every column of the Periodic Chart except Group V. Data are available[13] on three high-temperature species in this group, P_4, As_4, and Sb_4, but they have not yet been analyzed by us.†

II. ATOMS, INTENSITIES, ANGULAR DISTRIBUTIONS, AND ALL THAT

Since the only volatile species that are atomic are the noble gases, the high-temperature techniques will ultimately prove invaluable in accumulating detailed information for atomic species throughout the periodic chart. To date UPS data have only been reported for Zn, Cd, and Hg vapor[14-17] ‡ in the high-temperature domain, though unpublished data exist for Ba and Yb.[4]

As we shall try to show below, theoretical calculations of photoionization cross-sections and angular distributions of photoelectrons, and their dependence upon photon energy, cannot yet be carried out with confidence on systems other than atoms. Hence, it is particularly important to pursue this aspect of high-temperature UPS in the near future. Berkowitz[18] has discussed many of the atomic systems, their vapor pressures, and anticipated spectra in a review article.

The existing data on energetics can be summarized in short order. Zinc, cadmium, and mercury have an $nd^{10}(n+1)s^2$ valence orbital structure. Single-electron removal is the dominant photoionization process, yielding a $^2S_{\frac{1}{2}}$ state when an s electron is ejected, and $^2D_{\frac{5}{2}}$, $^2D_{\frac{3}{2}}$ states when a d electron is removed. The peaks occur at the predicted energies known from spectroscopy.[19] Some two-electron processes have been observed in Cd[17] and Hg,[16] with intensities relative to the one-electron processes of about 1% (also *see* note‡ added in proof). However, the data on Ba imply much more significance to the two-electron process in that system. In this context, see the note below on the three new Ba references. In Cd and Hg, the matrix

† *Note added in proof.* Spectra have recently been reported on $SbCl_3$, $SbBr_3$, and SbI_3 in the 25–115 °C region [see D. G. Nicholson and P. Rademacher, *Acta Chemica Scandinavica*, **A28**, 1136 (1974)]. *See also* discussion of PCl_3 and PBr_3 and spectra by P. A. Cox, S. Evans, A. F. Orchard, N. V. Richardson, and P. J. Roberts, *Disc. Far. Soc.* **54**, 26 (1972).

‡ *Note added in proof.* UPS spectra have recently been reported for Pb [S. Süzer, M. S. Banna and D. A. Shirley, *J. Chem. Phys.* **63**, 3473 (1975)]; Bi [S. Süzer, S. T. Lee, and D. A. Shirley, *ibid* **65**, 412 (1976)]; and the group II metal vapours of Ca, Sr, Ba, Zn and Cd (S. Süzer, S. T. Lee and D. A. Shirley, *Phys. Rev.* **A13**, 1842 (1976)]. For Ba, there now exist three independent studies: B. Brehm and K. Höfler, *Int. J. Mass Spectrom. Ion Phys.* **17**, 371 (1975); H. Hotop and D. Mahr, *J. Phys.* **B8**, L301 (1975); S. T. Lee, S. Süzer, E. Matthias, and D. A. Shirley, *Chem. Phys. Lett.* (submitted for publication).

elements describing the two-electron process have been inferred to result from configuration interaction in the initial state, and from "conjugate shake-up". The interested reader should see ref. 16 for a detailed description.

In the course of our experiments[14] on Zn, Cd, and Hg, using 584 Å radiation, we observed that the ratio of electron intensities corresponding to the spin–orbit partners $^2D_{\frac{5}{2}} : {}^2D_{\frac{3}{2}}$ was dependent on the atom, and in every instance was *greater* than or equal to the statistical value of $3:2$. This aroused curiosity since many experiments on the noble gases Ar, Kr, and Xe (see, for example, Samson and Cairns[20]) had shown that the ratio of spin–orbit partners $^2P_{\frac{3}{2}} : {}^2P_{\frac{1}{2}}$ was *less* than or equal to the statistical value.

In an attempt to rationalize these observations, Walker[14, 21] performed some calculations of photoionization cross-sections, using Dirac–Slater relativistic wavefunctions for both the bound and continuum states. From these calculations has emerged a general understanding of branching ratios and their departure from statistical weight predictions, which is a generalization of Cooper's calculations.[22] Cooper had previously noted that although the dipole selection rules permitted $\Delta l = \pm 1$, the $l+1$ channel usually dominates the photoionization cross-section.

Ionization from s or p orbitals rises step-like at the ionization threshold (see Figs. 2 and 3 for He, Ar), but for orbitals with $l \geqslant 2$ the term in the Hamiltonian of the form $[l(l+1)]/r^2$ gives rise to a substantial potential

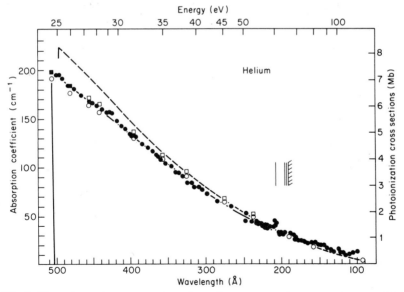

Fig. 2. Photoabsorption cross-section of He (from J. A. R. Samson, *Adv. Atom. Mol. Phys.* **2**, 177 (1966).

barrier for the ejection of an electron. As a consequence, the photo-ionization cross-section at the threshold for ionization from such an orbital is small and rises slowly with excess photon energy (see Fig. 4 for the *d*-orbital ionization of Hg). The theoretical explanation for this behavior is that at very low photoelectron energies, the first major maximum of the

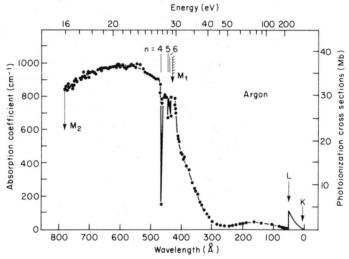

Fig. 3. Photoabsorption cross-section of Ar (from Samson).

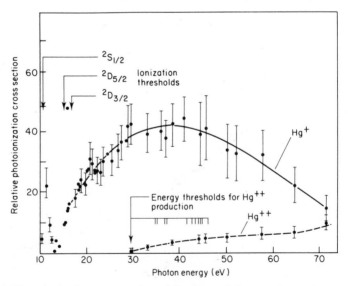

Fig. 4. Photoionization cross-section of Hg (from Cairns *et al.*, *J. Chem. Phys.* **53**, 96 (1970)).

continuum orbital of the f channel will lie outside the region of space occupied by the bound d orbital. As the photoelectron energy increases, this maximum moves nearer the nucleus, and the dipole matrix element increases. When the photoelectron energy increases still further, the first major maximum of the continuum orbital begins to overlap the nodes of the bound orbital and the dipole matrix element decreases and finally changes sign, giving rise to the "Cooper minimum" in the partial cross-section. The Cooper minimum therefore depends upon the presence of at least one node in the bound orbital. Note that Fig. 2 (He) displays no Cooper minimum whereas Fig. 3 (Ar) does.

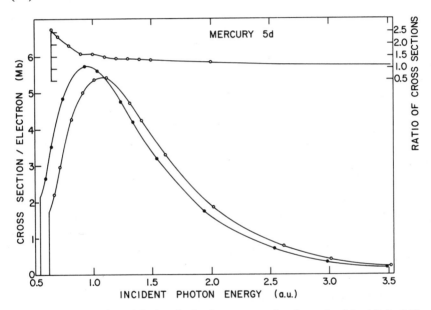

Fig. 5. Calculated partial photoionization cross-section from the $5d$ orbital of Hg (from T. E. H. Walker *et al.*, *Phys. Rev. Lett.* **31**, 678 (1973)).

Now consider what happens when spin–orbit coupling is introduced. This causes the $j = l - \frac{1}{2}$ component of the bound orbital, for which the spin–orbit potential is attractive, to be drawn slightly closer to the nucleus than the $j = l + \frac{1}{2}$. As a result, when the photoelectron energy increases from zero, the continuum orbital will have a greater overlap with the $j = l + \frac{1}{2}$ component of the bound state, and the ratio $\sigma(l+\frac{1}{2})/\sigma(l-\frac{1}{2})$ will be greater than the statistical value of $(l+1)/l$. Similarly, the continuum orbital will overlap the first node of the $j = l + \frac{1}{2}$ before that of the $j = l - \frac{1}{2}$, and in this part of the spectrum $\sigma(l+\frac{1}{2})/\sigma(l-\frac{1}{2}) < (l+1)/l$. Figure 5 illustrates the calculated behavior for Hg. The same considerations hold after the Cooper minimum is passed. Thus we can make the generalization that if the *partial* cross-section is rising, the ratio of cross-sections is greater than statistical, while if the partial cross-section is falling, the ratio will be less than statistical. For the noble gases Ar, Kr, and Xe, the cross-section is falling at 21·2 eV, while for Zn, Cd, and Hg it is rising.

In order to further test this prediction, we have measured[15] the ratio of intensities of $^2D_{\frac{5}{2}} : {}^2D_{\frac{3}{2}}$ for Hg at several photon energies spanning the wavelength region in which the cross-section for the d shell is rising, reaches a maximum, and begins to decline. The results, plotted in Fig. 6, provide a convincing verification of the theoretical predictions.

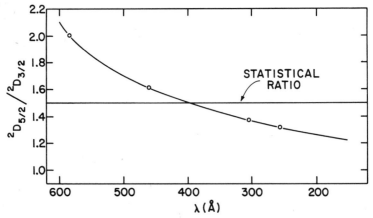

Fig. 6. $^2D_{\frac{5}{2}} : {}^2D_{\frac{3}{2}}$ ratio in Hg as a function of photon energy (from J. L. Dehmer and J. Berkowitz, *Phys. Rev.* **A10**, 484 (1974)).

III. PHOTOELECTRON ANGULAR DISTRIBUTIONS

The above discussion presupposes that the relative intensities being measured are meaningful. A possible discriminating factor in these measurements could arise if the angular distribution of photoelectrons was not the same for the transitions being compared. The angular distribution of photoelectrons for unpolarized incident light on an nl-subshell is given in the dipole approximation (which is excellent for low-energy photons) by[23]

$$\frac{d\sigma_{nl}(\varepsilon)}{d\Omega} = \frac{\sigma_{nl}(\varepsilon)}{4\pi}[1 - \tfrac{1}{2}\beta P_2(\cos\theta)]$$

where ε is the outgoing photoelectron energy, $\sigma_{nl}(\varepsilon)$ is the total subshell photoionization cross-section, and $\beta(\varepsilon)$ is the asymmetry parameter. The angle θ is between the photoelectron velocity and incoming photon direction and $P_2(\cos\theta) = \tfrac{1}{2}(3\cos^2\theta - 1)$. Using one-electron wave functions and LS coupling, it can be shown[23] that

$$\beta(\varepsilon) = \frac{\{l(l-1)\,R_{l-1}{}^2(\varepsilon) + (l+1)\,(l+2)\,R_{l+1}{}^2(\varepsilon) - 6l(l+1)\,R_{l-1}(\varepsilon)\,R_{l+1}(\varepsilon)\cos\,[\delta_{l+1}(\varepsilon) - \delta_{l-1}(\varepsilon)]\}}{(2l+1)\,[lR_{l-1}{}^2(\varepsilon) + (l+1)\,R_{l+1}{}^2(\varepsilon)]}$$

with the dipole matrix elements

$$R_{l\pm1}(\varepsilon) = \int_0^\infty P_{nl}(r)\,r P_{\varepsilon,l\pm1}(r)\,\mathrm{d}r$$

where $r^{-1}P_{nl}(r)$ and $r^{-1}P_{\varepsilon,l\pm1}(r)$ are the radial wavefunctions of the initial and final states of the photoelectron. Here the $\delta_{l\pm1}(\varepsilon)$ are the phase shifts of the continuum $l \pm 1$ partial waves with respect to free waves.

When $P_2(\cos\theta) = 0$, which occurs at $\theta = 54°\ 44'$, the various $\mathrm{d}\sigma_{nl}(\varepsilon)/\mathrm{d}\Omega$ are independent of β. The analyzer used in our experiments selects electrons close to this "magic angle", and hence the relative intensities discussed above are meaningful.

Manson[24] has calculated the angular distribution parameter β of the photoelectrons from the $5d$ shell of Hg as a function of energy within the

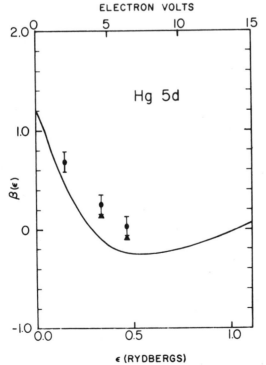

Fig. 7. Angular distribution parameter β as a function of photon energy in the photoionization of the $5d$ shell of Hg from Mansom (ref. 24).

LS approximation. The results are plotted in Fig. 7, together with experimental data of Harrison[25] and Niehaus and Ruf.[26] The agreement with experiment is fairly good, indicating among other things that the β values of the two spin–orbit components in Hg are not very different, since points from both states are plotted. A theory of photoelectron angular distributions taking into account different angular-momentum states has been developed by Fano and Dill.[27]

Harrison[25] has presented experimental evidence which indicates that a large difference exists between the β's associated with the $^2D_{\frac{5}{2}}$ and $^2D_{\frac{3}{2}}$ channels in the ionization of Cd $4d$ electrons. As Manson[24] notes, "This is rather puzzling since it implies that the spin–orbit interaction produces a major effect in Cd, but not in Hg where the effect would be expected to be even greater". It is clearly necessary to repeat this experiment.

IV. ALKALI HALIDES

The classical examples of molecules with ionic bonding are the alkali halides. Since they represent the opposite extreme from the oft-studied volatile and covalently bound systems and also because a number of *ab initio* calculations have been performed on this class of molecules, they were attractive candidates for photoelectron spectroscopic investigation. At this time, at least four laboratories[3, 5, 28, 29] have reported some results on the 584 Å PES of alkali halides. The cesium halide spectra reported by these laboratories are shown in Figs. 8–11. Only the threshold region has been reported in all cases† but one (CsI) where Berkowitz *et al.*[28] recorded another band in the 17–19 eV region.

The spectra published by the four groups demonstrate similarities, but also significant differences which are probably attributable to the method of generating the vapor, as well as the analyzer used for electron energy analysis. Our approach here will be first to compare the various results and the alternative interpretations and then to attempt to explain the differences in terms of experimental parameters.

There exists a rather extensive literature (*see*, for example, Rittner[30] and Berkowitz[31]) on the utilization of a classical ionic model with polarization for the calculation of various properties, such as dissociation energies, frequencies of vibration, and internuclear distances for alkali halide molecules. The noteworthy agreement of such calculations with experiment provided strong evidence for an M^+—X^- structure; recent extensive *ab initio* calculations[32] have verified this conclusion.

---

† Note added in proof. The He II photoelectron spectra of alkali halides have recently been reported. See A. W. Potts and T. A. Williams, *J. Chem. Soc. Far. Trans. II* (in press).

To translate this model into molecular orbital terminology, we note that atomic Cs has the configuration ...$(5p)^6 6s$. The halogens have a p^5 valence shell. A large degree of transfer of the Cs $6s$ will occur, tending to fill the halogen valence shell. In this $Cs^+—X^-$ picture, we are left with atomic-like molecular orbitals localized around either the Cs or X nucleus. The ionization potential of X^- in free space (i.e. the electron affinity (E.A.) of X) is by now a well-established quantity. The values for F, Cl, Br, and I are

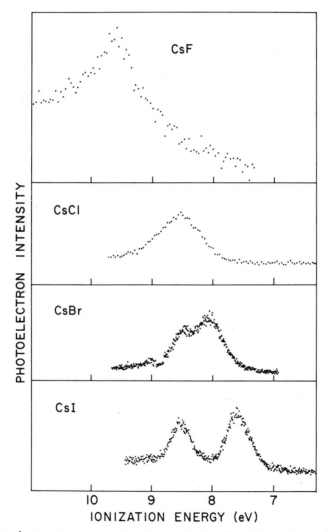

Fig. 8. Å photoelectron spectra of cesium halides from Berkowitz *et al.* (ref. 28)

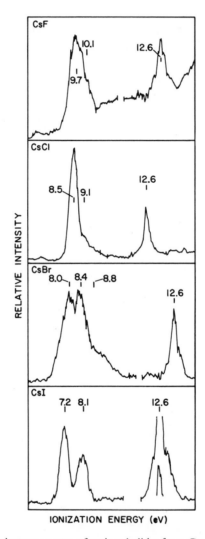

Fig. 9. 584 Å photoelectron spectra of cesium halides from Goodman *et al.* (ref. 29).

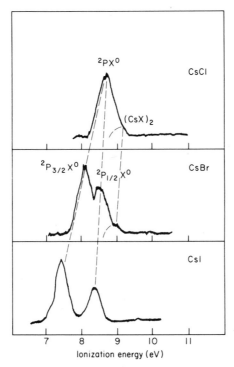

Fig. 10. 584 Å photoelectron spectra of cesium halides from Potts *et al.* (ref. 5).

3·448, 3·613, 3·363, and 3·063 eV, according to Berry.[33] However, the ionization potential of X^- in the field of Cs^+ is considerably larger, as is readily seen by a glance at Fig. 12. The reference point for ionization of X^- is depressed by the quantity D_{0i}, the energy for dissociation into ions. A typical ionic model calculation[34] of D_{0i} yields 5·55 eV for CsF, 4·77 eV for CsCl, 4·55 eV for CsBr, and 4·29 eV for CsI. If the resulting CsX^+ were unbound, we could calculate the first ionization potential of CsX as the sum $D_{0i}(CsX) + \text{E.A.}(X)$. However, Cs^+ and X^0 are weakly bound by a charge-induced dipole interaction $(\alpha e^2/2r^4)$ which leads to $D_0(CsX^+)$ in the neighborhood[5, 35] of 0·1–0·2 eV. One can readily form the identity

$$\text{I.P.}(CsX) + D_0(CsX^+) = D_{0i}(CsX) + \text{E.A.}(X)$$

With the previously cited values of D_{0i} and E.A.(X), we estimate I.P.(CsX) = 8·90, 8·28, 7·71, and 7·15 eV for CsF, CsCl, CsBr, and CsI. These values, together with our best reading of the adiabatic values from the graphs published by the aforementioned groups, are summarized in Table I.

The values of Goodman *et al.*[29] appear significantly lower for CsI, and Timoshenko and Akopyan[3] show a rather high value for CsBr, but apart

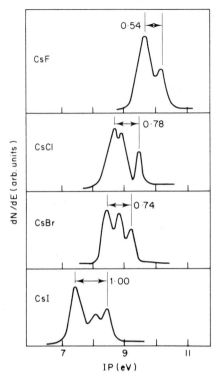

Fig. 11. 584 Å photoelectron spectra of cesium halides from Timoshenko and Akopyan (ref. 3).

TABLE I

Adiabatic First Ionization Potentials of CsX Molecules

	CsF	CsCl	CsBr	CsI
Berkowitz *et al.*	$\sim 8 \cdot 8_0{}^a$	$7 \cdot 8_4$	$7 \cdot 4_6$	$7 \cdot 1_0$
Potts *et al.*	—	$8 \cdot 2$	$7 \cdot 7$	$7 \cdot 2$
Goodman *et al.*	$9 \cdot 1$	$7 \cdot 7_5$	$7 \cdot 2_5$	$6 \cdot 5_5$
Timoshenko and Akopyan	$9 \cdot 2_5$	$8 \cdot 2$	$8 \cdot 2_5$	$7 \cdot 1_5$
Calculated	$8 \cdot 90$	$8 \cdot 28$	$7 \cdot 71$	$7 \cdot 15$

^{a} Sloping background taken into account.

from these discrepancies there is general agreement with the values based on this simple ionic model.

Now let us consider the possible higher ionization potentials. The binding energy of Cs $5p$ is 18–19 eV. (The other alkalis having p orbitals will have

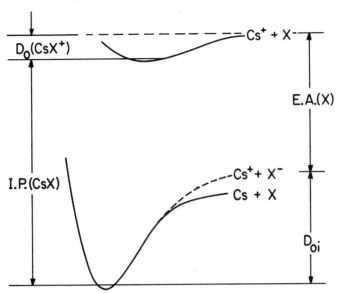

Fig. 12. Potential energy diagram for CsX and CsX^+.

still larger binding energies.) The halogen s orbitals lying just below the valence p orbitals have binding energies in excess of 21·2 eV. Hence the structure of the bands shown in Figs. 8–11 must somehow be associated with the halogen-like p orbitals. These p orbitals are split in both atomic and molecular systems by spin–orbit interaction into $j = \frac{3}{2}$ and $j = \frac{1}{2}$ subshells. The axial molecular field in a linear molecule introduces an additional splitting, analogous to Stark-field splitting, which separates the atomic p orbital into π and σ components according to the projection of the orbital angular momenta upon the internuclear axis. As we shall see below, these splittings are of comparable magnitude, and hence there is no preferred order for their evaluation in perturbation theory. One can begin with the spin–orbit split states, in the (atomic) zero field, and then show the influence of weak and strong axial fields (*see* Fig. 13 from Potts *et al.*[5]). Alternatively, one can start such a calculation with π–σ separations deduced from *ab initio* calculations, then introduce the spin–orbit interaction. In either case, one is left with a 2×2 secular equation to solve for the final effective splittings. Our first task is to estimate the spin–orbit and π–σ splitting parameters for the various alkali halides.

Ab initio calculations are available for 9 of the 20 alkali halides.[32] They demonstrate a monotonic decrease of π–σ splitting from 0·591 eV for LiF to 0·026 eV for RbF as the metal radius and internuclear distance increase (and hence the axial field decreases), and a monotonic increase (0·591 eV

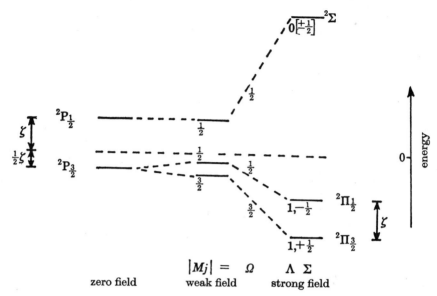

Fig. 13. Correlation of the various energies of ionized states formed by Stark splitting of the X^0 ^2P state from Potts *et al.* (ref. 5).

for LiF to 0·876 eV for LiBr; 0·362 eV for NaF to 0·472 eV for NaBr) as the halogen size and its polarizability increase. From this pattern, it is possible to estimate the π–σ splitting of the remaining alkali halides. Table II summarizes these values. Of immediate interest are the relatively small values for the cesium halides.

The spin–orbit parameters are dominated by the appropriate atomic halogen spin–orbit parameter for the valence p orbital, since the molecular orbital in the LCAO expansion is predominantly this atomic p orbital. Slight departures from this spin–orbit constant can be estimated from the

TABLE II

Separation of the π- and δ-Orbitals derived from the Uppermost Occupied (Halogen p-like) Orbitals (eV) (the π is least bound in all cases)[a]

	F	Cl	Br	I
Li	0·591	0·754	0·770	(0·8)
Na	0·362	0·453	0·472	(0·5)
K	0·215	0·248	(0·27)	(0·3)
Rb	0·026	(0·1)	(0·15)	(0·19)
Cs	(0)	(0·03)	(0·05)	(0·1–0·14)

[a] Values have been obtained assuming Koopmans' Theorem. The quantities in parentheses are estimated from the trends observed in the calculated values.

magnitudes of the contributions of other orbitals in the LCAO molecular wavefunction, and their spin–orbit parameters. Table III lists the spin–orbit coupling constants of the free halogen atoms, together with their modified

TABLE III

Diagonal Spin-orbit Coupling Constants of the Uppermost Occupied π-Orbitals of the Alkali Halides (eV)[a]

	F	Cl	Br	I
Atom	0·033	0·073	0·302	0·628
Li	0·028	0·063	0·263	0·530
Na	0·028	0·063	0·263	0·530
K	0·029	0·063	0·264	0·531
Rb	0·037	(0·068)	(0·273)	(0·546)
Cs	(0·043)	(0·073)	(0·298)	(0·628)

[a] The quantities in parentheses are estimated from the trends observed in the calculated values.

values in the particular alkali halides. For a more detailed description of this procedure, the reader is referred to the original article.[28] With the relevant parameters in hand we shall now illustrate the calculational method, using CsBr as our example. We begin by treating π–σ and spin–orbit splittings independently. Thus, for CsBr[+],

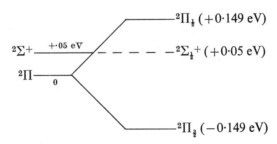

It is now necessary to take into account the off-diagonal elements corresponding to the Ω–Ω interaction of $^2\Pi_{\frac{1}{2}}$ and $^2\Sigma_{\frac{1}{2}}$. This matrix is of the form

	$^2\Pi_{\frac{1}{2}}$	$^2\Sigma_{\frac{1}{2}}$
$^2\Pi_{\frac{1}{2}}$	$\varepsilon(\Pi_{\frac{1}{2}})$	ε_{12}
$^2\Sigma_{\frac{1}{2}}$	ε_{21}	$\varepsilon(\Sigma_{\frac{1}{2}})$

where $\varepsilon(\Pi_{\frac{1}{2}}) = +0{\cdot}149$ eV, $\varepsilon(\Sigma_{\frac{1}{2}}) = +0{\cdot}05$ eV, and

$$\varepsilon_{12} = \varepsilon_{21} = \left\langle {}^2\Pi_{\frac{1}{2}} \left| \frac{Z_{\text{eff}}}{r^3}(l_x s_x + l_y s_y) \right| {}^2\Sigma_{\frac{1}{2}} \right\rangle$$

summed over both atoms. We transform $l_x s_x + l_y s_y = \tfrac{1}{2}(l^+ s^- + l^- s^+)$. But

$$l^+ s^- = \sqrt{[(l-\Lambda)(l+\Lambda+1)(s+\Sigma)(s-\Sigma+1)]}$$
$$l^- s^+ = \sqrt{[(l+\Lambda)(l-\Lambda+1)(s-\Sigma)(s+\Sigma+1)]}$$

where we have defined Λ and Σ as the projections m_l and m_s along the inter-nuclear axis of the electron's orbital and spin angular momenta. For $s = \tfrac{1}{2}$ and $\Sigma = \tfrac{1}{2}$, the second term drops out; for the first term operating on the Σ function, $\Lambda = 0$ and we are left with

$$l^+ s^- = \sqrt{[l(l+1)]}.$$

For a predominant "p" contribution, this becomes $\sqrt{2}$. In addition,

$$\left\langle \Pi \left| \frac{Z_{\text{eff}}}{r^3} \right| \Sigma \right\rangle \approx \left\langle \Pi \left| \frac{Z_{\text{eff}}}{r^3} \right| \Pi \right\rangle$$

in this case, as may be directly verified by evaluating these integrals for those cases where *ab initio* calculations are available. For our purposes, the off-diagonal elements are approximately

$$\tfrac{1}{2}(\sqrt{2}) \left\langle \Pi \left| \frac{Z_{\text{eff}}}{r^3} \right| \Pi \right\rangle$$

or $1/\sqrt{2}$ of the corresponding spin–orbit coupling constants. For the case of CsBr$^+$, our matrix becomes

	$^2\Pi_{\frac{3}{2}}$	$^2\Sigma_{\frac{1}{2}}$
$^2\Pi_{\frac{3}{2}}$	0·149	0·298/$\sqrt{2}$
$^2\Sigma_{\frac{1}{2}}$	0·298/$\sqrt{2}$	0·05

The roots of the secular equation are $+0\cdot316\,\text{eV}$ and $-0\cdot117\,\text{eV}$, so that the final energy level scheme becomes

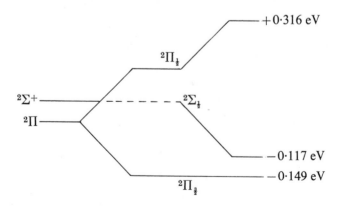

The $^2\Pi_{\frac{3}{2}}$ and $^2\Sigma_{\frac{1}{2}}^+$ states are separated by only 0·032 eV, while $^2\Pi_{\frac{1}{2}}$ is separated from the mean of the other states by ~ 0.45 eV. Both Potts et al.[5] and Berkowitz et al.[28] display two partially resolved peaks in the CsBr spectrum, the former reporting a splitting of 0·41 eV, the latter estimated from the graph to be 0·4–0·45 eV. In both experiments the lower energy peak is more intense; the calculation indicates that it has two components, the higher energy peak only one.

Note that this splitting is very nearly that of the bromine atom, not two-thirds of this value as in HBr or CH_3Br.[1] This is because the bromine in CsBr is almost spherically symmetric (i.e. the π–σ splitting is small compared with the spin–orbit splitting). Conversely, if the experimental molecular spin–orbit splitting is very near to the atomic value, it implies a small π–σ separation. The inherent width of the peaks prevents separation of the two lower components for CsBr although calculations on other alkali halides predict that three peaks should be observable in some cases.

Taking into account the peak widths, the corresponding calculation for CsI predicts a splitting into two well-resolved peaks, with a separation of ~ 0.95 eV, and for CsCl and CsF, a single peak, in good agreement with the data of Potts et al.[5] and Berkowitz et al.[28] However, Goodman et al.[29] and Timoshenko and Akopyan[3] present evidence for an additional peak to higher energy in almost every instance. Both these groups identify the higher-energy peak as the "missing" $^2\Sigma^+$ peak. Timoshenko and Akopyan go so far as to calculate the $^2\Pi_{\frac{3}{2}}$–$^2\Sigma_{\frac{1}{2}}^+$ separation, and report fair agreement with experiment. It is instructive to examine their calculation in detail.

They begin with the atomic spin–orbit split states $^2P_{\frac{1}{2}}$, $^2P_{\frac{3}{2}}$. Then a Stark field is applied (the axial field of the molecular ion). This field should only shift states with $J = 0, \frac{1}{2}$, but will split and shift higher J states. In particular, the $^2P_{\frac{3}{2}}$ state is supposed to split into $^2\Pi_{\frac{3}{2}}$ and $^2\Sigma$. (Note that this is not the splitting behavior shown in Fig. 13, taken from Potts et al. but $^2\Pi_{\frac{1}{2}}$ and $^2\Sigma_{\frac{1}{2}}$ can interact and lose their distinct identity.) Timoshenko and Akopyan estimate the Π–Σ splitting by making use of "known" polarizability values for the free halogen atoms.

The polarizability $\alpha(n)$ of an atom in the state n can be obtained from the relation[36]

$$\frac{\partial}{\partial \mathscr{E}}(\Delta E_n) = -\alpha(n)\,\mathscr{E}$$

where ΔE_n is the shift of the level in the electric field $\mathscr{E}$. Conventionally one speaks of the polarizability of an atom with reference to its ground state.

In our case, the halogen ground state $^2P_{\frac{3}{2}}$ will be shifted by an amount

$$\Delta E_1 = \tfrac{1}{2}\alpha_1 \mathscr{E}^2$$

where ΔE_1 and α_1 refer to the ground state. By arguments that are not entirely clear, they[3] then state that $^2P_{\frac{3}{2}}$ will also split into $^2\Pi_{\frac{3}{2}}$ and $^2\Sigma_{\frac{1}{2}}$, the upward and downward *splittings* each being equal to ΔE_1, and hence the separation between $^2\Pi_{\frac{3}{2}}$ and $^2\Sigma_{\frac{1}{2}}$ equalling $2\Delta E_1$. Sobelman[36] has shown that the general effect of a Stark field on an atom in quantum states γJM is

$$\Delta E_{\gamma JM} = \mathscr{E}^2\{A_{\gamma J} + B_{\gamma J} M^2\},$$

the first term representing a shift, the second a splitting into different $|M| = J, J-1, \ldots$. Sobelman notes that the asymmetry of the splitting (i.e. the dependence on M) is a special feature of this equation. In the present instance we are dealing with $|M| = \frac{3}{2}$ and $\frac{1}{2}$ and it is not clear why the splitting should be equal for these two states. A further confusion arises regarding the applicability of the quadratic Stark effect. When the unperturbed levels are strictly degenerate, the splitting depends linearly on $\mathscr{E}$ for infinitesimally small values of $\mathscr{E}$ if the states mix (i.e. if the dipole matrix element connecting these states is non-vanishing). This is the case for $n > 1$ in atomic hydrogen, where the l states are degenerate in zero field. In the present case, the $^2P_{\frac{3}{2}}$ state is degenerate in zero field, but its components ($^2\Pi_{\frac{3}{2}}$ and $^2\Sigma_{\frac{1}{2}}$) do not mix. The interaction that causes Stark shifts must then come from higher levels. Since the energy gap to these higher levels is larger than the magnitude of the Stark shift, the quadratic Stark effect is indeed applicable.

The change in energy due to the Stark effect, $\Delta E_{\gamma JM}$, is given by second-order perturbation theory as

$$\Delta E_{\gamma JM} = \sum_{\gamma' J'} \frac{|\langle \gamma JM | D_z | \gamma' J' M \rangle|^2}{E_{\gamma J} - E_{\gamma' J'}} \mathscr{E}^2 \tag{A}$$

Note that the dipole operator D_z will connect states differing by one unit of J, but M is unchanged.

The summation has contributions from all higher states that connect with the state in question, e.g. $^2\Pi_{\frac{3}{2}}$ and $^2\Sigma_{\frac{1}{2}}$. The $^2\Pi_{\frac{3}{2}}$ is the ground state, and hence the shift in its energy from the zero field value is in fact determined by the conventional dipole polarizability, α_1. However, the shift in the $^2\Sigma_{\frac{1}{2}}$ state should be determined by mixing with another class of states, namely those with $M = \frac{1}{2}$. Among these states is the heretofore neglected $^2P_{\frac{1}{2}}$ atomic state. It would be expected that the $^2\Sigma_{\frac{1}{2}}$ state will also be shifted downward in energy, and hence the net $^2\Pi_{\frac{3}{2}}$–$^2\Sigma_{\frac{1}{2}}$ splitting should be less than ΔE_1 for a homogeneous field.

We can make an assessment of the magnitude of this splitting in two special cases. Stevens and Billingsley[37a] have calculated the polarizabilities of the lowest $^2\Pi$ and $^2\Sigma$ states in atomic fluorine by considering their respective shifts in energy when a point charge approaches. They report

$\alpha(^2\Pi) = 0.48 \times 10^{-24}$ cm^3 and $\alpha(^2\Sigma) = 0.44 \times 10^{-24}$ cm^3. By considering the field due to a point charge at the equilibrium internuclear distance of CsF, we obtain $\Delta E_\Pi = -0.114$ eV and $\Delta E_\Sigma = -0.105$ eV, or $\Delta E_{\Pi-\Sigma} = 0.009$ eV. This is to be compared with an estimate of 0.0 eV by Berkowitz *et al.* and a contribution to splitting of 0.184 eV by Timoshenko and Akopyan. Stevens[37b] has made a calculation for atomic chlorine comparable to that for fluorine, and obtains $\alpha(^2\Pi) = 2.155 \times 10^{-24}$ cm^3, $\alpha(^2\Sigma) = 1.925 \times 10^{-24}$ cm^3. For CsCl at its equilibrium internuclear distance, this leads to $\Delta E_\Pi = -0.217$ eV, $\Delta E_\Sigma = -0.194$ eV and $\Delta E_{\Pi-\Sigma} = 0.023$ eV. Berkowitz *et al.* estimate 0.03 eV for this splitting, whereas the comparable value from Timoshenko and Akopyan is 0.462 eV.

It should be pointed out that this type of calculation is most valid at large internuclear separations, where the test charge does not penetrate the halogen charge cloud. Obviously, at short distances one is dealing with an HX$^+$ ion, which is quite different from an alkali-halide ion. At such short distances major molecular effects and internuclear repulsions enter, which are not considered in a Stark field model. It can readily be shown that the above calculations underestimate the Π–Σ separation more seriously as one goes to lighter metal chlorides and fluorides, which have shorter internuclear distances. However, for the cesium halides the calculation presented above does appear to be a reasonable approximation.

Timoshenko and Akopyan make a transition from Eq. (A) to the two-state Eq. (B)

$$\Delta E_{JM} = \frac{|\langle JM|D_z|J'M'\rangle|^2}{\Delta\varepsilon}\mathscr{E}^2 \tag{B}$$

omitting the summation, allowing for $M \neq M'$ and not specifying $\Delta\varepsilon$. If indeed the $^2\Pi_{\frac{3}{2}}$ and $^2\Sigma_{\frac{1}{2}}$ could mix and were initially separated by $\Delta\varepsilon$, this equation would be valid and the downward shift of $^2\Pi_{\frac{3}{2}}$ would be equal to the upward shift of $^2\Sigma_{\frac{1}{2}}$, so that the net splitting would be $2\Delta E_{JM}$. However, we have already shown that they do not, and we see no justification for Eq. (B).

The π–σ splittings which we have previously deduced[28] from extrapolation of *ab initio* calculations on alkali halides to the cesium halides (0.0, 0.03, 0.05, and 0.12 eV for CsF, CsCl, CsBr, and CsI, respectively) are an order of magnitude smaller than those calculated by Timoshenko and Akopyan. This would imply that the $^2\Pi_{\frac{3}{2}}$ and $^2\Sigma_{\frac{1}{2}}$ are primarily shifted downward together, and not in opposite directions. This conclusion agrees with our inference regarding the behavior of these states under the influence of a Stark field.

What, then, is the origin of the high-energy peaks observed by both Goodman *et al.* and by Timoshenko and Akopyan? We believe that the additional high-energy peak is due to a dimer, an M_2X_2 species. There exists an extensive literature on the relative abundance of dimers in various alkali halides (*see*, for example, ref. 38). One compilation[34] lists dimer : monomer ratios varying from 0·02 to 0·22 among the cesium halides, as much as 0·35 for potassium halides, and values of about unity for sodium and lithium halides. Do we have reason to believe that a dimer peak will occur at the observed energy, i.e. $1 \pm 0·5$ eV above the monomer band?

Photoionization measurements on NaI[39] and some rubidium and cesium halides[40] indicate that the first ionization potentials of monomer and dimer are very close to one another, probably within 0·2 eV. Electron impact measurements of ionization potentials[38, 41] provide further evidence for this conclusion. CNDO, INDO, and extended Hückel calculations on lithium and sodium halide dimers, based on a rhombic structure of known inter-nuclear distance, indicate six closely spaced levels in the threshold region (*see* Figs. 14, 15). In addition, an *ab initio* calculation has been performed[42]

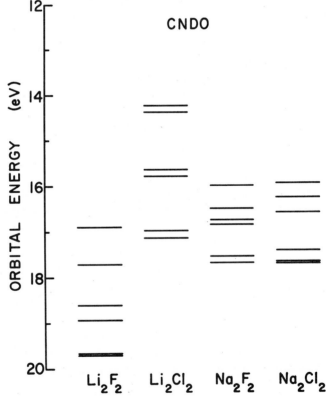

Fig. 14. Diagram of orbital energies of valence levels of $(LiF)_2$, $(LiCl)_2$, $(NaF)_2$, and $(NaCl)_2$ obtained from CNDO calculations.

on Li_2F_2, and also indicates six closely spaced levels, as shown in Fig. 16. For lithium halide dimers, these six levels may span ~ 3 eV, but for sodium halides this range is $\leqslant 2$ eV. Note that these calculations do not include spin–orbit coupling. Whereas the absolute binding energies are not given accurately by the semi-empirical calculations and Koopmans' Theorem must be invoked, the relative values can be expected to be much better according to the preponderance of evidence in photoelectron spectroscopy. Since the monomer spectrum dominates the first eV in the threshold region, one can anticipate that the photoelectron spectrum will be most sensitive to the dimer in the second eV above the lowest ionization limit.

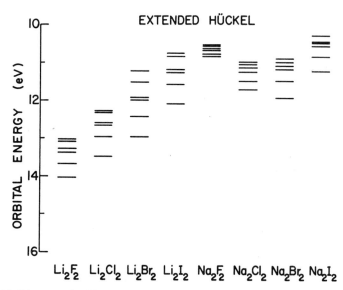

Fig. 15. Diagram of orbital energies of valence levels of $(LiX)_2$ and $(NaX)_2$ obtained from extended Hückel calculations (courtesy of L. C. Cusachs).

From our previous observation about the relative importance of dimers, we can anticipate that the high-energy peak will become relatively more intense as one progresses from cesium to rubidium to potassium and sodium halides. Figure 17, taken from the results of Potts *et al.*, shows just such a progression.

We are now in a position to examine why the results of the four groups of workers differ on the cesium halides. Berkowitz *et al.* and Potts *et al.* used a more-or-less conventional resistively heated oven to produce the vapor, and a selector analyzer to generate the photoelectron spectrum. Goodman *et al.* used a laser-heated crucible with a selector analyzer. It is quite conceivable that the laser-heated system operated at higher temperature, and hence yielded a higher dimer : monomer ratio. (Note that this does *not* imply superheating of the vapor). The experiments of Berkowitz *et al.* required the vapor to pass through a long, heated tube after sublimation, which could have the effect of diminishing the dimer : monomer ratio. Timoshenko and Akopyan used a quartz oven, presumably of conventional design, but their analyzer was a planar retarding field. The spectrum they

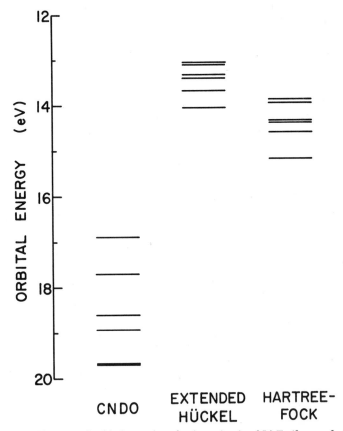

Fig. 16. Diagram of orbital energies of valence levels of Li_2F_2 (from ref. 41).

report is smoothed and differentiated. Their high energy peak may be real and due to dimer, indicating a fairly high vaporization temperature, but there is evidence of some spurious structure in their spectrum as well which may have been introduced by the smoothing and differentiating procedure. We refer here in particular to the splitting of the monomer peak in CsCl, and the odd structure in CsI, when compared to the results of the other investigators.

A summary of the observed splittings in the alkali halides, and the corresponding values calculated by our prescribed method, is given in a table by Potts et al.[5] The agreement is excellent. A more complete table of calculated values is given by Berkowitz et al.[28] Of particular note is the observation that in NaI, the $\pi-\sigma$ and spin–orbit separations are both large enough to suggest that a third peak should be barely resolvable in the monomer, and this shows up in the data of Potts et al.

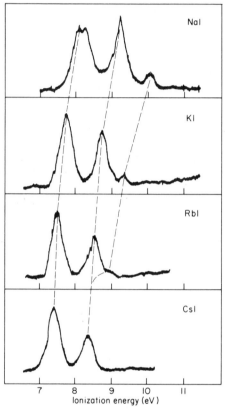

Fig. 17 The 584 Å photoelectron spectra of NaI, KI, RbI, and CsI (from ref. 5).

The only spectrum reported thus far[28] that displays higher ionization potentials is that of CsI, where there is a large band in the 17–19 eV region. This band is undoubtedly due to ionization from the cesium end of the molecule, with the attendant $\pi-\sigma$ and spin–orbit interactions, but a detailed interpretatoin is not possible at this time†. The implications of these studies for the interpretation of the fragmentation pattern (mass spectrum) of the alkali halides are dicussed by Bekowitz et al.[28]

† See footnote on p. 364.

Finally, it is necessary to take issue with a conclusion reached by Potts *et al.* in a note at the end of their paper. They infer that the ionization energy of the outermost electron increases as we go from monomer to dimer to polymer. We have already seen that the high-energy peak is a measure of the presence of dimer, and Potts *et al.* place significant emphasis on its higher energy as being representative of a higher first I.P. for the dimer. However, our earlier discussion has also shown that this is the high end of a 1–2 eV range of orbital energies for the dimer. Experimental evidence from photoionization studies cited earlier indicates that the first I.P. of monomer and of dimer differ by at most 0·2 eV. Hence, we feel that the difference in ionization energies of outermost electrons in dimer and monomer, listed by Potts *et al.*[5] as varying from 0·40 eV for CsCl to 0·96 eV for NaCl, is not really a variation in first ionization potential, but a variation in the span of the cluster of six orbitals in the valence region of the dimer spectrum. It will be recalled that the trend was toward a smaller span of energies as the size of the molecule increased, a result which is to be expected as the atomic orbitals interact less strongly.

V. GROUP II HALIDES

A. Group IIa Halides

The Group IIa alkaline earth dihalides have largely been avoided in experimental PES, and have only begun to be examined by *ab initio* calculations.[43-45] The lack of experimental data can be explained readily.

(a) The beryllium halides are dangerous materials, and the vapors tend to form dimer species, Be_2X_4.

(b) The other alkaline earth halides require vaporization temperatures at or beyond the limit of the current high temperature PES instruments.

By contrast, the Group IIb dihalides (i.e. the Zn, Cd, and Hg dihalides) have been investigated in varying degrees of completeness by five different groups. There is general agreement on some points, but significant discrepancy on others. There exists one *ab initio* calculation[46] (on ZnF_2) and semiempirical extended Hückel calculations[47] have been performed on the others. All are known to be linear.

Eland[48] first examined the most volatile members of this class, $HgCl_2$, $HgBr_2$, and HgI_2. These could be studied with a more or less conventional photoelectron apparatus, by heating the entire inlet system to temperatures between 80 °C and 140 °C. He showed that the orbital sequence in the valence region was $...\sigma_g^2 \sigma_u^2 \pi_u^4 \pi_g^4$, the π_g being most weakly bound. The π_g and π_u orbitals are additionally split by spin–orbit interaction. These valence orbitals, composed largely of atomic halogen p orbitals, are quite similar to those of the diatomic halogens, whose sequence is $...\sigma_u^2 \sigma_g^2 \pi_u^4 \pi_g^4$.

As a result of interaction with the outer "*s*" orbital of the central atom, the σ_g orbital lies relatively deeper than in the diatomic halogens. This observation was made by Eland, and has been corroborated by *ab initio* calculations on ZnF_2[46] and semi-empirical calculations on all of the Group IIb dihalides.[47] Subsequent results on $ZnCl_2$, $ZnBr_2$, and ZnI_2 by Cocksey *et al.*[2] and on the zinc and corresponding cadmium halides by Boggess *et al.*[49] confirmed that the orbital sequence was the same for these related compounds in the valence region. Somewhat higher resolution data of Berkowitz[50] on the zinc and cadmium halides are also in agreement with the general picture drawn by Eland. A summary of the results in the valence region is given in Table IV.

The most general observation that can be made about the variation of ionization potentials in Table IV is that for corresponding states they are highest for the zinc halides, decrease for the cadmium halides, then increase again or remain the same for the mercuric halides. The non-bonding π_g orbital has a higher binding energy in the ionic Group IIb dihalides than in the corresponding diatomic halogens; the bonding π_u and σ_u orbitals have lower binding energies in MX_2 than in X_2.

In his mercuric halide paper, Eland noted that there were additional peaks at higher energy, which he attributed to ionization from inner *d*-like orbitals of Hg. In atomic Hg, ionization from the $(5d)^{10}$ shell gives rise to $^2D_{\frac{5}{2}}$ and $^2D_{\frac{3}{2}}$ peaks, at 14·840 eV and 16·705 eV, respectively. In the axial field of the molecule, the *d*-orbital will split further into σ, π, and δ components. Hence, one is faced again with a combination of axial electric field and spin–orbit splittings, of comparable magnitude. Ionization from inner *d*-like orbitals has now been reported for the zinc halides by Cocksey *et al.*[2] and Orchard and Richardson[6] and on zinc and cadmium halides by Berkowitz.[50] Detailed analysis depends upon some knowledge of the spin–orbit parameters and ligand-field splittings of the various molecules. The spin–orbit separation for corresponding *d* levels is 0·337, 0·699, and 1·865 eV for Zn, Cd, and Hg, respectively. The only *a priori* handle we have on the ligand-field splitting comes from the *ab initio* calculation of Yarkony and Schaeffer[46] on ZnF_2. Pertinent results from this calculation are summarized in Table V. We note that ionization from the zinc 3*d* region primarily spans the range from 23·41 eV to 23·70 eV. It is of interest that while the σ_g sub-level of this *d* complex lies highest, it is followed by δ_g and then π_g. Orchard and Richardson have analyzed their data on $ZnCl_2$, $ZnBr_2$, and ZnI_2 in terms of an orbital sequence $...\delta\pi\sigma$, the σ lying highest. The separations are small, and violation of Koopmans' Theorem might bring the two alternative calculations into harmony. Orchard and Richardson find $\Delta_{\sigma\pi} = 0.17$ eV, $\Delta_{\sigma\delta} = 0.01$ eV, and ζ_d (spin–orbit parameter) = 0·135 eV for $ZnCl_2$. The axial field splitting parameter and spin–orbit

TABLE IV

Vertical Ionization Energies (eV) of Zinc, Cadmium and Mercury Halides[a]

State	$ZnCl_2$	$ZnBr_2$	ZnI_2	$CdCl_2$	$CdBr_2$	CdI_2	$HgCl_2$	$HgBr_2$	HgI_2
$^2\Pi_{\frac{3}{2},g}$	11·85[b]	10·90	9·76	11·42[b]	10·59	9·53	11·37	10·62	9·50
$^2\Pi_{\frac{1}{2},g}$		11·28$_5$	10·32		10·96$_5$	10·07	11·50	10·96	10·16
$^2\Pi_{\frac{3}{2},u}$	12·41[c]	11·46	10·40	11·92[c]		10·21	12·13[c]	11·20	10·00
$^2\Pi_{\frac{1}{2},u}$		11·62$_5$	10·57$_5$		11·31			11·54	10·40
$^2\Sigma_u$	13·09	12·33	11·53	12·46	11·85	11·20	12·74	12·09	11·29
$^2\Sigma_g$	14·13	13·55	12·80	13·29	12·84	12·27	13·74	13·39	12·85

[a] The values for Zn and Cd halides are from Berkowitz (ref. 50); those for Hg halides are from Eland, (ref. 48).
[b] Values for both the $^2\Pi_{\frac{3}{2},g}$ and $^2\Pi_{\frac{1}{2},g}$ states.
[c] Values for both the $^2\Pi_{\frac{3}{2},u}$ and $^2\Pi_{\frac{1}{2},u}$ states. A value listed between two states indicates that the two were not resolved, and both states probably contribute to the peak.

TABLE V

Atomic Composition and Binding Energies of Orbitals in $ZnF_2{}^a$

Binding energy	Type	Zinc atomic orbitals (%)			Fluorine atomic orbitals (%)	
		s	p	d	s	p
15·38	$2\pi_g$			3·3		96·7
15·38	$2\pi_g$			3·3		96·7
15·83	$5\sigma_u$		6·6		1·2	92·2
15·83₅	$3\pi_u$		1·8			98·2
15·83₅	$3\pi_u$		1·8			98·2
15·83₆	$7\sigma_g$	17·2		7·9	1·1	73·8
23·41	$6\sigma_g$	0·4		88·2	2·0	9·4
23·65	$1\delta_g$			100·0		
23·65	$1\delta_g$			100·0		
23·70	$1\pi_g$			96·9		3·1
23·70	$1\pi_g$			96·9		3·1

^{a} From the *ab initio* calculations of D. R. Yarkony and H. F. Schaefer (ref. 46).

splitting parameter for Zn $3d$ are very nearly the same. From our study of the alkali halides, we might expect the axial field splitting to diminish as the internuclear distance increases. However, Orchard and Richardson find a larger orbital splitting parameter for ZnI_2 than for $ZnCl_2$ and $ZnBr_2$. They argue that this "... can be understood in terms of more 'exposed' d orbitals in ZnI_2, the $3d$ radial functions being somewhat expanded as a result of greater s/p covalency (which increases the shielding of the d electrons)". In any event, this effect is likely to be small compared to the increase in spin–orbit coupling as one proceeds to Cd, and finally Hg halides. For the latter cases, one can anticipate that spin–orbit coupling will be dominant. Eland has analyzed his mercuric halide data on this assumption, and concludes that the $^2D_{\frac{5}{2}}$ spin–orbit state should be split into three components of equal statistical weight, $\pm\frac{5}{2}$, $\pm\frac{3}{2}$, and $\pm\frac{1}{2}$, while the $^2D_{\frac{3}{2}}$ should be split likewise into $\pm\frac{3}{2}$ and $\pm\frac{1}{2}$. He concludes that the sharp line observed in each of his spectra of the mercuric halides corresponds to $\pm\frac{5}{2}$, and is sharp because this state involves electrons lying in a plane perpendicular to the molecular axis, which are hardly broadened by the ligand field (although stabilized by it). A second band represents the $\pm\frac{3}{2}$ and $\pm\frac{1}{2}$ components of $^2D_{\frac{5}{2}}$, while the third band represents $\pm\frac{3}{2}$ and $\pm\frac{1}{2}$ components of $^2D_{\frac{3}{2}}$. The separation between the third band and the sharp $\pm\frac{5}{2}$ line ranges from 1·92 eV to 1·95 eV, close to the spin–orbit splitting in atomic mercury. In gross terms, one can talk about the spin–orbit split states $^2D_{\frac{5}{2}}$ and $^2D_{\frac{3}{2}}$ of Hg shifted in energy and modified in shape by ligand field splitting, but retaining about the same separation they have in atomic mercury. The only data on cadmium halides,[49] while wanting for higher resolution, appear to be interpretable in the same manner. The zinc halides appear more complex, presumably because of the closely comparable spin–orbit and ligand field splittings. If one slightly "blurs" the higher resolution data of Orchard and Richardson on $ZnCl_2$, one can infer two bands at 19·10 eV and 19·45 eV; this may be compared with the values reported by Berkowitz at 19·23 eV and 19·51 eV. The separation of $\sim 0·3$ eV is close to the spin–orbit splitting of $^2D_{\frac{5}{2}}-^2D_{\frac{3}{2}}$ in atomic zinc. A similar analysis of $ZnBr_2$ yields $\sim 18·80$ eV and 19·12 eV from the data of Orchard and Richardson, while Berkowitz reports 18·89 eV and 19·19 eV, again resulting in an average separation of $\sim 0·3$ eV, with the data of Berkowitz being about 0·08 eV larger than the averaged values of Orchard and Richardson. Hence, even these zinc halides are crudely interpretable by the same model. The ZnI_2 may be too complex to be incorporated in the simple model, however. The signal of Cocksey et al.[2] is barely detectable in this region, the data of Berkowitz are insufficiently resolved, and cannot readily be compared with those of Orchard and Richardson, which show at least five distinct peaks. It is of interest to note that Eland had initially identified the lowest energy component as a sharp, $\pm \frac{5}{2}$ peak. By the time one gets to ZnI_2, the first peak is the broadest one, according to the spectrum of Orchard and Richardson.

B. Energy Shifts and the Electronegativity Scale

Whereas there was general agreement among the various investigators regarding the identification of peaks and the corresponding ionization energies in the valence region, this is not the case for the d-ionization region. Boggess et al.[49] do not report on this region at all. Berkowitz,[50] and Orchard and Richardson[6] are in fairly good agreement on the zinc halides, if account is taken of the different resolution in the two experiments. The data of Berkowitz correspond to an average shift in the d-orbital binding energies to higher values by $2·04_5$, $1·69_5$, and 1·23 eV for $ZnCl_2$, $ZnBr_2$, and ZnI_2, respectively. For the mercuric halides, Eland gives corresponding shifts of 1·87, 1·56, and 1·15 eV for $HgCl_2$, $HgBr_2$, and HgI_2. Hence, the zinc halides are seen to show a *larger* shift in each instance than the mercuric halides. On the other hand, Cocksey et al.[2] report shifts of 1·72, 1·39, and 1·09 eV for $ZnCl_2$, $ZnBr_2$, and ZnI_2. In each instance, they report a *smaller* shift for the zinc halides than for the corresponding mercuric halides. This difference in observed energies leads directly to opposite interpretations regarding the relative electronegativities of zinc and mercury.

A simple electrostatic model was introduced by the Uppsala group[51] to account for chemical shifts in inner shells. It attributes the shift to a different distribution of charges of the atom in question in the chemical environment of the molecular ground state, i.e. relaxation phenomena in the final ionized states are implicitly assumed to play a minor role. Eland[48] has also used this model, which proceeds as follows.

A displacement of charge q in an outer valence shell of radius r will produce a change in all inner shell orbital energies ΔE proportional to q. For a particular shell, there is a specific proportionality constant k_n such that

$$\Delta E_n = k_n q/r \qquad (C)$$

We can evaluate k_n from tables of atomic energies[19] if we know ΔE_n corresponding to $q = 1$. In the cases at hand, if we know the d-orbital ionization energy in Hg^0, and Hg^+ (where Hg^+ refers to removal of a $6s$ electron), then k_n is determined for a reasonable choice of r. Eland had some success with the choice of $r = 1.2$ Å for the $6s$ shell of mercury. With this choice, and $\Delta E_n = 9.427$ eV for the $5d$ orbital in Hg, $k_n = 11.3124$. It so happens that 1.20 Å corresponds to the outer peak of the $6s$ wavefunction in an accurate relativistic calculation[52] of the Hg atom.

We have chosen to maintain this same pattern of calculation for Zn and Cd. Thus for Zn the outermost peak of the $4s$ orbital is calculated to be at 1.30 Å, ΔE_n for the $3d$ shell is 10.592 eV and $k_n = 13.7670$. For Cd, the outermost peak of the $5s$ orbital is calculated to be at 1.275 Å, $\Delta E_n = 9.445$ for the $4d$ shell and $k_n = 12.0463$.

The modification of Eq. (C) for the case where a charge q is not removed to infinity, but rather to the mean position of the nearby atom which attracts this charge, is[51]

$$\Delta E_{cs} = k_n\left(\frac{1}{r} - \frac{1}{R}\right)q \qquad (D)$$

where ΔE_{cs} is the chemical shift in inner orbital energy due to a shift of charge q from the neutral atom to its neighbor, whose mean position is R units away, i.e. the internuclear distance R. The internuclear distances of the pertinent Group IIb dihalides are well known.[53] From the experimentally determined ΔE_{cs} and k_n, r and R determined in the prescribed way, we have calculated q for each of these cases and included these values in Table VI.

From the charge q, it is possible to calculate the per cent ionic character, χ (see ref. 50). These quantities are also included in Table VI. The average values of χ are $1.77_4 \pm 0.02_5$ for Zn, $1.77_0 \pm 0.01_7$ for Cd and $1.91_2 \pm 0.01_1$ for Hg. It seems clear from this analysis that $\chi_{Zn} \approx \chi_{Cd} < \chi_{Hg}$. This is virtually identical to the conclusion reached by McCoy and Allred[54] in a study of the NMR shifts of $(CH_3)_2Zn$, $(CH_3)_2Cd$, and $(CH_3)_2Hg$, and the trend is similar to a number of other electronegativity scales, in particular that of Pauling.

Cocksey et al. deduce an opposite trend between Zn and Hg, which is of course directly traceable to the different magnitude ΔE_{cs} that they report for the zinc halides. They conclude that their results, while in disagreement with the Pauling scale of electronegativities, do agree with (and hence favor) the Allred–Rochow scale.

TABLE VI

Chemical Shifts of d-Orbitals, Partial Ionic Charges and
Metal Electronegativities

		Cl	Br	I
Zn:	ΔE_{cs}	$2.04_5 \pm 0.002$ eV	$1.69_5 \pm 0.05$ eV	1.23 ± 0.03 eV
	q	0.528	0.388	0.256
	χ	1.71_3	1.80_4	1.80_6
Cd:	ΔE_{cs}	1.98	1.70	1.36
	q	0.495	0.389	0.288
	χ	1.78_0	1.80_1	1.73_0
Hg:	ΔE_{cs}	1.87^a	1.56^a	1.15^a
	q	0.417	0.330	0.228
	χ	1.94	1.93_6	1.87

^{a} These chemical shifts are taken from Eland (ref. 48). His criterion, which was the shift in the $^2D_{\frac{3}{2}}$ peak alone, was somewhat different from the criterion adopted here, which was the average of the shifts of the two main sharp peaks at 304 Å. The difference is slight. For Hg, the average χ using Eland's criterion is 1.91_5; the present criterion yields 1.91_2.

Allred and Rochow[55] introduced a method of calculating electronegativity based on the equation

$$F = \frac{e^2 Z_{eff}}{r^2} \tag{E}$$

where F is the force of attraction between the nucleus and an electron from a bonded atom, r is taken as the distance between the nucleus and an electron "at the covalent boundary", and "eZ_{eff} is the charge which is effective at the electron due to the nucleus and its surrounding electrons". The values of r were taken from Pauling[56] while Z_{eff} was evaluated by Slater's early screening rules.[57] The electronegativity scale derived from

Eq. (E) was found to be directly proportional to Pauling's thermochemical scale,[58] to a good approximation for the 31 elements examined. Little and Jones[59] extended the Allred–Rochow prescription to the heavier elements, and in this way arrive at an electronegativity for Hg which is smaller than that of Zn. This extension is often referred to as the Allred–Rochow scale, but it is one to which the latter authors can justifiably take exception. We shall try to show why this prescription must be modified for the heavier elements.

The Allred–Rochow scale (Eq. E) is based on two input parameters, Z_{eff} obtained from theory and r derived from experiment. We have compared the screening constants obtained from Slater's rules with corresponding values from the more recent and more precise calculations of Clementi and Raimondi,[60] Clementi et al.,[61] and Froese-Fischer.[62] In Table VII, we note that Slater's rules offer an adequate approximation through the second complete row, ending with Ar. However, with the onset of the first transition element group the deviation of screening constant using Slater's rules from the more recent calculations becomes apparent, and even more pronounced through the second transition element group. Thus, Allred and Rochow give Z_{eff} (i.e. $Z - \sigma$) as 4·00 for Zn and Cd, and extension of this approach to Hg would also give $Z_{eff} = 4·00$. In the conventional application of Slater's rules for the outermost electron (rather than an additional test charge, as used by Allred and Rochow) the value of Z_{eff} becomes 4·35 in all three cases. This is compared with Z_{eff} from Clementi et al. and Froese-Fischer in Table·VIII. The absolute values obtained for Z_{eff} in the more recent calculations are much larger, and a clear trend is evident in the Zn–Cd–Hg sequence which is absent in the Slater-based values. This is evidence that for heavier atoms the outermost electron penetrates the diffuse outer charge cloud much more than we would infer by application of Slater's rules. Also shown in Table VIII are values of Z_{eff}/r^2 for the three sets of computations. Since the atomic radius r given by Pauling increases somewhat in the Zn–Cd–Hg sequence, the ratio Z_{eff}/r^2 diminishes monotonically in the Slater-based computation, and would suggest that the corresponding electro-negativities also decrease. However, the ratio Z_{eff}/r^2 increases slightly from Zn to Cd, and then increases more dramatically for Hg, for both sets of computations based on more recent values of the screening constant. Although these numbers cannot be directly placed on the Allred–Rochow scale, since Z_{eff} has been obtained from the outermost electron rather than an additional test charge, the general features of this trend should be un-affected. A reasonable conclusion from these computations is that $\chi_{Zn} \approx \chi_{Cd} < \chi_{Hg}$, which agrees with the results of the present experiments. The modified Allred–Rochow scale is also seen to be in line with the Pauling scale, at least for the systems considered here.

TABLE VII

A Comparison of Screening Constants (σ) for the Outermost Electron

Atom	Slater's rules (1930)[a]	Clementi *et al.*[b]	Froese-Fischer[c]
He	0·30	0·313	0·382
Li	1·70	1·721	1·451
Be	2·05	2·088	1·735
B	2·40	2·579	2·732
C	2·75	2·864	3·131
N	3·10	3·167	3·544
O	3·45	3·547	3·965
F	3·80	3·900	4·391
Ne	4·15	4·242	4·820
Na	8·80	8·493	7·792
Mg	9·15	8·693	7·850
Al	9·50	8·934	9·360
Si	9·85	9·715	9·516
P	10·20	10·114	9·723
S	10·55	10·518	9·959
Cl	10·90	10·884	10·214
Ar	11·25	11·236	10·483
K	16·80	15·505	14·423
Ca	17·15	15·602	14·311
Sc	18·00	16·368	14·939
Ti	18·85	17·183	15·627
V	19·70	18·019	16·346
Cr	20·55	18·867	17·083
Mn	21·40	19·717	17·834
Fe	22·25	20·566	18·596
Co	23·10	21·424	19·367
Ni	23·95	22·289	20·144
Cu	24·80	23·158	20·928
Zn	25·65	24·035	21·718
Ga	26·00	24·778	24·283
Ge	26·35	25·220	24·080
As	26·70	25·551	24·019
Se	27·05	25·713	24·038
Br	27·40	25·972	24·108
Kr	27·75	26·231	24·215
Rb	34·80	32·016	30·341
Sr	35·15	31·930	29·906
Y	36·00	32·743	30·279
Zr	36·85	33·555	30·804
Nb	37·70	34·304	31·402
Mo	38·55	35·024	32·047
Tc	39·40	35·774	32·725
Ru	40·25	36·548	33·430
Rh	41·10	37·357	34·154

TABLE VII (*cont.*)

Atom	Slater's rules (1930)[a]	Clementi et al.[b]	Froese-Fischer[c]
Pd	41·95	38·163	34·895
Ag	42·80	38·972	35·650
Cd	43·65	39·808	36·417
In	44·00	40·530	39·339
Sn	44·35	40·898	38·892
Sb	44·70	41·006	38·635
Te	45·05	41·192	38·487
I	45·40	41·389	38·410
Xe	45·75	41·576	38·388
Cs	52·80	48·637	46·436
Ba	53·15	48·425	45·727
La	54·15	47·688	46·571
Ce	55·15	47·204	47·430
Pr	56·15	51·253	48·300
Nd	57·15	50·693	49·175
Pm	58·15	51·605	50·056
Sn	59·15	53·988	50·940
Eu	60·15	54·878	51·828
Gd	61·15	55·785	52·719
Tb	62·15	56·700	53·612
Dy	63·15	57·656	54·507
Ho	64·15	58·561	55·404
Er	65·15	59·524	56·302
Tm	66·15	60·416	57·203
Yb	67·15	61·407	58·105
Lu	68·00	62·196	58·320
Hf	68·85	62·836	58·727
Ta	69·70	63·475	59·225
W	70·55	64·146	59·782
Re	71·40	64·884	60·382
Os	72·25	65·677	61·015
Ir	73·10	66·433	61·675
Pt	73·95	67·249	62·357
Au	74·80	68·062	63·058
Hg	75·65	68·847	63·776
Tl	76·00	68·746	67·501
Pb	76·35	69·607	66·668
Bi	76·70	69·660	66·102
Po	77·05	69·779	65·692
At	77·40	69·837	65·386
Rn	77·75	69·924	65·161

[a] J. C. Slater (ref. 57).
[b] E. Clementi and D. L. Raimondi (ref. 60). E. Clementi et al. (ref. 61).
[c] C. Froese-Fischer (ref. 62).

TABLE VIII

Alternative Values of Z_{eff} and Z_{eff}/r^2 for Zn, Cd and Hg

	Z_{eff}			Z_{eff}/r^2		
	Zn	Cd	Hg	Zn	Cd	Hg
Slater's rules	$4 \cdot 35^a$	$4 \cdot 35^a$	$4 \cdot 35^a$	$2 \cdot 78_8$	$2 \cdot 17_9$	$2 \cdot 09_8$
Clementi et al.[b]	$5 \cdot 965$	$8 \cdot 192$	$11 \cdot 153$	$3 \cdot 82_4$	$4 \cdot 10_3$	$5 \cdot 38_0$
Froese-Fischer[c]	$8 \cdot 282$	$11 \cdot 583$	$16 \cdot 224$	$5 \cdot 30_9$	$5 \cdot 80_1$	$7 \cdot 82_4$

[a] This value differs from that (4·00) of Allred and Rochow (ref. 55), who applied Slater's rules to an additional test change. The value used here is more appropriate for comparison with the other calculations quoted.
    [b] Refs. 60 and 61.   [c] Ref. 62.

C. Intensity Variations with Photon Energy

In Figs. 18, 19, and 20, the 584 Å photoelectron spectra of $HgCl_2$, $HgBr_2$, and HgI_2, taken from Eland,[48] are reproduced. In Fig. 21, the corresponding 304 Å spectra, obtained by Berkowitz,[50] are shown. Before embarking upon a detailed examination of these spectra, two points must be made.

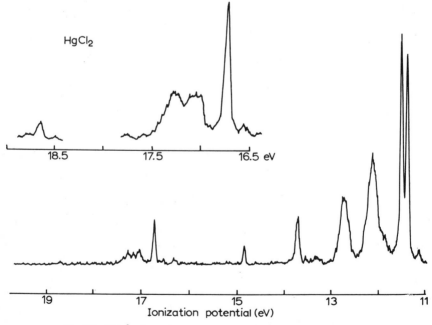

Fig. 18. 584 Å photoelectron spectrum of $HgCl_2$ (from ref. 47).

(1) Although the resolving power (expressed as $\Delta E/E$) was comparable in the two instruments, the resolution width ΔE is substantially poorer in the 304 Å experiment because of the higher kinetic energy of the electrons. A retarding field could have improved the resolution, but might have influenced the transmission of the apparatus, a situation which we tried to avoid for this comparison.

(2) Eland measured photoelectrons ejected at 90° with respect to the photon propagation direction; in the experimental arrangement of Berkowitz, this angle was ~60°. Angular effects can influence relative intensities, but in this case (see ref. 50), this effect is not marked.

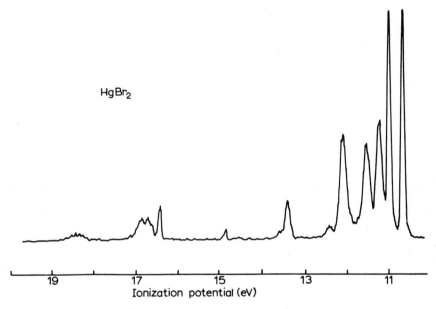

Fig. 19. 584 Å photoelectron spectrum of $HgBr_2$ (from ref. 47).

A glance at the two sets of spectra reveals immediately that the cross-section for ionization from the p-like valence orbitals is much larger than that for ionization of the inner d-like orbitals at 584 Å, whereas exactly the opposite is true at 304 Å. Orchard and Richardson[6] also show much more intense valence bands than d bands for ZnI_2 at 584 Å whereas the opposite is true at 304 Å. Looking more carefully, one notes that within the valence shell the deepest-lying σ_g level, which was the weakest of these peaks at 584 Å, has become markedly more prominent at 304 Å. We had noted earlier that this σ_g level is stabilized by interaction with the metal s orbital.

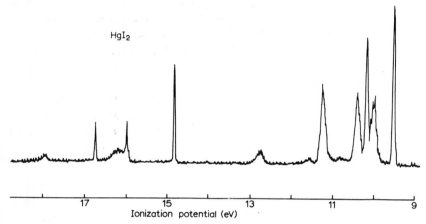

Fig. 20. 584 Å photoelectron spectrum of HgI_2 (from ref. 47).

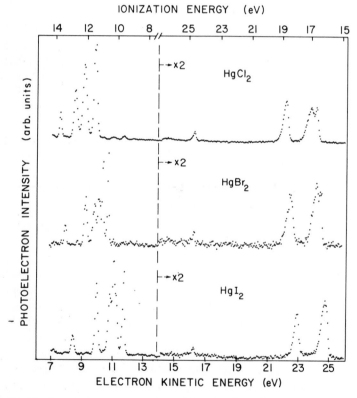

Fig. 21. 304 Å photoelectron spectra of $HgCl_2$, $HgBr_2$ and HgI_2 (from ref. 49).

Since s orbitals have more probability density in the vicinity of the nucleus than do p-type orbitals, the cross-section for ionization from s orbitals does not decline as rapidly with increasing photon energy as it does for p-type orbitals. This phenomenon has been discussed in some detail by Price et al.[63] We shall not focus our attention here on the absence of peaks that correspond to ionization from the σ_g and σ_u orbitals composed of the outermost occupied halogen s orbitals (see Berkowitz[50] for a brief discussion of this point). A final observation concerns the relative intensity of the various components resulting from ionization of the d-like orbitals. In the mercuric halides, one can distinguish two major components split by approximately the atomic spin–orbit separation. The lower energy branch itself has a sharp component (which Eland has identified as $\pm\frac{5}{2}$) and a broader companion region. (Note that this latter separation is about 0·4 eV in $HgCl_2$, and diminishes as one goes to $HgBr_2$ and HgI_2. It is a measure of ligand field splitting, and decreases as one goes to larger internuclear distance and hence weaker field, as found for the alkali halides, but unlike the recent observation of Orchard and Richardson on ZnI_2.)

A dramatic change in relative intensity of the sharp $\pm\frac{5}{2}$ component and the "$^2D_{\frac{3}{2}}$" component is evident in comparing Eland's spectrum at 584 Å with the 304 Å spectrum. In the 584 Å spectrum the "$^2D_{\frac{3}{2}}$" component is barely detectable, whereas the $\pm\frac{5}{2}$ component is prominent. In the 304 Å spectrum, the two components are of about the same intensity. This variation is qualitatively, if not quantitatively, the same as has been described for atomic Hg[14, 15] and its rationalization is very likely similar to that already discussed for the atomic case (see Section II).

With the aid of some additional experimental considerations (see ref. 50) it is possible to construct a crude partial cross-section curve for these metal halides. A representative curve of this type for $HgCl_2$, is shown in Fig. 22. The decline in partial cross-section of the p-like valence bands and the increase in cross-section for the atomic d-like orbitals with increasing photon energy is dramatically illustrated in this figure. Obviously if one were measuring a total cross-section the variation would be slight, and one would be unaware of the changes taking place.

The decline in p cross-section can be readily understood from our earlier discussion. The p-like orbitals are extended in space, with no probability density in the vicinity of the nuclei. The wavefunction for the departing electron acquires higher frequency oscillation as the kinetic energy of this electron (and hence the photon energy) increases. The matrix element describing photoionization, in the one-electron approximation, connects the bound orbital wavefunction with the departing-electron wavefunction. As the oscillations increase in frequency, there is more cancellation and the cross-section decreases. For d-like orbitals (and, in general, higher l orbitals) one must take into account a term in the Hamiltonian of the form

$$+\frac{l(l+1)}{r^2}$$

which combines with the attractive (negative) term in the potential to form an effective potential. The consequence of this effective potential is to give rise to a potential barrier for the escape of a d-like (or higher l) electron.[22] At threshold, this barrier may be only weakly penetrated, but as the kinetic energy increases, the probability of passing through the barrier increases. The effect is readily seen in atomic mercury (Fig. 4) and the molecular behavior shown in Fig. 22 is apparently a manifestation of the same phenomenon.

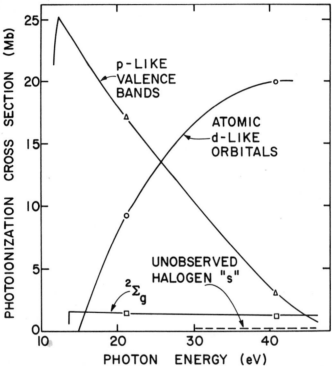

Fig. 22. A crude estimate of the variation of partial cross-section with photon energy for $HgCl_2$, based on data obtained with 584 and 304 Å radiation.

VI. GROUP III HALIDES

A. Group IIIa Halides

This class of systems is characterized by a metal valence orbital sequence $...s^2 p$. The metals may exhibit monovalent or trivalent behavior, the relative tendencies often being described in terms of the stability of the s^2 as a lone pair. As the metal gets larger and its electronegativity decreases, the trend is from trivalent to monovalent behavior. Thus, boron, aluminum, and gallium most readily form trihalides, indium can be produced as monohalide and trihalide, and thallium halide vapors have only been produced as monohalides. An additional complication in the study of these vapors is their tendency to dimerize. This is not serious for the boron trihalides, but the aluminum trihalides are largely dimeric (with the exception of AlI_3). Of the gallium halides, only the trichloride causes difficulty in this respect, and the problem is still less severe with the indium trihalides. The monohalides are largely monomeric, with the exception of thallium fluoride, whose dimer poses sufficient experimental and theoretical problems to warrant special treatment (see below). In summary, the monohalides most susceptible to study are those of thallium and indium; the trihalides of boron, gallium, and indium are most readily produced free of dimer; and the aluminum and gallium trihalides are the best candidates for studying M_2X_6 molecules.

Historically, the thallium halides were the first truly high temperature species studied by photoelectron spectroscopy.[64] They and the other IIIa monohalides continue to be the best examples to illustrate the difference between ionic and covalent bonding upon photoelectron spectra. In Fig. 23, the 584 Å photoelectron spectra of InCl and TlCl are displayed, and in Fig. 24, those of InBr and TlBr. In each instance, we note a sharp peak corresponding to the first vertical I.P., and a broad second peak. We know that sharp peaks correspond to transitions from a neutral to an ionized state in which the potential energy curves have not altered greatly, and, particularly, in which the equilibrium internuclear distance has not shifted significantly. The first peak in the photoelectron spectra of the hydrogen halides is very sharp, and it corresponds to ionization from a halogen lone-pair orbital. However, in hindsight we already know that ionization from an orbital localized on the halogen in alkali halides does not lead to a sharp peak. In fact, there are no really sharp peaks in the valence region of the alkali halides comparable to the sharp peak in the indium and thallium halides. Another confusing feature is the absence of any obvious spin–orbit splitting, which is transparently obvious in the hydrogen halides, and (as we have seen) can be inferred from the alkali halides. A covalent bonding picture such as that invoked to explain the features of HX molecules cannot possibly explain the observed behavior. This point is discussed in some detail by Berkowitz.[64] However, the ionic picture also presents an apparent paradox. If we visualize the neutral ground state as being Tl^+—Cl^- (for example), then our naïve expectation would be that the most easily removable electron would come from the Cl^-, as was seen to be the case for the alkali halides.

(It will be recalled that in the cesium halides, about 10 eV separated the region of X^- ionization from the region of M^+ ionization.) However, removal of an electron from Cl^- should disrupt the ionic bond, forming a weakly-bound Tl^+—Cl^0 molecular ion, and a broad peak should appear in the photoelectron spectrum. This is, of course, contrary to the observation of a sharp first peak.

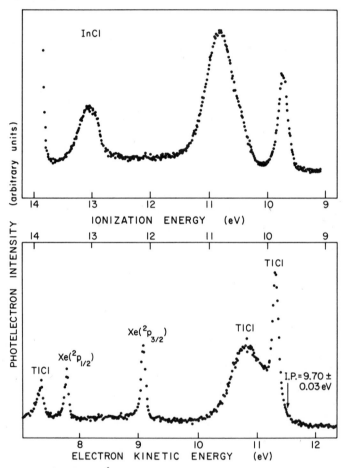

Fig. 23. 584 Å photoelectron spectra of InCl and TlCl.

The whole problem becomes trivially simple (as problems usually do when one understands them) when calculations are available to interpret the experiment. But *ab initio* calculations on heavy element-containing molecules such as indium and thallium halides are still not tractable, and even the semi-empirical CNDO and extended Hückel programs are usually not parametrized for these heavy element systems. However, one such program of the extended Hückel type was available and provided the insight necessary to understand the spectra of Figs. 23 and 24. Hastie and Margrave[65] had utilized such a program to compute the orbital energies of a number of metal halides, including TlF, TlCl, InF, and InCl. They noted that in each of these cases, the uppermost occupied orbital had predominantly metal s character, i.e. Tl $6s$ and In $5s$. If this were so, then ionization (in the ionic model representation) would correspond to $Tl^+—Cl^- \rightarrow Tl^{++}—Cl^-$. Why would

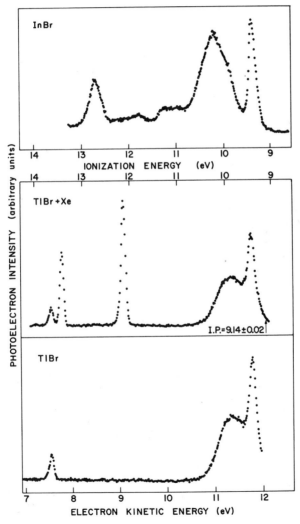

Fig. 24. 584 Å photoelectron spectra of InBr and TlBr.

this appear as the first ionization in the Group IIIa halides, and at a much higher ionization potential for the alkali halides? This ionization energy represents a balance between the energy expended to go from M^+ to M^{++}, and the binding energy recovered from the Coulombic $M^{++}-X^-$ interaction. If both CsCl and TlCl are assumed to have complete charge separation in their neutral ground states, then ionization of a Cs $5p$ electron requires 25·1 eV, that of a Tl $6s$ electron, 20·42 eV.[43] The Coulombic energy recovered at the respective equilibrium distances[66] of CsCl and TlCl are 9·90 eV and 11·58 eV. Hence, in this crude model one would predict an ionization potential from the metal end of the molecule to be 15·2 eV for CsCl and 8·84 eV for TlCl. Both calculations give too low a value, the CsCl being low by about 2 eV, the TlCl by about 1 eV, but the difference between the thallium and cesium cases is clear, and the value obtained for TlCl readily explains why ionization from the thallium end of the molecule can give rise to the first I.P. in this case. The sharpness of the first peak in the thallium and indium halide spectra is due to the fact that r_e changes slightly, to a *smaller* value. The system acts as if a somewhat *anti-bonding* electron has been removed. The broad second peak results from removal of an electron from an orbital centered on X^-, and this acts as the bonding orbital in the ionic M^+-X^- molecule.

Once this general pattern was understood, it was possible to extract some detailed variations among the Group IIIa monohalides from a comparison of extended Hückel and *ab initio* calculations. Table IX is a set of extended Hückel calculations obtained by Cusachs on the Group IIIa monochlorides. In every case, the orbital sequence is the same as that already discussed. The separation between the first two peaks increases monotonically in the sequences Tl–In...B. Experimentally, we note that this is indeed the behavior in Figs. 23 and 24. From the eigenvectors listed in Table IX, we can observe an increase in ionic character (charge separation) in the sequence $B \rightarrow Tl$, owing to the decrease in the electronegativity of the metal. This is corroborated by dipole-moment trends. Although the sequence of dipole moments for Group IIIa chlorides is not well established, the dipole moments[67] of AlF, GaF, InF, and TlF are 1·53, 2·45$_5$, 3·40, and 4·2282$_8$ D, respectively.

Our failure to detect spin–orbit splitting is also understandable, since only the π orbital can exhibit spin–orbit splitting, and it is broader than the anticipated splitting of Cl (0·08 eV) and even Br ($\sim 0·35$ eV), although there is a hint of asymmetry in this peak in both InBr and TlBr. We might expect to see this splitting more clearly in InI and TlI, but in Fig. 25 we observe a rather different character to the spectra in the threshold region. Instead of having a sharp first peak and a broad second one, both are approximately of equal half-width, the first peak having about double the intensity of the second. From our previous experience with circumstances in which π–σ

TABLE IX

Extended Hückel Calculations for Group IIIa Monochlorides

| Orbital | Type of AO | LCAO expansion coefficients | | | | |
		BCl	AlCl	GaCl	InCl	TlCl
$^1\sigma$	M (s)	0·6562	0·7771	0·7420	0·7941	0·8019
	M ($p\sigma$)	−0·6788	−0·5094	−0·4617	−0·4370	−0·4188
	Cl (3s)	0·0023	−0·0844	−0·0739	−0·0849	−0·0881
	Cl ($3p\sigma$)	0·4263	0·4766	0·5954	0·5210	0·5428
Eigenvalue (eV)		−10·964	−8·836	−9·170	−8·682	−8·502
$^1\pi$	M ($p\pi$)[a]	0·2380	0·1544	0·1514	0·1448	0·1412
	Cl ($3p\pi$)[a]	0·9075	0·9527	0·9562	0·9619	0·9645
Eigenvalue (eV)		−14·292	−12·145	−12·003	−11·409	−11·090
$^1\sigma$	M (s)	−0·4142	−0·4123	−0·5247	−0·4596	−0·4626
	M ($p\sigma$)	0·0603	0·0485	0·0748	0·0455	0·0425
	Cl (3s)	0·1885	0·1562	0·1821	0·1489	0·1443
	Cl ($3p\sigma$)	0·7768	0·8113	0·7237	0·7889	0·7918
Eigenvalue (eV)		−15·581	−13·429	−13·736	−12·681	−12·271

[a] $C_{p\pi} = (C_{p_x}^2 + C_{p_y}^2)^{\frac{1}{2}}$.

and spin–orbit splitting are comparable (*see* alkali halides and Group II dihalides) we are in a much better position to understand these spectra than was the author when he first encountered them. The situation in TlI is in fact very similar to that of CsI, where the spin–orbit splitting is ∼0·65 eV and the π–σ splitting (which decreases in the sequence F—Cl—Br—I) is estimated to be ∼0·3 eV. The CsI interaction was solved in ref. 28, and was seen to describe correctly the valence photoelectron spectrum, which is seen to be very similar to that of TlI. For InI, an extrapolation from the observed π–σ separation in InCl (∼1·1 eV) and InBr (∼0·8 eV) would suggest a value of ∼0·6 eV for InI, comparable to the spin–orbit separation. Hence, we might expect a splitting into three peaks, and indeed a shoulder is distinctly visible on the high-energy side of the first peak.

The extended Hückel calculations summarized in Table IX clearly display the similarity in ordering of orbital energies and in their eigenvectors as one progresses through a column of the Periodic Chart, a feature I have referred to as the principle of chemical similarity. If this behavior is accepted, we can explore the Group IIIa monohalides in further detail by taking advantage of the accuracy which one can obtain from high-quality *ab initio* calculations on the lighter members of the group. Attermeyer *et al.* have performed such a calculation on the neutral ground state of AlCl, and also for the first three states of AlCl⁺. The potential energy curves derived from these calculations are shown in Fig. 26. The $^2\Pi$ state of AlCl⁺ is seen to be almost completely repulsive, demonstrating the destruction of bonding that results when an electron is removed from the Cl⁻ end of the molecule. A contour map of the electron density of this 3π orbital (Fig. 27) clearly indicates the accumulation of charge density around the chlorine nucleus. The ground state of AlCl⁺ would be expected to behave as Al⁺⁺Cl⁻, and it does indeed behave this way near its equilibrium internuclear distance, which is shorter than in AlCl. However, at large r this $^2\Sigma$ state begins to interact with other $^2\Sigma$ states before it can reach its high asymptotic limit of Al⁺⁺+Cl⁻, resulting in an

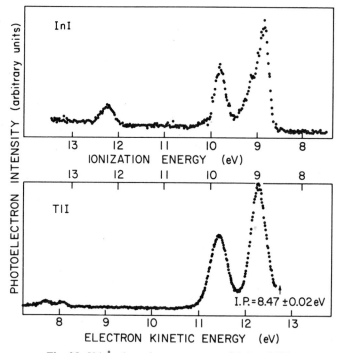

Fig. 25. 584 Å photoelectron spectra of InI and TlI.

avoided crossing. There are two such $^2\Sigma$ states, one having $Al^+(3s^2\,^1S)+Cl(3p^5\,^2P)$ as its asymptote, whereas the other has $Al^+(3s3p\,^3P)+Cl(3p^5\,^2P)$ as its diabatic limit. One consequence of this interaction is the appearance of a potential maximum in the lowest $^2\Sigma$ state; another is the odd double minimum in the first excited $^2\Sigma$ state. This double minimum may be the explanation for the bimodal character of the third peak in the photoelectron spectrum of TlI, although it is not clear why it is manifested in only this member of the Group IIIa monohalides. A contour map of electron density for the uppermost occupied orbital in AlCl(9σ) is reproduced in Fig. 28. Here, it is apparent that the charge density is concentrated on the Al center.

Another relevant parameter is the change of molecular dipole moment as a result of removing an outermost σ or π electron. The electric dipole moment of neutral AlCl ($X^1\Sigma$) is calculated[68] to be 2 D (and an experimental estimate[69] gives 1–2 D). Removal of an electron from the uppermost σ

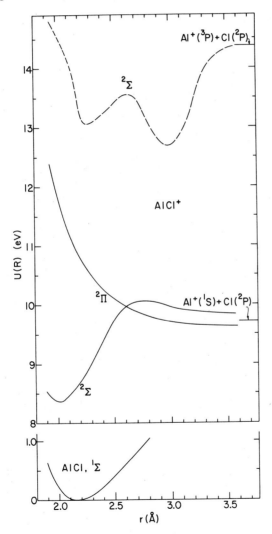

Fig. 26. Potential energy curves for $X^1\Sigma$ state of AlCl and the $X^2\Sigma$, $^2\Pi$ and upper $^2\Sigma$ states of AlCl.[+]

orbital yields AlCl⁺(X²Σ) having $\mu \approx 6.5$ D, whereas removal of an electron from the uppermost π orbital yields AlCl⁺(²π) having $\mu \approx 0$ D. Therefore, removal of the uppermost σ shortens the internuclear distance, significantly increases the dipole moment, and can be viewed as producing a molecular ion described as Al⁺⁺—Cl⁻; whereas removal of an electron from the uppermost π orbital produces a repulsive state which has almost zero dipole moment and can be described as Al⁺—Cl⁰.

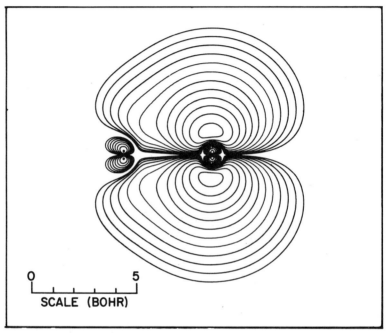

Fig. 27. Contour map of electron density of the 3π orbital of AlCl at $R = 4.252A_0$ ($A_0 = 0.529$ Å). The aluminum atom is on the left.

Finally, we take note of some recent theoretical developments which hold great promise for the computation of molecular properties of systems containing heavy elements. Rosén and Ellis[70] have used the Dirac–Slater Hamiltonian (which includes relativistic effects) and a technique they call the discrete variational method (DVM) to generate eigenfunctions and orbital energies for molecules containing heavy, as well as light elements. Good agreement is obtained with experiment when the theoretical binding energies are calculated by a transition-state procedure.

In Fig. 29, various types of calculations on AlCl are compared. HF refers to the aforementioned Hartree–Fock calculation of Attermeyer *et al.* The extended Hückel and CNDO notations are obvious. The next columns

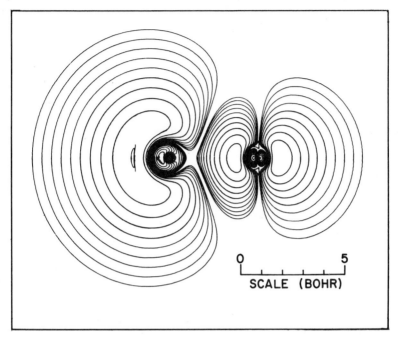

Fig. 28. Contour map of electron density of the 9σ orbital of AlCl at $R = 4\cdot252A_0$
The aluminum atom is on the left.

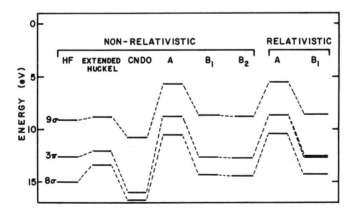

Fig. 29. Comparison of various calculations on the orbital energies of AlCl (from
ref. 70).

represent the calculations of Rosén and Ellis, both non-relativistic and relativistic, without and with the transition-state procedure. The last column (relativistic, and with transition-state procedure) is seen to be in good concordance with the Hartree–Fock calculation. In Fig. 30, the experimental orbital energies for InCl, InBr, and InI are compared with their relativistic transition state calculation, and in Fig. 31 a similar comparison is made between experiment and theory for the Group IIIa monohalides. Note that spin–orbit splitting is now a parameter that derives directly from the calculation. The agreement is quite satisfactory, not only in the ordering of the orbital energies but their absolute magnitude. The calculations at this

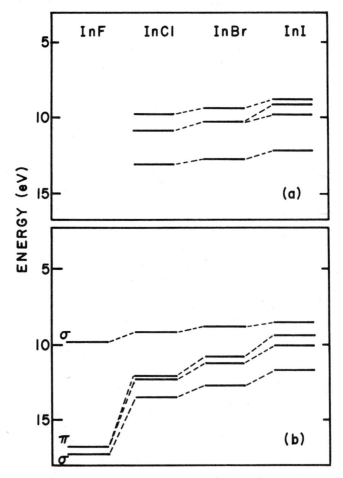

Fig. 30. Comparison of experimental orbital energies in InCl. InBr and InI with relativistic transition state calculations (from ref. 70).

stage seem to place the π level closer to the upper σ than experiment indicates. For details of the calculation, the reader is referred to the papers of Rosén and Ellis.[70]

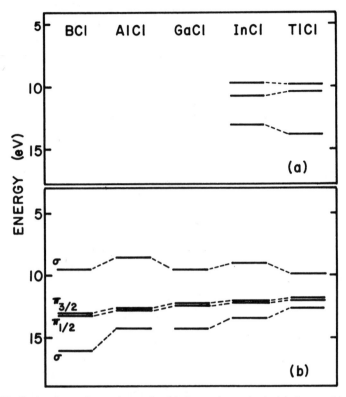

Fig. 31. Comparison of experimental orbital energies and relativistic transition state calculations for the Group III a monochlorides (from ref. 70).

B. The Strange Saga of Thallous Fluoride Dimer, Tl_2F_2

It has been known for some time that the equilibrium vapor pressure above TlF consists of dimers as well as monomers, the dimer composition exceeding 50%. By contrast, the other thallous monohalides have $\sim 1\%$ dimer in their equilibrium vapor. On the basis of vapor pressure, transpiration, and mass spectrometric studies of Tl_2F_2, Keneshea and Cubicciotti[71] concluded that Tl_2F_2 has a linear symmetrical F—Tl—Tl—F geometry. This structure was chosen over the planar rhombic structure of the alkali halide dimers[72] since, according to these authors, the total entropy of the linear configuration was in better agreement with their measurements.

More recently, Brom and Franzen[73] have measured the infrared spectrum of matrix-isolated thallous fluoride and have observed two infrared-active bands attributable to the dimer. This is consistent with the linear $D_{\infty h}$ symmetry, whereas three bands would be observed were the dimer planar rhombic (D_{2h}) or bent (C_{2v}).

There were weaknesses in the above arguments, however. Brom and Franzen pointed out that the entropy argument of Keneshea and Cubicciotti failed to provide sound proof for the linear structure, since their own entropy calculations were unable to distinguish linear and rhombic structures. Furthermore, their own conclusion was based on the assumption that their failure to detect a third band in the infrared spectrum was not due merely to weak intensity of an infrared active mode.

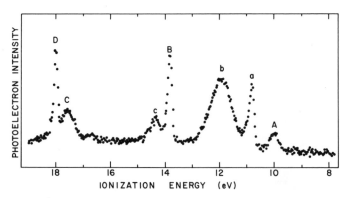

Fig. 32. 584 Å photoelectron spectrum of thallous fluoride. The dimer peaks are labelled A–D, the monomer peaks a–c.

Our approach was to obtain the photoelectron spectrum of thallous fluoride vapor, sort out the peaks attributable to the monomer, and assign the remainder to the dimer. Thereafter, calculations could be obtained of orbital energies for various possible structures, and a selection made according to the best match with experiment. The photoelectron spectrum obtained is shown in Fig. 32. The peaks labelled a, b, and c were selected on various grounds as attributable to monomer. The only calculations available to us were those of Cusachs,[74] using the extended Hückel method. The comparisons (Fig. 33) of experiment with a symmetric linear and a rhombic configuration appeared to favor the linear one, primarily because the span of orbital energies for the rhombic structure appeared too compact.

The problem has since been pursued by other techniques. Fickes et al.[75] studied the recoil of a thermal molecular beam of thallous fluoride when scattered by low (< 16 eV) energy electrons. Their ability to distinguish structures is based on the argument that polar molecules cause scattering which peaks at very small angles, while scattering from a non-polar molecule should show at most a $\cos^2 \theta$ dependence. Earlier experiments by this group had demonstrated the feasibility of detecting the polarity of alkali halide molecules. After correcting for the contribution of monomer TlF to their scattering function, these authors came to the surprising, but ". . . cautious

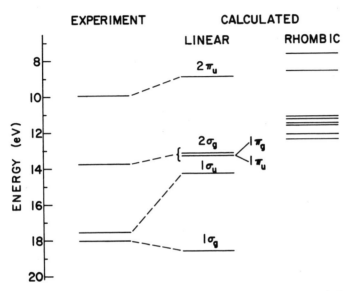

Fig. 33. Comparison of the experimental binding energies and calculated eigenvalues of Tl_2F_2.

conclusion . . . that the dimer scatters electrons as if it has an instantaneous permanent dipole moment comparable to that of the monomer", which we have noted is $4 \cdot 2282_8$ D. Such a dipole moment cannot easily be rationalized with either linear symmetric or rhombic structures and visions of

F—Tl—F—Tl or $\overset{Tl-Tl}{\underset{F}{\diagup}}\,F$ began to appear. Shortly thereafter, Muenter[76] performed a molecular beam electric deflection experiment on Tl_2F_2. He observed no significant focussing as would exist for a polar dimer, although he could observe focussing effects from peaks due to monomer. "Therefore", he concluded, "all observations are consistent with the polar monomer and a non-polar dimer structure." This brings us back to linear symmetric or planar rhombic structures.

The most recent, and perhaps conclusive, contribution to this problem is an electron diffraction study by Solomonik et al.[77] which finds good concordance between the diffraction pattern produced by a planar rhombic structure and their observed diffraction pattern. They deduce a Tl—Tl distance of 3.678 ± 0.003 Å, and a Tl—F distance of 2.290 ± 0.004 Å, equivalent to a F—Tl—F angle of $73 \pm 2°$. The large Tl—Tl distance, which should be the most prominent part of an electron scattering study, is difficult to attribute to any other structure.

If they are correct, there are a number of experiments left unexplained in the wake of the research on this problem. Where is the missing band in the infrared spectrum? Why did the molecular beam deflection-electron scattering study lead to a polar dimer structure? Why did the photoelectron spectroscopic study show preference for a linear structure? While casually discussing this latter spectrum, Dr. D. W. Turner and this author were struck by the similarity between the peaks B, C, and D in Fig. 32, and the peaks in the photoelectron spectrum of CO_2. It now seems quite likely that a CO_2 impurity was present in this experiment, and that the other dimer peaks may be hidden under peaks b and c of the monomer, which would be consistent with a rhombic structure. We plan to repeat this experiment in the near future.†

C. Group IIIa Tridalides: Monomers

This class of molecules has generally been assumed to be planar, with D_{3h} symmetry, although the evidence is far from complete. Measurements on BF_3,[78] BCl_3, GaF_3, and GaI_3 [79] have been interpreted to infer planarity, but the observation[80] of a symmetric stretch frequency in the infrared spectrum of matrix-isolated $AlCl_3$ has led to the conclusion that this molecule is pyramidal.‡

The boron trihalides are relatively volatile and can be investigated by more or less conventional photoelectron spectroscopic techniques. Spectra of the entire set of boron trihalides have been reported by Potts et al.[81] and by King et al.[82] The spectra of aluminum and gallium trihalides, which require temperatures in the range 50–250 °C, have been presented by Lappert et al.[83] while the gallium trihalides and the higher temperature indium trihalides have been investigated by Dehmer et al.[84] Many peaks are involved, and though there is general agreement among the groups of workers about most of the assignments, there are some differences. It is instructive to plot the experimentally observed orbital energies for a given halogen system as the central atom is varied, and we have done this in Figs. 34, 35, and 36 for the chlorides, bromides, and iodides, respectively. Several features

† Note added in proof. Dr. D. G. Streets and the author have repeated the TlF experiment, and have confirmed that the peaks B, C, and D in Fig. 32 are due to CO_2 impurity. The peaks we do observe are consistent with a planar rhombic structure. A note detailing this work appears in Chem. Phys. Lett., **38**, 475 (1976).

‡ Note added in proof. Two recent infrared and Raman spectral studies have concluded that $AlCl_3$ is planar. See I. R. Beattie, H. E. Blayden, and J. S. Ogden, J. Chem. Phys. **64**, 909 (1976); J. S. Shirk and A. E. Shirk, ibid. **64**, 910 (1976).

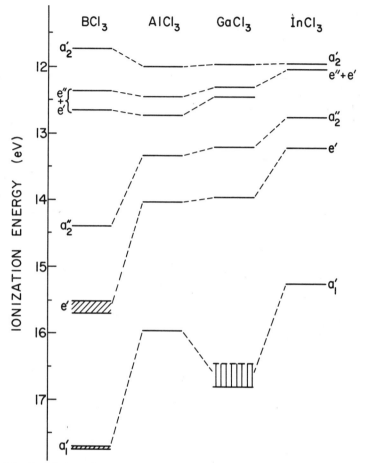

Fig. 34. Correlation diagram of the experimentally observed vertical ionization potentials of the Group III a trichlorides.

immediately become apparent, but before we discuss them we note that the identification of levels has been deduced from extended Hückel calculations,[84] which predict $a_2' < e'' \lesssim e' < a_2'' < e' < a_1'$. This general ordering is agreed upon by the several groups of investigators.

The deepest lying orbital, a_1', has a large metal "s" contribution according to the extended Hückel calculation. In all three figures, the variation of binding energy of this a_1' orbital with the central atom roughly parallels the ionization energy of the singly ionized atoms B^+, Al^+, Ga^+, and In^+, which are 25·149, 18·823, 20·51, and 18·86 eV, respectively.[19] The non-monotonic progression of this orbital's binding energy with variation of central atom would seem to be explained.

The remaining orbitals in the valence region are primarily combinations of halogen p orbitals conforming to the symmetry of the molecule. The uppermost occupied orbital, a_2', decreases slightly between boron and aluminum and then remains essentially constant at a value between that of atomic and diatomic halogen's first I.P. The other generalization that can

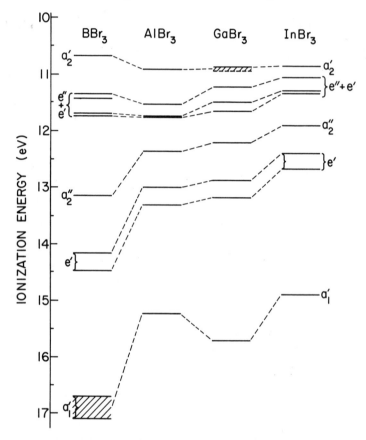

Fig. 35. Correlation diagram of the experimentally observed vertical ionization potentials of the Group III a tribromides.

be made is that the span of valence orbitals becomes more compressed as the metal atom increases or, perhaps more significantly, as the size of the molecule increases. The halogen orbitals are thereby farther removed from one another, the interaction between them decreases, and hence the molecular splitting decreases. This is also borne out by the extended Hückel calculations.

We are left with four orbitals to account for, three of which are doubly degenerate. The deepest-lying e' orbital appears easiest to discuss. It manifests a rather constant spin–orbit splitting of ~0·3 eV for the entire class of tribromides, and a corresponding splitting of ~0·5 eV for all the tri-iodides, both reasonable values from our knowledge of spin–orbit splitting in halide-containing molecules. The splitting in the chlorides is too small to detect, but in BCl_3 (and BF_3)—see ref. 81—the splitting observed in the e' level is attributed to Jahn–Teller interaction. The binding energy of the non-degenerate a_2'' orbital progresses monotonically to lower values in the sequence $B \to In$. The major problem in the entire assignment now arises— distinguishing the ordering and splitting of the e'' and upper e' orbitals.

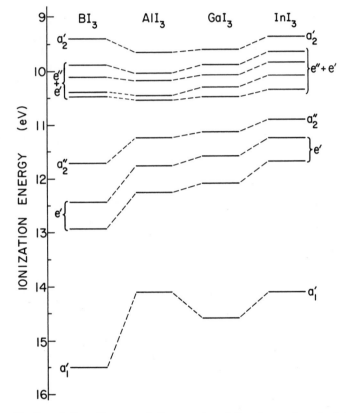

Fig. 36. Correlation diagram of the experimentally observed vertical ionization potentials of the Group III a tri-iodides.

Potts *et al.* have argued that the e' orbital may be expected to show spin–orbit splitting but the e'' orbital should exhibit neither spin–orbit nor Jahn–Teller splitting. On this basis, we should expect to be left with three un-assigned peaks in this energy range. However, four are clearly observed in the spectrum of GaI_3 (Fig. 37) where they are labelled B, C, D, and E. Lappert *et al.* also indicate four peaks for AlI_3 as well as GaI_3 in this energy region. The spectrum of InI_3, though not as well resolved, can also be interpreted as consisting of four peaks in the same region. In addition, Potts *et al.* also report four ionization energies, although only three peaks are clearly evident in their BI_3 spectrum in the relevant energy range. In the BI_3 spectrum of King *et al.*[82] all four peaks are evident. The relative intensities in the spectrum of GaI_3 (Fig. 37) suggest that the peaks B and E belong to one spin–orbit pair, and the peaks C and D to some other pair. The B–E energy gap (0·6 eV) is about what one expects for spin–orbit splitting in iodides. It is larger than the splitting (0·5 eV) in the deeper e' orbital, but the extended Hückel calculations[84] indicate that the ratio of the squared coefficients of the iodine $5p$ atomic orbitals in the upper e' to those in the lower e' is 1·37, which is reasonably close to the ratio of spin–orbit splittings, 1·20.

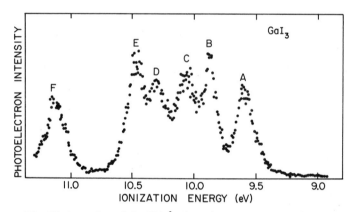

Fig. 37. A portion of the 584 Å photoelectron spectrum of GaI_3.

On this basis, Dehmer *et al.*[84] had concluded that the two components of e'' are surrounded, as it were, by the two e' components. The e'' then is seen to have a more-or-less constant splitting across the entire iodide series. (As has been noted earlier, the BI_3 spectrum displays only three prominent peaks in this region. We have taken the one of highest energy to represent components of both e' and e'', a conclusion which is mildly supported by Table 1 of Potts *et al.* and the spectrum of King *et al.* which records a fourth ionization potential 0·1 eV to higher energy.) In the tribromides, it is possible to pick out a spin–orbit spacing that is reasonably constant across the series, but the assignment must be considered less certain in this group. Four peaks are reported[81] for BBr_3 in the relevant energy range, but a glance at the spectrum suggests that one cannot be certain that four peaks are observable.

The spectra of the iodides are more conclusive on this point and clearly demonstrate that the e'' orbital is split, contrary to the conclusion of Potts et al. that neither spin–orbit nor Jahn–Teller splitting is possible. First-order spin–orbit coupling splits the degeneracy of e' levels, but not those of e'' symmetry. However, this conclusion is no longer valid when one considers the double group formed as the direct product of the irreducible representations of D_{3h} with the representation of the spin functions, $e_{\frac{1}{2}}$. The irreducible representations of the double-group D_{3h}' become[85]

$$a_1' \times e_{\frac{1}{2}} = a_2' \times e_{\frac{1}{2}} = e_{\frac{1}{2}}$$

$$a_2'' \times e_{\frac{1}{2}} = e_{\frac{3}{2}}$$

$$e' \times e_{\frac{1}{2}} = e_{\frac{3}{2}} + e_{\frac{5}{2}}$$

$$e'' \times e_{\frac{1}{2}} = e_{\frac{1}{2}} + e_{\frac{3}{2}}$$

It is evident in this representation that e'' can mix with e', a_1', and a_2'. In particular, e'' and e' have been seen to be very close in energy, and hence the interaction should be strong. King et al.[82] had demonstrated that this interaction would give rise to a splitting in e'' (as well as e'). Manne et al. have elaborated on this treatment for the boron trihalides[86] and more recently Wittel and Manne[87] have extended these calculations to GaI_3.

The three sets of calculations differ somewhat in their conclusions, although they agree that four separate levels should result from e' and e''. For BI_3, King et al.[82] conclude that the levels derived from e' surround those from e'', as do Manne et al., but the latter authors reverse the ordering of the $e_{\frac{5}{2}}$ and $e_{\frac{3}{2}}$ derived from e'. On the other hand, the corresponding calculation for GaI_3 by Wittel and Manne has the orbital energies derived from e' and e'' alternating in order. The correlation shown in Fig. 36 suggests that there should be a smooth progression between the energy ordering in BI_3–AI_3–GaI_3–InI_3. Perhaps it is no longer meaningful to retain the D_{3h} irreducible representations. In that event, the sequence in the double-group representation that may be appropriate in Fig. 36 is $e_{\frac{3}{2}} < e_{\frac{3}{2}} < e_{\frac{1}{2}} < e_{\frac{5}{2}}$. As pointed out by Wittel and Manne,[87] the above analysis makes it unnecessary to invoke the hypothesis of deviations from planar geometry to explain the presence of four peaks in this energy region.

Finally, a comparison of the Group IIIa monohalides and trihalides is in order. The best examples are the indium halides which can be prepared and studied in both valence states. It will be recalled that the first I.P. of InCl and InBr corresponded to removal of a metal s-like electron, whereas the deepest I.P. reported for $InCl_3$ and $InBr_3(a_1')$ corresponds to removal of a metal s-like electron. This dramatic re-ordering can be rationalized in the following way:

(1) The three halogens around the indium withdraw more charge than does a single halogen. Hence, an indium $5s$ orbital is screened less from its nucleus and has a higher binding energy.

(2) We have previously shown how the Tl^{++}—X^- structure (and hence also the In^{++}—X^- structure) strongly stabilizes this state of the ion. For the trihalides, although the total withdrawal of charge from the indium may be greater, the separation of charge between indium and a given halogen may not be as great, and hence the stabilization of the corresponding *ionic* state may be a weaker effect.

It may be difficult to separate the effects of (1) and (2) but it is noteworthy that the binding energy of a_1' diminishes in the sequence $InCl_3$–$InBr_3$–InI_3, as the degree of ionicity, and hence charge separation, decreases.

D. Group IIIa Trihalides: Dimers

This class of molecules is known[79] to have an interesting halogen-bridged struture of D_{2h} symmetry. Lappert *et al.*[83] have presented photoelectron spectra for Al_2Cl_6, Al_2Br_6, and Ga_2Cl_6 (the latter also presented inadvertently by Dehmer *et al.*)[84] thus far without assignment of the bands.

VII. GROUP IV HALIDES

Of the members of this group that are high temperature vapors, the only ones investigated at the time this review was undertaken were the lead dihalides. Representative spectra of $PbCl_2$ and PbI_2 are shown in Fig. 38. An interpretation of these and more recent spectra is presented below.

The valence orbital configuration of Pb is $...(6s)^2(6p)^2$; that of Hg is $...(6s)^2$. The lead dihalides are bent[88] (C_{2v}) with a bond angle of 95°, whereas the mercuric dihalides are linear. There should be some correlation between the orbitals of HgX_2, in $D_{\infty h}$ symmetry, and those of PbX_2, in C_{2v} symmetry. A Walsh diagram describing such a correlation for the isoelectronic valence system of SiF_2 is reproduced[89] in Fig. 39.

The orbitals $(4\sigma_g)(3\sigma_u)(1\pi_u)(1\pi_g)$ represent the correct sequence for the mercuric halides, as determined by Eland.[48] Extended Hückel calculations (together with Eland's experimental values in parentheses) for $HgCl_2$ and HgI_2 are shown in Fig. 40(a, b). Corresponding calculations for $PbCl_2$ and PbI_2 are shown on the left-hand sides of these diagrams. Lines are drawn connecting orbitals of appropriate symmetry, as prescribed by the Walsh diagram of Fig. 39. (More properly, one should use the irreducible representations of the double group, which are obtained as direct products of the species of the molecular orbitals and of the spin functions. For the $D_{\infty h}$

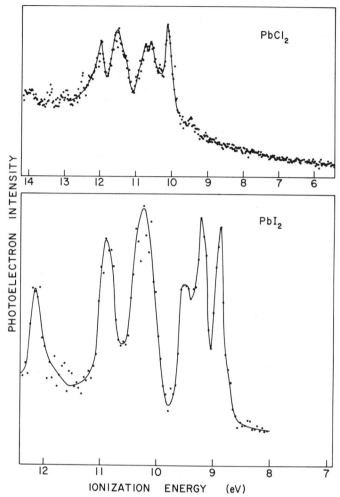

Fig. 38. 584 Å photoelectron spectra of (a) $PbCl_2$ and (b) PbI_2.

symmetry of HgX_2, this gives rise to $e_{g,\frac{3}{2}}$, $e_{g,\frac{1}{2}}$, $e_{u,\frac{3}{2}}$ and $e_{u,\frac{1}{2}}$, but for the C_{2v} symmetry of PbX_2 all orbitals reduce to $e_{\frac{1}{2}}$, and hence all correlation is lost.)[90–91]

We shall utilize the principle of chemical similarity as a guide in the assignment of peaks. The prototype of the Group IV dihalides is CF_2. Recent *ab initio* calculations[44, 92] and experimental values[93] for the orbital energies of this molecule are summarized in Table X. The agreement between both calculations and the experimental results is very good, and clearly establishes that $6a_1$ is the uppermost occupied orbital, followed successively by the orbitals in C_{2v} symmetry derived from π_g and π_u. For the next member of the group, SiF_2, photoelectron spectra[94] and *ab initio* calculations[89, 95] appear to agree upon the sequence $...(7a_1)^2(2b_1)^2(1a_2)^2(5b_2)^2(8a_1)^2$. Again, the uppermost occupied orbital has the irreducible representation a_1 (derived from σ_g in $D_{\infty h}$ symmetry) and the corresponding molecular orbital has a large contribution from the a.o. of the central atom. The successively deeper orbitals are derived from π_g and π_u. In Thomson's[95] calculation the next deeper orbitals are $4b_2$ and $6a_1$, correlating with σ_u and σ_g, respectively, in $D_{\infty h}$ symmetry.

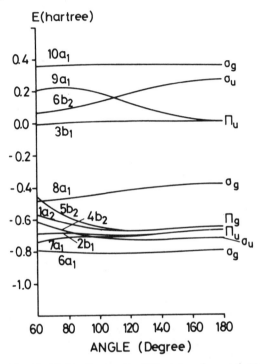

Fig. 39. Walsh diagram for SiF_2, taken from ref. 85.

The more readily studied molecule SO_2 is isoelectronic with SiF_2, and the latest assignment[96] of orbitals in this molecule is $...(6a_1)(4b_2+7a_1+2b_1)$ $(1a_2+5b_2)(8a_1)$, essentially the same as in SiF_2. Extended Hückel calculations of Hastie and Margrave[65] indicate that the uppermost occupied molecular orbital in CF_2, SiF_2, GeF_2, SnF_2, PbF_2, $SiCl_2$, $GeCl_2$, and $PbCl_2$ has a substantial, if not dominant, central atom component, which implies that it is similar in nature to the aforementioned $6a_1$ orbital in CF_2 and $8a_1$ orbital in SiF_2. Hence, all the calculations and experimental deductions mentioned above agree that the uppermost occupied orbital in this iso-electronic sequence is a_1, which correlates with σ_g in $D_{\infty h}$ symmetry. We there-fore assign the first sharp band in the photoelectron spectra of both $PbCl_2$ and PbI_2 to ionization from such an a_1 orbital.

TABLE X

Orbital Energies of CF_2 deduced from *ab initio* Calculations and from PES Experiments

	Calculations		Experiment
	Rothenberg and Schaefer[a]	Snyder and Basch[b]	Dyke et al.[c]
$6a_1(\sigma_g)$	12·94	13·39	12·27
$4b_2(\pi_g)$	18·65	18·94	16·40
$1a_2(\pi_g)$	19·03	19·26	17·4
$1b_1(\pi_u)$	21·19	21·50	20·83
$5a_1(\pi_u)$	21·75	22·01	19·2
$3b_2(\sigma_u)$	22·49	22·44	22·2
$4a_1(\sigma_g)$	26·21	26·41	24·0

[a] S. Rothenberg and H. F. Schaefer III, *J. Am. Chem. Soc.* **95**, 2095 (1973).
[b] L. C. Snyder and H. Basch (ref. 92).
[c] J. M. Dyke *et al.* (ref. 93).

Another factor that may be helpful in identification of the bands is the covalent or ionic nature of the compound. We have seen that the Group III monohalides can be treated as ionic molecules. The Group III and Group IV elements have very nearly the same electronegativities. According to the extended Hückel calculations, the uppermost occupied orbital for each of the Group IV dihalides has most of its charge density around the central atom. Ionization from this orbital may be expected to give a sharp peak (*see* Section VI-A, Group III halides), which is seen to occur as the first I.P. in both $PbCl_2$ and PbI_2 (and also in $SnCl_2$, $SnBr_2$, and $PbBr_2$, *vide infra*[97]). In Section VI-A we noted that the energy gap between this "metal-like" first ionization band and the next inner band (which is halogen-like) was large (3–4 eV) for the lighter elements (B, Al) and gradually diminished toward In and Tl. The same behavior can be inferred from our limited knowledge of the Group IV dihalides, the corresponding gap in CF_2 and SiF_2 being ~ 4 eV, and much less for the tin and lead dihalides.

In the course of preparation of this manuscript, an article by Evans and Orchard[97] appeared, presenting new data on $SnCl_2$, $SnBr_2$, and $PbBr_2$, together with new assignments. Unfortunately, these authors have chosen a different frame of reference for the C_{2v} symmetry, so one must establish the proper correlations, shown below.

$D_{\infty h}$	C_{2v} (present)	C_{2v} (Evans and Orchard)
σ_g	a_1	a_1
σ_u	b_2	b_1
π_g	$a_2 + b_2$	$a_2 + b_1$
π_u	$a_1 + b_1$	$a_1 + b_2$

The orbital sequence chosen by Evans and Orchard for their Group IV dihalides, in their reference frame, is (b_1) (two mixed $\pi_g + \pi_u$ bands) (a_1) (a_1), the b_1 corresponding to the first ionization energy. In our reference frame, the b_1 changes to b_2, the a_1 does not change, and the second and third bands

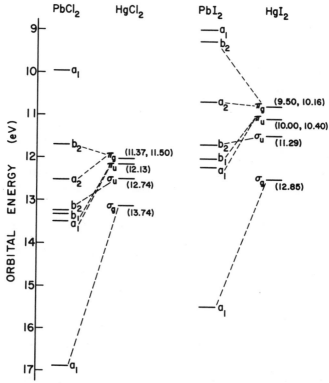

Fig. 40. Correlation between orbital energies of HgX$_2$ and PbX$_2$, from extended Hückel calculations. Values in parentheses are experimental values. (a) PbCl$_2$ and HgCl$_2$. (b) PbI$_2$ and HgI$_2$.

are still mixtures of π_g and π_u components. It is evident that this sequence differs markedly from the one we had deduced earlier. However, interchanging b_1 and a_1 in their frame of reference almost recovers the orbital sequence presented above for CF_2 and SiF_2, the mixing of the π_g and π_u components then being the only discrepancy. Our reasons for selecting a_1 as the uppermost occupied orbital in this isoelectronic sequence have been stated earlier in some detail, and we can find no reason for the alternative selection made by Evans and Orchard.

In $PbCl_2$, the second band appears as a broad peak, perhaps with incipient splitting, while the counterpart in PbI_2 is a distinct doublet, with ~ 0.3 eV separation. The $PbBr_2$ spectrum of Evans and Orchard appears rather similar to that of $PbCl_2$. In $D_{\infty h}$ symmetry, this band would correspond to ionization from a π_g orbital (see Fig. 39, Walsh diagram). Spin–orbit interaction may split the π_g degeneracy in $D_{\infty h}$ symmetry, giving rise to $^2\Pi_{g,\frac{3}{2}}$ and $^2\Pi_{g,\frac{1}{2}}$ states, as we have seen in the mercuric dihalides. In C_{2v} symmetry, the π_g orbital splits into a_2 and b_2 components, themselves non-degenerate. To a large extent, the spin–orbit splitting is washed out in C_{2v} symmetry.[91, 98] It is recovered as one goes to principal axes C_n with $n \geqslant 3$, since for these point groups there remains at least a doubly degenerate (E) irreducible representation. This is then split when one forms the direct product of the species of the molecular orbitals and the spin functions. In other words, a group containing E or T irreducible representations may exhibit further (spin–orbit) splitting, but for a group such as C_{2v} this is no longer possible. In pictorial terms, C_{2v} represents a major perturbation to local cylindrical symmetry. As $n \to \infty$ for C_{nv}, we finally attain the unperturbed cylindrical symmetry of a diatomic (or, in general, linear) molecule.

We have noted that the splitting in this second band in PbI_2 is ~ 0.3 eV. This is about the splitting[81] between b_2 and a_2, and between b_1 and a_1 in CH_2I_2. As we have seen (Section VI-A, Group III halides) this spin–orbit splitting is largely restored to its diatomic value of ~ 0.63 eV in the C_3 system of $BI_3 \to InI_3$, and Green et al.[99] have shown that this trend continues in the C_4 systems AB_4^+.

Two strong bands remain to be assigned in the photoelectron spectra of the lead and tin halides, and Evans and Orchard have succeeded in detecting a weak peak at ~ 16 eV for $SnCl_2$, $SnBr_2$, and $PbBr_2$. Of the two strong bands, the one having lower ionization energy is substantially more intense and somewhat broader than the other.

From the calculations and arguments given above, we assign the larger (and lower ionization) band to a composite of $^2A_1$ and $^2B_1$, which is derived from π_u in $D_{\infty h}$ symmetry. We have shown that spin–orbit splitting should be quenched, and the extended Hückel calculations indicate that splitting in C_{2v} symmetry is very small. The higher ionization band is assigned to

$^2B_2$ derived from σ_u in $D_{\infty h}$ symmetry, both because it is weaker and narrower.

The innermost valence orbital from which ionization is expected is a_1 derived from σ_g in $D_{\infty h}$ symmetry. This peak is already very weak in HgI_2, and the weak peak observed by Evans and Orchard at ~ 16 eV in $SnCl_2$, $SnBr_2$, and $PbBr_2$ is assigned to such a $^2A_1$ state.

VIII. THE GROUP VI HOMONUCLEAR DIATOMICS

One of the major simplifications emerging from the many experimental studies of molecular photoelectron spectroscopy has been the similarity in spectra exhibited by molecules from the same column of the Periodic Chart. I have referred to this as the principle of chemical similarity. Among diatomic molecules, good examples are provided by the homonuclear diatomic halogens,[100, 101] the hydrogen halides,[102] and, as we have already seen, the Group IIIa halides and alkali halides. Among the Group VI homonuclear diatomics we observe an apparent departure from this principle of chemical similarity, which can be rationalized in terms of the dynamics of the photoionization process and a transition to a different Hund's coupling scheme.

In this group, O_2 is a permanent gas and has been studied by a number of investigators. Probably the most detailed spectrum and analysis is that of Edqvist et al.[103] The other members of this group, S_2, Se_2, and Te_2, must all be prepared by high-temperature techniques. We[104] have recently reported results on S_2 and Te_2, and are currently obtaining some data on Se_2.

If ordinary orthorhombic sulfur is vaporized, it produces a complex vapor mixture containing, most prominently, S_8, S_7, and S_6. By vaporizing a sulfide which decomposes and has a low decomposition pressure, it is possible to shift the equilibrium to simpler species. Previous studies with Knudsen cells[105] had demonstrated that ZnS and CdS would generate the metal atoms and S_2 almost quantitatively, but the temperature necessary for vaporizing these systems was just beyond the instrumental capability at that time. A compromise candidate was found in HgS, which generates Hg atoms, S_2 with about 90% of the sulfur molecular flux, and smaller amounts of S_6 and S_5. The spectrum obtained is shown in Fig. 41(b). Vaporization of selenium produces similar problems to that of sulfur, but Se_2 can be generated rather cleanly by vaporizing CdSe. We are currently pursuing this study with a high-temperature modification of our apparatus. The molecule Te_2 can fortunately be generated in rather pure form by volatilization of crystalline tellurium. The spectrum obtained is shown in Fig. 41(c). Figure 41(a) is a spectrum of O_2 taken in this laboratory under similar conditions to those used for studying S_2 and Te_2, and is displayed for convenient comparison of the spectra.

A. S_2

In Fig. 41(b), the sharp peaks at 10·437, 14·840, and 16·704 eV, readily identified by their energies and their narrowness, are due to atomic Hg. The other major features in Fig. 41(b) can be assigned to the S_2 molecule. Returning for a moment to O_2, we note that its orbital configuration in the

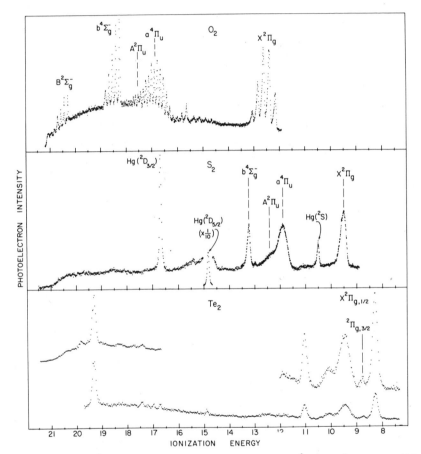

Fig. 41. (a) 584 Å photoelectron spectrum of O_2; (b) 584 Å photoelectron spectrum of S_2, obtained by sublimation of HgS_2; (c) 584 Å photoelectron spectrum of Te_2, obtained by sublimation of crystalline tellurium.

neutral ground state may be written $(\sigma_g 1s)^2 (\sigma_u 1s)^2 (\sigma_g 2s)^2 (\sigma_u 2s)^2 (\sigma_g 2p)^2$ $(\pi_u 2p)^4 (\pi_g 2p)^2$. In the normal Hund's coupling scheme, the π_g^2 configuration results in three possible states, $^1\Sigma_g^+$, $^3\Sigma_g^-$, and $^1\Delta_g$, of which $^3\Sigma_g^-$ is the lowest-lying (ground) state.

Removal of an electron from the $\pi_g 2p$ orbital leaves this orbital singly occupied, resulting in a $^2\Pi_g$ state. This is split by spin–orbit interaction into $^2\Pi_{g,\frac{3}{2}}$ and $^2\Pi_{g,\frac{1}{2}}$, of which the latter is the ground state. (Note that this ordering is opposite to that in the homonuclear diatomic halogen ions, and is a consequence of the π shell having only one electron or missing only one electron, resulting in the spin–orbit matrix element being positive or negative.) The spin-doublet splitting has been observed in photoelectron spectra by Edqvist et al.[103] to be 23 meV, and by more accurate optical spectroscopic measurements[106] to be 24.2 meV. The spin–orbit parameter for the $2p$ level (ζ_{2p}) of atomic oxygen has been calculated[62] to be 18.3 meV and can be inferred from experimental data[19] by application of the Landé interval rule to be 18.72 meV. Hence, to first order the molecular spin-orbit splitting is approximately equal to the appropriate atomic ζ, as has been found for the halogens[100, 101] and alkali halides (*vide supra*). In Table XI, we summarize atomic ζ parameters obtained from theory and experiment, their per cent deviation, and the corresponding molecular spin–orbit splittings (where observed) for the halogens and chalcogens.

From Table XI, we can infer that the $^2\Pi_{g,\frac{3}{2}}$–$^2\Pi_{g,\frac{1}{2}}$ splitting in S_2^+ should be ~ 50 meV. This is difficult to observe experimentally, because the vibrational separation in ground state S_2 is ~ 90 meV, kT at the experimental temperature is ~ 50 meV, and hence a few vibrational levels are populated. Each of these vibrational states has its Franck–Condon region, and the inherent instrumental resolution, together with rotational broadening, further blur the possibility of seeing structure. Despite this, there appears to be a shoulder on the high-energy side of the first photoelectron peak in Fig. 41(b), which is separated from the peak maximum by ~ 60 meV. If we identify the peak maximum as the approximate vertical ionization potential for formation of $^2\Pi_{g,\frac{1}{2}}$ (9.50 eV), and the shoulder as the corresponding value for formation of $^2\Pi_{g,\frac{3}{2}}$ (9.56 eV), then we note for subsequent discussion that the intensity of $^2\Pi_{g,\frac{3}{2}}$ is somewhat lower than that of $^2\Pi_{g,\frac{1}{2}}$. This contrasts with the observation of Edqvist et al.[103] on O_2, where the $^2\Pi_{g,\frac{3}{2}}$ was higher. (However, in the deconvoluted photoelectron spectrum of O_2 reported by Dromey et al.[107] the $^2\Pi_{g,\frac{1}{2}}$ intensity is somewhat larger than the $^2\Pi_{g,\frac{3}{2}}$ intensity.)

The broad, asymmetric peak between 11.5 eV and 13.0 eV in Fig. 41(b) is very similar to the envelope of transitions in O_2 between 16 eV and 18 eV (see Edqvist et al. for a better spectrum of this region), and is therefore directly assigned to an $\alpha\,^4\Pi_u$ (vertical I.P. = 11.88 eV) and an $A\,^2\Pi_u$ (vertical I.P. $\cong$ 12.33 eV). The sharp peak at slightly higher energy (vertical I.P. = 13.23 eV) is assigned as $b\,^4\Sigma_g$, by analogy with O_2.

The remaining band in the 584 Å photoelectron spectrum of O_2 is a $B\,^2\Sigma_g$, and the only evidence for the corresponding band in S_2 appears to lie under the $^2D_{\frac{5}{2}}$ peak of Hg at 14.84 eV. Note that the bulge at the base of

TABLE XI

Spin–orbit Parameters and Splittings in the Chalcogens and
Halogens (all energies are in meV)

| | A. Chalcogens | | | |
	O	S	Se	Te
Calculated ζ_{np}[a]	18·3	45·1	205	419
Landé interval experimental[b]	18·72	47·36	209·5	389
Deviation (%)	2·2	4·8	2·1	7·7
Diatomic molecule spin–orbit splitting[c]	23–24·2	—	—	—
CY_2^+ molecular spin–orbit splitting	19·8[g]	54·6[g]	260[h]	

| | B. Halogens | | | |
	F	Cl	Br	I
Calculated ζ_{np}[a]	33·0	68·0	275	534
$\frac{2}{3}$ of atomic splitting, experimental[d]	33·4	72·8	305	628
Deviation (%)	1·2	6·6	9·8	15·0
Molecular spin–orbit splitting	30[e]	80[e]	350[e]	650[e]
	41·8[f]	80·0[f]	350[f]	635[f]

[a] Ref. 62, with np referring to the valence p orbital in each instance.
[b] Ref. 19. The spin–orbit parameter ζ is obtained from the expression $\frac{3}{2}\zeta \cong {}^3P_0 - {}^3P_2$.
[c] Ref. 99.
[d] For the atomic halogens the splitting between $^2P_{\frac{3}{2}}$ and $^2P_{\frac{1}{2}}$ is $\frac{3}{2}\zeta_{np}$, and hence one must divide by $\frac{3}{2}$.
[e] Ref. 96.   [f] Ref. 97.
[g] G. Herzberg "Molecular Spectra and Molecular Structure. III. Electronic Spectra and Electronic Structure of Polyatomic Molecules" (D. Van Nostrand Co., Princeton, 1966).
[h] S. Cradock and W. Duncan *Mol. Phys.* **27**, 837 (1974).

the $^2D_{\frac{5}{2}}$ peak does not occur at the base of the $^2D_{\frac{3}{2}}$ peak of Hg at 16·70 eV, and hence it is not an instrumental effect.

The $a\,^4\Pi_u$ and $A\,^2\Pi_u$ states result from ejection of an electron from a $\pi_u\,2p$ orbital in O_2, and the $b\,^4\Sigma_g$ and $B\,^2\Sigma_g$ from the next inner $\sigma_g\,2p$ orbital. In S_2 the corresponding orbitals would be $\pi_u\,3p$ and $\sigma_g\,3p$. In the 304 Å photoelectron spectrum[103] of O_2 a peak appears at 24·6 eV, assigned as $c\,^4\Sigma_u$, presumably resulting from ejection of an electron from the $\sigma_u\,2s$ orbital. The absence of such a peak in the 584 Å photoelectron spectrum of S_2 implies that it has ionization energy $> 20·5$ eV (allowing for poor sensitivity in the 20·5–21 eV region).

In O_2, the orbital under discussion ($\sigma_u 2s$) can be described as the anti-bonding component resulting from the linear combination of the $O\,2s$ electrons. A crude estimate of the binding energy of $O\,2s$ in atomic oxygen can be obtained from the photoelectron spectrum of H_2O, where the $1a_1$ orbital[108, 109] which is largely $O\,2s$, has a binding energy of 32·2 eV. Siegbahn et al.[51] list values of 28 eV and 32 eV. From the photoelectron spectrum[109] of O_2 we can infer a weighted mean of 26·2 eV for the anti-bonding component ($\sigma_u 2s$) and 40·3 eV for the bonding component ($\sigma_g 2s$) of the interaction of $O\,2s$ electrons. The center of gravity of this pair (33·2 eV) is not far from the value of the $1a_1$ orbital in H_2O.

For the corresponding situation in sulfur, the orbital designated $1a_1$ in H_2S (which is largely $S\,3s$) has a binding energy[108, 109] of 22·2 eV. We can expect the anti-bonding component in S_2 to have a smaller binding energy. When combined with the lower limit implied by the present experiment, we can conclude with some confidence that the $c\,^4\Sigma_u^-$ state in S_2^+ should lie between 20·5 eV and 22·2 eV above ground state S_2. One might argue that the intensity of this peak may be low and difficult to detect, but this does not seem to be the case[103] in O_2.

Finally, we note that bands due to other sulfur molecular species, in particular the next most abundant one, S_6, are very weak or undetectable. In earlier photoionization studies on sulfur species,[110, 111] it was established that the adiabatic ionization potential of S_6 at 0 °K was 9·16 eV, and that it would appear still lower at the experimental temperature, perhaps at 9·00 eV. There is no evidence for any significant peak at such a low energy. If there is contamination by other sulfur species in Fig. 41(b), it is no larger than 5–10% relative to the major S_2 peaks, in agreement with our anticipated intensities based on equilibrium studies. The adiabatic and vertical ionization energies of S_2 are summarized in Table XII.†

---

† Note added in proof. J. M. Dyke, L. Golob, N. Jonathan, and A. Morris (J. Chem. Soc. Faraday Trans. II, **71**, 1026 1975) have recently reported the UPS of S_2 generated in a five- to ten-fold excess of He, and hence possibly cooled vibrationally. Their spectrum displays vibrational structure, and hence the vibrational frequencies of the various S_2^+ states can be deduced. Peaks at 14·62 eV and 15·58 eV, which are partially obscured in our spectrum because of the large Hg $^2D_{\frac{3}{2}}$ peak are clearly visible in their spectrum. The spin–orbit splitting they observe in the X $^2\Pi_g$ state is 58 meV, very close-to that (~ 60 meV) inferred from our spectrum. The intensity of the lower binding energy component is higher than that of the excited component, as in our spectrum, but their identification of the lower binding energy component as $^2\Pi_{g,\frac{3}{2}}$ disagrees with our assignment. In our view, the ordering of these levels should be regular, i.e. $^2\Pi_{g,\frac{1}{2}} < {}^2\Pi_{g,\frac{3}{2}}$, for states having a single electron in a π orbital, as in the present case, and inverted, i.e. $^2\Pi_{g,\frac{3}{2}} < {}^2\Pi_{g,\frac{1}{2}}$, for states having a single hole in a Π orbital, as in F_2^+, Cl_2^+, Br_2^+, and I_2^+.

In our view, the assignment of the weak bands observed by Dyke et al. in the 17·7–18·1 eV region to 2s-like orbitals ($^4\Sigma_u$ and $^2\Sigma_u$) of S_2 are questionable.

B. Te_2

From the arguments given earlier, we should expect the first band in this photoelectron spectrum to correspond to formation of a $^2\Pi_g$ state of $Te_2{}^+$. This state should be spin–orbit split (as observed in O_2, and possibly S_2). From Table XI, we would infer a splitting of about 0·5 eV between two peaks of approximately equal intensity, corresponding to $^2\Pi_{\frac{1}{2},g}$ and $^2\Pi_{\frac{3}{2},g}$. The expectation of approximately equal intensities for $^2\Pi_{\frac{3}{2}}$ and $^2\Pi_{\frac{1}{2}}$ is based on the fact that these states have equal degeneracies, or statistical weights, and is observed in the diatomic halogens[100, 101] and O_2.[103]

TABLE XII

Ionization Energies of S_2 and Te_2, in eV

	S_2		Te_2	
	Adiabatic	Vertical	Adiabatic	Vertical
$X^2\Pi_{g,\frac{1}{2}}$	$9\cdot36 \pm 0\cdot02^a$	9·50	$8\cdot29 \pm 0\cdot03^b$	8·30
	9·30		8·05	
$^2\Pi_{u,\frac{3}{2}}$		9·56		8·77
$a^4\Pi_u$	11·28	11·88	$9\cdot00^c$	$9\cdot44^c$
$A^2\Pi_u$		12·33		$10\cdot10^c$
$b^4\Sigma_g{}^-$	13·06	13·23	$10\cdot75^c$	$11\cdot04^c$
$B^2\Sigma_g{}^-$	$(14\cdot48)^d$	?	—	—
$c^4\Sigma_u{}^-$	—	—	$19\cdot13^c$	$19\cdot30^c$

[a] Photoionization: J. Berkowitz and C. Lifshitz (ref. 106).
[b] Photoionization: J. Berkowitz and W. A. Chupka, *J. Chem. Phys.* **50**, 4245 (1969).
[c] See text for alternate designation of these states, due to the different coupling mechanisms in Te_2.
[d] The position of this peak is uncertain because it is strongly masked by the $^2D_{\frac{3}{2}}$ peak of Hg.

In Fig. 41(c), however, the first peak at $\sim 8\cdot3$ eV has a width (FWHM) of only 0·2 eV, and the nearest peak is only one-tenth as intense, although it has roughly the expected energy separation (0·47 eV). Hence, at first glance there appears to be a departure from the principle of chemical similarity. In order to rationalize the observed peak structure, it is necessary to consider the nature of the neutral ground state and ionic states of Te_2 and related Group VI diatomics. R. F. Barrow and his co-workers[112, 113] have studied the neutral states of these systems in considerable detail in the past decade, and we shall quote liberally from their conclusions. We quote below from ref. 113.

"In the $^3\Sigma$ states of light molecules the triplet splitting, which arises partly from the spin–orbit interaction, is small, and the rotational energy levels are well represented by the (Hund's) case-b expressions

$$F_1(N) = B_v N(N+1) + B_v(2N+3) - \lambda_v - \gamma/2$$

$$= -\left[\lambda_v^2 - 2\lambda_v\left(B_v - \frac{\gamma}{2}\right) + (2N+3)^2\left(B_v - \frac{\gamma}{2}\right)^2\right]^{\frac{1}{2}}$$

$$F_2(N) = B_v N(N+1)$$

$$F_3(N) = B_v N(N+1) - B_v(2N-1) - \lambda_v - \gamma/2$$

$$+ \left[\lambda_v^2 - 2\lambda_v\left(B_v - \frac{\gamma}{2}\right) + (2N-1)^2\left(B_v - \frac{\gamma}{2}\right)^2\right]^{\frac{1}{2}},"$$

where

B_v = rotational constant

N = rotational quantum number

λ_v = a constant which measures the magnetic interaction of the spin–spin and polarization types.

γ = spin–rotation constant

"In these expressions, the F_1 levels have $J = N+1$, the F_2 levels have $J = N$ and for F_3, $J = N-1$; terms in the centrifugal distortion are omitted."

"As the atomic number increases, λ begins to assume values of the order of atomic spin–orbit coupling constants . . .. In the limit of strong interaction with states of different Λ and Σ, the coupling approximates to Hund's case-c. The spin–rotation constant γ is always small, so that for $^3\Sigma^-$ states when $\lambda \gg B_v$, the F_1 levels separate to form the levels of an $\Omega = 0^+$ state, while the F_2 and F_3 levels converge to form the levels of an $\Omega = 1$ state . . . the limiting case-c energies are

$$\Omega = 0^+, \quad F_1(J) = B_v J(+1) - 2\lambda$$

$$\Omega = 1, \quad F_{2,3}(J) = B_v J(J+1)."$$

Barrow and Yee[113] summarize the results of many experiments aimed at deducing 2λ. Of relevance to the present study are the following: O_2, $3 \cdot 9696$ cm^{-1}; S_2, $23 \cdot 68$ cm^{-1}; Se_2, $366 \cdot 6$ cm^{-1}; and Te_2, $\leqslant 2230$ cm^{-1}.

The rotational constants B_v are $1 \cdot 44566$ cm^{-1} for O_2, $0 \cdot 2956$ cm^{-1} for S_2, $0 \cdot 08977$ cm^{-1} for Se_2, and $0 \cdot 0396$ cm^{-1} for Te_2. Hence, Te_2 presents a marked contrast to the other Group VI homonuclear diatomics since even for high rotational quanta, the $^3\Sigma_g^-$ state is markedly into case-c, and splits into 0_g^+ and 1_g states, of which 0_g^+ is the ground state.[114] The splitting

($\sim$0·27 eV) is such that at the temperature of our experiment the 1_g state should represent only $\sim 1\%$ of the beam.

Now let us consider the nature of the excited (ionic) states permitted by electric-dipole allowed transitions. In transitions from $(\pi_g)^2$, $0_g{}^+$, the accessible Rydberg molecular orbitals must be $np\sigma$ or $np\pi$ and hence in the limit the continuum electron must also be $np\sigma$ of $np\pi$. If it is $np\sigma$, the coupling[106] of the Rydberg electron with the $^2\Pi_{g,\frac{3}{2}}$ ion core gives $\Omega = 2_u$ or 1_u; coupling with the $^2\Pi_{g,\frac{1}{2}}$ ion core gives $\Omega = 0_u{}^-$, $0_u{}^+$, and 1_u.

In case-c coupling, the selection rule $\Delta\Omega = 0$, ± 1 would permit transitions from the ground state $(0_g{}^+)$ to $0_u{}^+$ or 1_u, and hence to the continuum associated with the $^2\Pi_{g,\frac{1}{2}}$ and $^2\Pi_{g,\frac{3}{2}}$ core. In an extensive study of the related TeO molecule, which is also treated as case-c, Chandler *et al.*[112] were puzzled by the absence of $0^+ \to 1$ bands. They commented: "While it is impossible to affirm that these transitions are absent, it is certain that they are, at most, inconspicuous features of the spectrum." Hence, superposed upon the selection rule there appears to be a propensity rule which strongly prefers $\Delta\Omega = 0$ to $\Delta\Omega = \pm 1$. Since in the present case only the $^2\Pi_{g,\frac{1}{2}}$ core can give rise to $0_u{}^+$, this state is strongly enhanced in the photoelectron spectrum.

We must now carry through the analysis for an $np\pi$ Rydberg orbital, which is also permissible. Coupling of such a Rydberg electron with a $^2\Pi_{g,\frac{3}{2}}$ ion core gives $0_u{}^+$ and 1_u; with a $^2\Pi_{g,\frac{1}{2}}$ core one obtains 1_u (twice) and $0_u{}^+$. Thus an ionizing transition in which the departing electron is characterized as $np\pi$ would permit a $^2\Pi_{g,\frac{3}{2}}$ ionic state to be formed, even if $\Delta\Omega = 0$. If our observations are to be consistent with this analysis, we are forced to conclude that transitions to $np\pi$ are an order of magnitude weaker than those to $np\sigma$.

The remaining peaks in the 584 Å photoelectron spectrum of Te_2, while showing resemblances to those of $O_2{}^+$ and $S_2{}^+$, also exhibit differences. The double-humped band with peak maxima at 9·44 eV and 10·10 eV can be correlated with the $a\,^4\Pi_u$ and $A\,^2\Pi_u$ states in $O_2{}^+$ and $S_2{}^+$, both arising from $(\pi_u)^{-1}$. However, the peak at 11·04 eV, which one is tempted to identify with the corresponding $b\,^4\Sigma_g{}^-$ peak in $O_2{}^+$ and $S_2{}^+$, does not seem to have a prominent companion $B\,^2\Sigma_g{}^-$ peak, such as would be expected from $(\sigma_g)^{-1}$ and the consequent interaction of two unfilled shells.

The peak at 19·30 eV and its weak companion at 19·8 eV do not correspond to additional states of $Te_2{}^+$. Subsequent studies with another lamp have shown that these peaks are due to an impurity of hydrogen Lyman-α ($h\nu = 10\cdot2$ eV) photoionizing Te_2, and reproducing the region of the first ionization produced by He I radiation ($h\nu = 21\cdot2$ eV) on Te_2. They should appear 11·0 eV apart, and the spectrum indicates that this is indeed the case. It is noteworthy that the anomalous intensity ratio of $^2\Pi_{g,\frac{1}{2}} : ^2\Pi_{g,\frac{3}{2}}$ persists at $h\nu = 10\cdot2$ eV.

The absence or weakness of a companion peak to the 11·04 eV peak assigned as $b\,^4\Sigma_g^-$ may also be related to the case-c coupling. Since the neutral ground state is 0_g^+, the contribution of the outermost $(\pi_g)^2$ to the component of total angular momentum is zero. Ejection of an electron from an inner orbital leads to a net spin (and in the case of π orbitals, orbital) angular momentum contribution from this orbital. If there is substantial interaction between the angular momentum of the unfilled core orbital and the unfilled valence orbital, then splitting into quartets and doublets should occur, as is the case with O_2^+. If, on the other hand, this interaction is very weak, then only one state might be anticipated. This interaction is governed by the exchange integral between the pertinent core and valence levels and should diminish in the sequence O_2, S_2, Se_2, Te_2 since it is a matrix element of the operator e^2/r_{ij}, and the spatial extent of the orbitals is increasing with increasing Z. The strength of the exchange interaction must presumably be compared with the λ-interaction, which splits 0_g^+ and 1_g. We can estimate the relative strengths of these interactions by comparing the $^4\Pi_u$-$^2\Pi_u$ separations with the λ values for O_2 and S_2, and try to extrapolate to Te_2. In O_2, $2\lambda = 0·0005$ eV and the $^4\Pi_u$-$^2\Pi_u$ separation is $\sim 1·0$ eV; for S_2, $2\lambda = 0·003$ eV and the $^4\Pi_u$-$^2\Pi_u$ separation is $ca.$ 0·4 eV. For Te_2, we can expect an exchange splitting significantly smaller than 0·4 eV, and $2\lambda \simeq 0·27$ eV. If the exchange integral between $(\sigma_g\,5p)^{-1}$ and $(\pi_g\,5p)^{-2}$ has also diminished drastically in this sequence, this could account for the observation of only a single Σ_g^- peak. The fact that two peaks result from $(\pi_u\,5p)^{-1}$ can readily be explained by the intra-orbital spin–orbit splitting, which (as we have seen) has also been increasing in the sequence O_2, S_2, Se_2, Te_2. In fact, the magnitude of the splitting in Te_2^+ resulting from $(\pi_u\,5p)^{-1}$ is $\sim 0·66$ eV, in the range expected for spin–orbit splitting from Table X.

In summary, a comparison of the 584 Å photoelectron spectra of O_2, S_2, and Te_2 reveals greater differences than between F_2, Cl_2, Br_2, and I_2, suggesting that the principle of chemical similarity is being violated. (Preliminary data on Se_2† indicate that its behavior, particularly the relative intensity of $^2\Pi_{g,\frac{1}{2}}$ and $^2\Pi_{g,\frac{3}{2}}$, lies between that of S_2 and Te_2.) These differences are attributable to the dynamics of the photoionization process, rather than to the orbital sequence. The anomalous behavior of Te_2 is not attributable to its relatively large spin–orbit splitting. That of I_2 is larger, and it does not display anomalous behavior. It is related to the open-shell $(\pi_g)^2$ structure and to the large case-c λ splitting.‡

---

† D. G. Streets and J. Berkowitz, *Journal of Electron Spectroscopy* (in press) presents a photoelectron spectrum of Se_2, as well as more complete data on Te_2.

‡ S. T. Lee, S. Süzer, and D. A. Shirley, *Chem. Phys. Lett.* **41**, 25 (1976), have recently proposed an alternative interpretation for the anomalous $^2\Pi_{g,\frac{1}{2}}$-$^2\Pi_{g,\frac{3}{2}}$ ratios in the O_2, S_2, Se_2, Te_2 sequence.

The analysis given here leads to the conclusion that the formation of the ionic ground state proceeds primarily by way of a $\Delta\Omega = 0$ transition, and that the continuum corresponding to this state has a strong preference for $\varepsilon p\sigma$, rather than $\varepsilon p\pi$. Presumably, this conclusion could be tested by a study of the angular distribution of photoelectrons produced in the formation of $^2\Pi_{g,\frac{1}{2}}$, although (to the author's knowledge) such an analysis for the coupling described herein has not yet been carried out.

IX. CONCLUSION

High-temperature photoelectron spectroscopy is only about five years old, but in that time a number of new effects have been uncovered. The influence of ionic bonding upon photoelectron spectra, new ideas about bonding and antibonding orbitals, effects of photon energy upon spin–orbit pair intensities, chemical shifts in sub-valence shells and electronegativity and deviations from the principle of chemical similarity—all of these concepts have resulted from the introduction of high-temperature capability. The studies mentioned in this article have been limited to $\sim 1000\,°K$. Very few atoms have been investigated, and almost no high-temperature oxides or sulfides have been explored. Studies with different photon energies have just begun, and we are on the threshold of measuring the angular distributions of photoelectrons for these species at different wavelengths. Much remains to be done and, hopefully, we may be rewarded with new scientific insights.

REFERENCES

1. D. W. Turner, C. Baker, A. D. Baker, and C. R. Brundle, "Molecular Photoelectron Spectroscopy" (Wiley–Interscience, London, 1970).
2. B. G. Cocksey, J. H. D. Eland, and C. J. Danby, *J. Chem. Soc. Faraday Trans. II*, **69**, 1558 (1973).
3. M. M. Timoshenko and M. E. Akopyan, *Khim. Vys. Energii*, **8**, 211 (1974); *Engl. Transl. High Energy Chem.* **8**, 175 (1974).
4. K. Hoefler, private communication.
5. A. W. Potts, T. A. Williams, and W. C. Price, *Proc. Roy. Soc. Lond.* **A341**, 147 (1974).
6. A. F. Orchard and N. V. Richardson, *J. Electr. Spectr.* **6**, 61 (1975).
7. J. Berkowitz, *J. Chem. Phys.* **56**, 2766 (1972).
8. E. M. Purcell, *Phys. Rev.* **54**, 818 (1938).
9. V. V. Zashkvara, M. I. Korsunskii, and O. S. Kosmachev, *Soviet Phys. Tech. Phys.* (*Engl. Transl.*), **11**, 96 (1966).
10. H. Z. Sar-El, *Rev. Sci. Instr.* **38**, 1210 (1967), **42**, 1601 (1971). See also M. Karras, M. Pesa, and S. Arsela, *Ann. Acad. Scient. Fenn. Series A VI. Physica No.* 289 (1968) (Helsinki).
11. J. D. Allen, Jr., G. W. Boggess, T. D. Goodman, A. S. Wachtel, Jr., and G. K. Schweitzer, *J. Electr. Spectr.* **2**, 289 (1973).
12. J. L. Gardner and J. A. R. Samson, *J. Electr. Spectr.* **2**, 267 (1973).

13. J. Berkowitz, Photoelectron and photoion spectroscopy of molecules, "Vacuum Ultraviolet Radiation Physics" (Pergamon/Vieweg, 1974).
14. T. E. H. Walker, J. Berkowitz, J. L. Dehmer, and J. T. Waber, *Phys. Rev. Letters*, **31**, 678 (1973).
15. J. L. Dehmer and J. Berkowitz, *Phys. Rev.* **A10**, 484 (1974).
16. J. Berkowitz, J. L. Dehmer, Y.-K. Kim, and J. P. Desclaux, *J. Chem. Phys.* **61**, 2556 (1974).
17. S. Suzer and D. A. Shirley, *J. Chem. Phys.* **61**, 2481 (1974).
18. J. Berkowitz, *Proc. Intern. Symp. Synchrotron Radiation Users, Daresbury, Jan.* 1973 DNPL/R26, p. 151 (1973).
19. C. E. Moore, "Atomic Energy Levels", Natl. Bur. Std. (US) Circ. 467 (1958).
20. J. A. R. Samson and R. B. Cairns, *Phys. Rev.* **173**, 80 (1968).
21. T. E. H. Walker and J. T. Waber, *J. Phys.* **B7**, 674 (1974).
22. J. W. Cooper, *Phys. Rev.* **128**, 681 (1962).
23. J. Cooper and R. N. Zare, in: "Lectures in Theoretical Physics, Vol. 11C, Atomic Collision Processes", S. Geltman, K. Mahantharpa, and W. Brittin, eds. (Gordon and Breach, New York, 1969), p. 317.
24. S. T. Manson, *Chem. Phys. Lett.* **19**, 76 (1973).
25. H. Harrison, *J. Chem. Phys.* **52**, 901 (1970).
26. A. Niehaus and M. W. Ruf, *Z. Physik*, **252**, 84 (1972).
27. U. Fano and D. Dill, *Phys. Rev.* **A6**, 185 (1972).
28. J. Berkowitz, J. L. Dehmer, and T. E. H. Walker, *J. Chem. Phys.* **59**, 3645 (1973).
29. T. D. Goodman, J. D. Allen, Jr., L. C. Cusachs, and G. K. Schweitzer, *J. Electr. Spectr.* **3**, 289 (1974).
30. E. S. Rittner, *J. Chem. Phys.* **19**, 1030 (1951).
31. J. Berkowitz, *J. Chem. Phys.* **29**, 1386 (1958), **32**, 1519 (1960).
32. R. L. Matcha, "Compendium of Alkali Halide Wavefunctions", Chemical Physics Group, Univ. of Houston, 1 April 1970.
33. R. S. Berry, *Chem. Rev.* **69**, 533 (1969).
34. L. Brewer and E. Brackett, *Chem. Rev.* **61**, 425 (1961).
35. E. N. Nikolaev, *Khim. Vys. Energii*, **3**, 491 (1969).
36. I. I. Sobel'man, "An Introduction to the Theory of Atomic Spectra", T. F. J. Le Vierge. G. K. Woodgate, ed. (Pergamon Press, Oxford, 1972), p. 263 *et seq.*, 308.
37a. W. J. Stevens and F. P. Billingsley, *Phys. Rev.* **A8**, 2236 (1973).
37b. W. J. Stevens, private communication.
38. J. Berkowitz and W. A. Chupka, *J. Chem. Phys.* **29**, 653 (1958).
39. J. Berkowitz and W. A. Chupka, *J. Chem. Phys.* **45**, 1287 (1966).
40. J. Berkowitz, *Adv. in High Temp. Chem.* **3**, 123 (1971).
41. J. Berkowitz, H. A. Tasman, and W. A. Chupka, *J. Chem. Phys.* **36**, 2170 (1962).
42. C. P. Baskin, C. F. Bender, and P. A. Kollman, *J. Am. Chem. Soc.* **95**, 5868 (1973).
43. J. L. Gole, A. K. Q. Siu, and E. F. Hayes, *J. Chem. Phys.* **58**, 857 (1973).
44. S. Rothenberg and H. F. Schaefer III, *J. Ann. Chem. Soc.* **95**, 2095 (1973).
45. D. R. Yarkony, W. J. Hunt, and H. F. Schaefer III, *Mol. Phys.* **26**, 941 (1973).
46. D. R. Yarkony and H. F. Schaefer III, *Chem. Phys. Lett.* **15**, 514 (1972).
47. L. C. Cusachs, F. A. Grimm, and G. K. Schweitzer, *J. Electr. Spectr.* **3**, 229 (1974).

48. J. H. D. Eland, *Int. J. Mass Spec. Don. Phys.* **4**, 37 (1970).
49. G. W. Boggess, J. D. Allen, Jr., and G. K. Schweitzer, *J. Electr. Spectr.* **2**, 467 (1973).
50. J. Berkowitz, *J. Chem. Phys.* **61**, 407 (1974).
51. K. Siegbahn *et al.*, ESCA-atomic, Molecular and Solid-state Structure, studied by means of Electron Spectroscopy" (Almqvist and Wiksells, Uppsala, 1967), p. 79–82.
52. These relativistic wavefunctions, available from Dr. Y.-K. Kim, Argonne National Laboratory, were obtained from a program written by Dr. J. P. Desclaux.
53. P. A. Akishin and V. P. Spiridonov, *Kristallografia*, **2**, 472 (1957).
54. C. R. McCoy and A. L. Allred, *J. Inorg. Nucl. Chem.* **25**, 1219 (1963).
55. A. L. Allred and E. G. Rochow, *J. Inorg. Nucl. Chem.* **5**, 264 (1958).
56. L. Pauling, *J. Am. Chem. Soc.* **69**, 542 (1947).
57. J. C. Slater, *Phys. Rev.* **36**, 57 (1930).
58. L. Pauling, "The Nature of the Chemical Bond" (Cornell University Press, Ithaca, N.Y., 1939).
59. E. J. Little and M. M. Jones, *J. Chem. Educ.* **37**, 231 (1960).
60. E. Clementi and D. L. Raimond, *J. Chem. Phys.* **38**, 2686 (1963).
61. E. Clementi, D. L. Raimond, and W. P. Reinhardt, *J. Chem. Phys.* **47**, 1300 (1967).
62. C. Froese-Fischer, *Atomic Data*, **4**, 301 (1972); Atomic Data and Nuclear Data Tables, **12**, 87 (1973).
63. W. C. Price, A. W. Potts, and D. G. Streets, in: "Electron Spectroscopy", D. A. Shirley, ed. (North-Holland, Amsterdam, 1972), p. 187.
64. J. Berkowitz, *J. Chem. Phys.* **56**, 2676 (1972).
65. J. W. Hastie and J. L. Margrave, *J. Phys. Chem.* **73**, 1105 (1969).
66. B. Rosen, "Spectroscopic Data Relative to Diatomic Molecules" (Pergamon Press, Oxford, 1970).
67. The references for the dipole moments are as follows: For AlF, D. R. Lide, Jr., *J. Chem. Phys.* **42**, 1013 (1965); for GaF and Inf: J. Hoeft, F. J. Lovas, E. Tiemann, and T. Törring, *Z. Naturf.* **25**a, 1029 (1970); for TlF: R. V. Boeckh, G. Gräff, and R. Ley, *Z. Physik*, **179**, 285 (1964).
68. M. Attermeyer, G. Das, and A. C. Wahl, to be published.
69. D. Lide, *J. Chem. Phys.* **42**, 1013 (1965).
70. A. Rosén and D. E. Ellis, *J. Chem. Phys.* **62**, 3039 (1975).
71. F. J. Keneshea and D. Cubicciotti, *J. Phys. Chem.* **69**, 3910 (1965), **71**, 1958 (1976).
72. S. H. Bauer, T. Ino, and R. F. Porter, *J. Chem. Phys.* **33**, 685 (1960); P. A. Akishin and N. G. Rambidi, *Z. Physik. Chem.* **213**, 111 (1960).
73. J. M. Brom, Jr. and H. F. Franzen, *J. Chem. Phys.* **54**, 2874 (1971).
74. J. L. Dehmer, J. Berkowitz, and L. C. Cusachs, *J. Chem. Phys.* **58**, 5681 (1973).
75. M. G. Fickes, R. C. Slater, W. G. Becker, and R. C. Stern, *Chem. Phys. Lett.* **24**, 105 (1974).
76. J. S. Muenter, *Chem. Phys. Lett.* **26**, 97 (1974).
77. V. G. Solomonik, E. Z. Zasorin, G. V. Girichev, and K. C. Krasnov, *Khimia: Khimicheskaya Tekhnologia*, **17**, 136 (1974).

78. G. Herzberg, "Molecular Spectra and Molecular Structure. II. Infrared and Raman Spectra of Polyatomic Molecules" (D. Van Nostrand Co., Princeton, 1945).
79. P. A. Akishin, V. A. Naumov, and V. M. Talevskii, *Sov. Phys.-Crystallogr.* **4**, 174 (1959).
80. M. L. Lesiecki and J. S. Shirk, *J. Chem. Phys.* **56**, 4171 (1972).
81. A. W. Potts, H. J. Lempka, D. G. Streets, and W. C. Price, *Phil. Trans. Roy. Soc. Lond.* **A268**, 59 (1970).
82. G. H. King, S. S. Krishnamurty, M. F. Lappert, and J. B. Pedley, *Discuss. Faraday Soc.* **54**, 70 (1973).
83. M. F. Lappert, J. B. Pedley, G. J. Sharp, and N. P. C. Westwood, *J. Electr. Spectr.* **3**, 237 (1974).
84. J. L. Dehmer, J. Berkowitz, L. C. Cusachs, and H. S. Aldrich, *J. Chem. Phys.* **61**, 594 (1974).
85. G. Herzberg, "Molecular Spectra and Molecular Structure. III. Electronic Spectra of Polyatomic Molecules" (Van Nostrand Reinhold, New York, 1966).
86. R. Manne, K. Wittel, and B. S. Mohanty, *Mol. Phys.* **29**, 485 (1975).
87. K. Wittel and R. Manne, *J. Chem. Phys.* **63**, 1322 (1975).
88. K. C. Krasnov, "Molekularnie Postoyanie Neorganicheskikh Soedinenii" (Leningrad, 1968).
89. B. Wirsam, *Chem. Phys. Lett.* **22**, 360 (1973).
90. K. Wittel, *Chem. Phys. Lett.* **15**, 555 (1972).
91. K. Wittel, H. Bock, and R. Manne, *Tetrahedron*, **30**, 651 (1974).
92. L. C. Snyder and H. Basch, "Molecular Wave Functions and Properties" (Wiley, New York, 1972).
93. J. M. Dyke, L. Golob, N. Jonathan, A. Morris, and M. Okuda, *J. Chem. Soc. Faraday Trans. II*, **70**, 1828 (1974).
94. T. P. Fehlner and D. W. Turner, *Inorg. Chem.* **13**, 755 (1974).
95. C. Thomson, *Theor. Chem. Acta*, **32**, 93 (1973).
96. D. R. Lloyd and P. J. Roberts, *Mol. Phys.* **26**, 225 (1973).
97. S. Evans and A. F. Orchard, *J. Electr. Spectr.* **6**, 207 (1975).
98. F. Brogli and E. Heilbronner, *Helv. Chim. Acta*, **54**, 1423 (1971).
99. J. C. Green, M. L. H. Green, P. J. Joachim, A. F. Orchard, and D. W. Turner, *Phil. Trans. Roy. Soc. Lond.* **A268**, 111 (1970).
100. A. W. Potts and W. C. Price, *Trans. Faraday Soc.* **67**, 1242 (1971).
101. A. B. Cornford, D. C. Frost, C. A. McDowell, J. L. Ragle, and I. A. Stenhouse, *J. Chem. Phys.* **54**, 2651 (1971).
102. H. J. Lempka, T. R. Passmore, and W. C. Price, *Proc. Roy. Soc. Lond.* **A304**, 53 (1968).
103. O. Edqvist, E. Lindholm, L. E. Selin, and L. Åsbrink, *Physica Scripta*, **1**, 25 (1970).
104. J. Berkowitz, *J. Chem. Phys.* **62**, 4074 (1975).
105. J. Berkowitz and J. R. Marquart, *J. Chem. Phys.* **39**, 275 (1963).
106. G. Herzberg, "Molecular Spectra and Molecular Structure. I. Spectra of Diatomic Molecules" (D. Van Nostrand Co., Princeton, 1950).
107. R. G. Dromey, J. D. Morrison, and J. B. Peel, *Chem. Phys. Lett.* **23**, 30 (1973).
108. A. W. Potts and W. C. Price, *Proc. Roy. Soc. Lond.* **A326**, 181 (1972).
109. K. Siegbahn *et al.*, "ESCA applied to Free Molecules" (North-Holland Publishing Co., Amsterdam, 1969), pp. 84, 87.

110. J. Berkowitz and C. Lifshitz, *J. Chem. Phys.* **48**, 4346 (1968).
111. P. M. Guyon and J. Berkowitz, *J. Chem. Phys.* **54**, 1815 (1971).
112. G. G. Chandler, H. J. Hurst, and R. F. Barrow, *Proc. Phys. Soc.* **86**, 105 (1965).
113. R. F. Barrow and K. K. Yee, *Acta Phys. Acad. Scient. Hung.* **35**, 239 (1974).
114. R. P. Du Parcq and R. F. Barrow, *Chem. Comm.* 270 (1966).
115. J. P. Desclaux, *Atomic Data and Nuclear Data Tables*, **12**, 311 (1973).

8

Two-parameter Coincidence Experiments

M. E. GELLENDER

and

A. D. BAKER

Queens College, Department of Chemistry,
The City University of New York,
Flushing, New York 11367

I. INTRODUCTION

Electron spectroscopy primarily involves the measurement of electron kinetic energy spectra resulting from particular processes of photoionization, inelastic scattering, or excitation. Considerable information about the energy levels of molecules and ions has thus been derived from photoelectron, electron impact, and Auger spectra.

Scanning an electron kinetic energy spectrum may thus be referred to as a first-order experiment since only one parameter, the electron kinetic energy, is measured. This type of experiment is generally characterized by a high attainable resolution and signal/noise ratio. However, the extent of the information which may be derived is limited, for while the detection of an electron of known energy reveals the state of the ion or molecule from which it was dispatched, it can yield little information about the fate of the resulting excited species. It is only after the initial process of ionization or excitation that the resulting species will embark on a chemical adventure of unimolecular fragmentation into ions, reaction with other molecules, or the emission of radiation.

As techniques of energy analysis and data processing have become more refined, there has been an expanding interest in two-parameter coincidence experiments. Although these experiments are inherently more difficult to perform and provide much lower rates of data collection, they are the only unambiguous means of examining the fates of excited ions and molecules.

Underlying all electron spectroscopy experiments is a general trade-off between the resolution of electron energies and the detected signal count rate. The energy resolution, which characterizes the extent of useful spectral information, may be enhanced if the experimenter is willing to accept a sacrifice of signal count rate. However, the number of electrons detected each second, S, is also a major consideration, as it determines the signal/noise ratio which may be achieved within a given period of data collection time, t. The maximum attainable signal/noise ratio in an ideal one-parameter experiment (in which the only source of noise is the inherent statistical fluctuations in the number of electrons arriving at the detector) is

$$(S/N)_{\text{optimal}} = S^{\frac{1}{2}} t^{\frac{1}{2}} \tag{1}$$

Thus, from a theoretical viewpoint, any desired resolution and signal/noise ratio can be achieved if sufficient time is allocated for the collection of data. Any discussion of the feasibility and practicality of an experimental technique must therefore consider both the resolution and rate of data collection which may simultaneously be attained.

A. Signal Limitations of One-parameter Experiments

A basic one-parameter electron spectrometer is depicted in Fig. 1. Electrons photoejected or scattered from a sample molecule enter an energy analyzer, whose function is to transmit electrons of kinetic energy within an energy range ΔE, while discriminating against all electrons of other energies. Of the total number of electrons produced per second, N, only a small fraction have the proper energy, and enter the analyzer with the proper orientation, to be transmitted by the analyzer and detected. The resulting signal due to electrons within energy range ΔE is

$$S = Ngf_e \Delta E \text{ electrons/second} \qquad (2)$$

where g is the fraction of all electrons within energy range ΔE, and where f_e is a constant determined by the focussing properties of the energy analyzer (for the $127°$ sector analyzer commonly employed in early electron spectrometers, and for all similar first-order focussing analyzers, it may be shown[1] that

$$f_e = \frac{1}{2\pi E}.$$

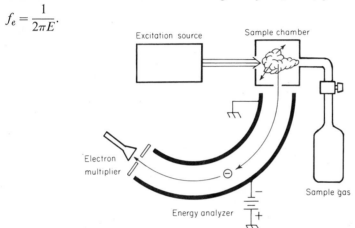

Fig. 1. Schematic diagram of a one-parameter experiment.

The subsequent discussion will demonstrate that two-parameter coincidence experiments are characterized by inherently inferior data-collection rates.

II. ELECTRON–ION COINCIDENCE SPECTROSCOPY

Considering the extensive work done in electron impact and photo-ionization[2] mass spectrometry, it is curious how poorly understood are the ionization and fragmentation processes of molecules. Since the mass spectrum of a compound contains the cumulative ion fragmentation products of all possible excitations and ionizations of the molecule, it is virtually impossible to infer the ion products resulting from dissociation of a particular parent ion state. Although a theoretical analysis of the fragmentation of ions is possible using the Quasi-equilibrium Theory of Mass Spectra,[3,4] agreement with experimental data is often only semiquantitative and in some cases glaring discrepancies exist.

Whether electron impact or photoionization is employed, the processes of ionization and fragmentation may be visualized as occurring in two distinct successive steps. Initially, molecules are ionized into various excited states of the parent ion, which then subsequently undergo fragmentation into the various ions characteristic of the mass spectrum of the compound.

Once a molecule is ionized, the dissociation into fragments may be direct, whereby the breaking of a specific bond within the molecule follows as an immediate consequence of the removal of a strongly bonding electron. An alternate fragmentation process is predissociation, in which the nature of the original molecular orbital from which ionization has occurred has little bearing upon the fragmentation products; even ionization of "non-bonding" lone-pair electrons may well result in dissociation. Predissociation occurs when the electronic excitation energy of the parent ion is converted into vibrational energy, which will become redistributed until eventually sufficient vibrational energy is localized within one portion of the ion to rupture a bond.

By the use of photoelectron–photoion coincidence spectroscopy, the fragmentation products of a particular ion state can be directly observed. In essence the technique allows the experimenter to select one particular ionization process, and then observe the resulting mass spectrum. Photoelectron–photoion coincidence spectroscopy has allowed ion fragmentation pathways to be determined for CO_2,[5] SO_2,[6] CS_2,[7] N_2O,[8] COS,[8] CH_4,[9] C_2H_6,[9] CH_2Cl_2,[10] CH_2Br_2,[11] CH_2I_2,[11] C_2F_6,[12, 13] and C_6H_6[14] (not including earlier[15, 16, 17] coincidence spectra). For these compounds then, the origin of each ion in the photoelectron mass spectrum has been completely determined.

Electron impact on molecules, as well as forming the same parent ion states as are encountered in photoionization, will lead to excited neutral molecules as well, which may then undergo autoionization. Therefore, a full understanding of electron-impact mass spectra will require autoionization coincidence experiments to be performed, although, as of this writing, this type of experiment has yet to be successfully undertaken. The characteristics and limitations of such a technique would be very similar to photoelectron–photoion coincidence spectroscopy, which although difficult to conduct, has been demonstrated as a viable technique.

A. Limitations

The basic experimental apparatus for a photoelectron–photoion coincidence experiment is depicted in Fig. 2. As a source for the photoionization of the sample gas within the ionization region, monochromatic radiation (usually He I radiation of 21·21 eV energy) is commonly used. Some experiments,[9, 11] however, have utilized a continuous source of radiation with a vacuum-uv monochromator, providing a monochromatic source of variable energy. Although the use of a monochromator results in substantial attenuation of the photon intensity (and adds considerable cost), it nonetheless allows the photoejection of electrons of low kinetic energy. Since the collection efficiency of energy analyzers is inherently superior for electrons originating with low kinetic energy,[1] a substantial gain in performance may be realized by the use of a variable energy source.

In any event, the photoelectrons produced enter an energy analyzer, which transmits a fraction $f_e\Delta E$ of electrons of a preselected energy (corresponding to a particular ionization process of interest). Upon detection of an electron, indicating that the selected ionization process has occurred, the arrival of the resulting ion is anticipated at the ion detector.

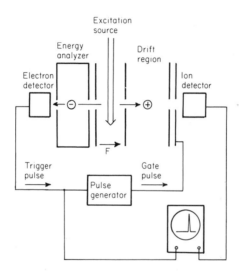

Fig. 2. Schematic diagram of a photoelectron–photoion coincidence experiment.

Within the ionization region, a constant electric field F will accelerate the resulting ion out of the ionization region, whereupon it will drift toward the ion detector, the expected arrival time depending upon the mass of the ion.

Consider an ion of mass m resulting from the detected photoionization process. Since the time of the ionization process is specified almost exactly by the detection of the photoelectron (the time of flight of the electron through the analyzer is usually negligible, and can be compensated for), it is tempting to believe that the arrival time of an ion of mass m could be predicted exactly. However, ions formed with initial velocities in the direction of the accelerating field will arrive at an earlier time than those initially moving opposite the field, which must be retarded to a standstill and then reaccelerated. Consequently, as a result of the initial kinetic energy E_I of the ion, the arrival time of an ion of mass m at the detector cannot be specified exactly, but only within a time range t_m, the "turnaround time" of an ion.

$$t_m = (2{\cdot}9 \times 10^{-6})\,(mE_I)/F \text{ seconds} \tag{3}$$

While parent ions are formed only with thermal energy (averaging about 0·040 eV at room temperature), fragment ion kinetic energies are often considerable, averaging up to 1 eV.

The uncertainty in the exact arrival time of the ion of mass m will be an inherent limitation of the instrument to distinguish between the ion formed as a result of the detected photoionization and any other ion which by random chance happens to strike the ion detector during time range t_m.

Consider the case where I ionizations occur each second in the ionization region, of which only a fraction $g(f_e \Delta E)$ of the photoelectrons are of the proper energy and will enter the analyzer with the proper orientation to be transmitted and detected. Consequently, $I g f_e \Delta E$ electrons are detected each second and trigger a time-of-flight spectrum of the resulting ion.

The detection of each electron will be followed by the successful observation of the resulting ion if ideal ion collection is assumed. Of the ions resulting from the selected ionization process, a fraction p_m will be of mass m. Therefore, the number of ions of mass m detected in coincidence with photoelectrons of known energy is

$$S = (I g f_e \Delta E) p_m \qquad (4)$$

However, during each time range t_m corresponding to the detection of an ion of mass m in each mass scan, $I t_m$ ions will arrive at the ion detector by random chance. These ions comprise a constant average background of "false coincidences" which will be detected during the entire duration of each time-of-flight spectrum. This background itself will present no difficulty, as it can be predicted and subtracted from the resulting spectrum. Statistical fluctuations, however, will inevitably be present in this background, which can neither be predicted nor eliminated, and will result in noise.

Such noise degrades the information content of the detected signal, and can only be diminished by increasing the time period t during which data are collected. At an optimal ionization rate (such that a maximum signal is present without saturation of the detectors) the signal/noise ratio for ions of mass m formed in coincidence with the selected ionization will be limited to a maximum value given by[18]

$$(S/N)_{\text{optimal}} = (g f_e \Delta E p_m^2 / t_m)^{\frac{1}{2}} t^{\frac{1}{2}} \qquad (5)$$

The signal/noise ratio determines the actual rate at which useful data of real coincidences is collected, and may be thought of as arising from an effective signal S_{eff} consisting only of real coincidences. As all ions resulting from the selected ionization process will be detected, the total rate at which data is accumulated will be the summation of the effective signal S_{eff} for each ion. By analogy with Eq. (2), the total effective signal count rate for the detection of all ions in coincidence with the selected photoionization is

$$S_{\text{eff}} = N_{\text{eff}} \, g f_e \, \Delta E \text{ coincidences/second} \qquad (6)$$

where

$$N_{\text{eff}} = \sum_{\text{all ions}} P_m{}^2 / t_m$$

The rate of effective ionization per second, N_{eff}, will depend primarily upon the initial energy of the ions, the width of the ionization region, and the electric field within the ionization region. Typically, N_{eff} will be of the order of 10^5 ionizations/second, as compared to a typical ionization rate in a one-parameter experiment of about 10^8 ionizations/second.

The most convenient way to increase N_{eff}, and hence the rate of data accumulation, is by increasing the electric field within the ionization region, and in doing so, decreasing the ion "turnaround time". However, this electric field results in an energy spread of the photoelectrons and so, degrades the electron energy resolution which may be achieved. Thus, the rate of data collection will be enhanced with a concurrent sacrifice of electron energy resolution. The overall performance of the instrument, including both the resolution and data collection rate, ultimately depends upon the initial ion kinetic energy and width of the ionization region.

Lest the reader be left exasperated with the prospect of poor data collection rates in coincidence experiments, several possibilities may be exploited in decreasing the required data collection time. Restricting the width of the ionization region or making use of high-transmission analyzers (such as the cylindrical mirror analyzer) will allow, according to Eq. (6), a higher rate of data collection to be achieved. Furthermore, considerable opportunity exists for the incorporation of multichannel array detectors and multichannel electron energy analysis. Substantial improvement in resolution and data collection rates can therefore be envisioned due to the increasing availability, at reasonable cost, of complex data-handling electronics.

B. Results of Photoelectron–Photoion Coincidence Spectra

For most molecules hitherto studied by photoelectron–photoion coincidence spectroscopy, agreement is usually found between the general behavior predicted by the quasi-equilibrium theory and actual experimental results. Typically, at the onset of ionization only the parent ion is observed, as insufficient energy exists for the formation of other possible ion products. At higher ionization energies however, as other ion products become energetically accessible, the ion fragments begin to appear in increasing abundance as the energy above the threshold for their appearance is exceeded. Parent ions and other ions requiring low energy for formation will then be formed in decreasing numbers at higher energies.

Thus if the quasiequilibrium theory (QET) was applicable to all molecules, it would be a simple matter to predict the general characteristics of the fragmentation processes of a molecule. However, quantitative agreement between experimental results and QET predictions often does not exist, and some molecules may exhibit major disparities between the experimentally determined fragmentation process and that predicted theoretically. One notable example is C_2F_6.

The photoelectron spectrum of C_2F_6, given in Fig. 3, contains three bands located approximately at 14·5 eV, 16·1 eV, and 17·6 eV. Within this range of ionization energy, three ions are energetically possible and appear in the photoionization mass spectrum, CF_3^+, $C_2F_5^+$, and CF^+. No parent $C_2F_6^+$ ions appear in the photoionization or electron impact mass spectrum.

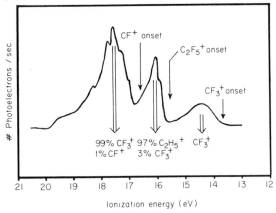

Fig. 3. Photoelectron spectrum of C_2F_6, and ions resulting from photoionization.

Ionization from the first band in the photoelectron spectrum produces only CF_3^+, being below the energy required for the formation of other ions. The second band, however, is above the appearance potential of $C_2F_5^+$, and not surprisingly, CF_3^+ is formed in decreasing abundance while $C_2F_5^+$ becomes the dominant ion product.

At higher ionization energies, in the third photoelectron band, it would be expected that CF^+ and $C_2F_5^+$ would become virtually the only ion products as progressively less CF_3^+ would be present. This in fact is not observed. Rather, the photoelectron–photoion coincidence spectrum shows that above 17·5 eV, the CF_3^+ ion once again becomes the dominant ion product, with small quantities of $C_2H_5^+$ and CF^+ ions. Above 18·4 eV, CF_3^+ is formed with the total exclusion of other ions.

Such behavior for the fragmentation of C_2F_6 cannot be accounted for on a theoretical basis, and exposes the limited applicability of the current theoretical approach in the understanding of fragmentation mechanisms.

C. Fragmentation of Multiply Charged Ions

The earliest application of a coincidence technique to the study of chemical problems was the investigation of the outcome of multiple ionization processes.

At the electron impact energies commonly employed in mass spectrometry in excess of 50 eV, sufficient energy is present to result in double, or even higher, ionization of the sample molecule. The resulting multiply charged ion will dissociate into several singly charged fragments, and possibly neutral fragments as well. Such fragmentation products of multiply charged ions are typically responsible for 2–4% of the total ionization products of the molecule (at high impact energies). Since considerable Coulombic repulsion will be present between the dissociating ion fragments, dissociation products of multiply charged ions will be characterized by high kinetic energies.

In order to investigate which ions are formed simultaneously as the outcome of a multiple ionization, it is necessary to utilize an experimental apparatus employing coincidence techniques. Sample molecules within the ionization region are excited by a high-energy (1 keV) electron beam, whereupon the occurrence of a single or multiple ionization process is indicated by the detection of a low-energy electron ejected out of the initial beam path. An electric field within the ionization region extracts the resulting ions into a drift region towards an ion detector so that a time-of-flight separation allows identification of the ion products. In this manner, a time-of-flight spectrum is scanned of the ions resulting from the initial ionization process, noting pairs of ions detected within the same mass scan.

TABLE I

Dissociation Products of Multiply Ionized CF_4

	Approx. total ionization (%)
$CF_4 \longrightarrow CF_4^{+n} \longrightarrow F^+ + CF^+ + \text{fragments}$	1·6
$\longrightarrow F^+ + CF_2^+ + \text{fragment}$	1·0
$\longrightarrow F^+ + CF_3^+$	0·8
$\longrightarrow F^+ + F^+ + \text{fragments}$	Suspected

n = primarily 2 (double ionization); higher ionization may contribute

Initial experiments[19] were performed on CO_2, CF_4, CH_4, and C_3H_6, and results of the CF_4 spectra are briefly presented here in Table I. All dissociations resulting from multiple ionization of CF_4 produce F^+ ions, and it is estimated that these comprise at least 25% of the total yield of F^+ ions produced by electron impact. As is expected of multiple ionizations, consider-able kinetic energy is imparted to the ion products. Kinetic energies of F^+ ions as high as 5 eV may be inferred from the coincidence mass spectrum.

Interestingly, even larger kinetic energies are imparted in the dissociation of multiply ionized hydrocarbons. Since protons are of much lower mass than other fragments, kinetic energy will be preferentially distributed among the ion products so that protons receive the bulk of the available energy. Dissocia-tion of multiply charged methane produces H^+ ions with 7–10 eV of kinetic energy.

The interpretation of such multiple ionization coincidence spectra leaves yet to be explained whether double (or higher) ionization results from direct removal of electrons, or by autoionization of a highly excited molecule. How the multiply charged parent ion fragments into neutral radicals and product ions has also to be determined.

D. Photoion Kinetic Energy Analysis

Ionization from a molecular orbital may well result in the ion possessing energy in excess of that required to dissociate the ion into fragments. This excess energy will appear in the form of kinetic or vibrational energy of the product fragments (it is generally anticipated that fragmentation will not lead to significant rotational excitation).

Consider the general case of ionization from a molecular orbital of ionization potential P resulting in fragmentation leading to an ion m_1^+ and a neutral fragment m_2. If A is the threshold energy required to form m_1^+ (i.e. the sum of the dissociation energy of the m_1—m_2 bond and the ionization potential of the fragment m_1), then energy $(P-A)$ will ultimately appear as the total kinetic and vibrational energy of the fragments. Consequently, the total kinetic energy of both fragments, $E_1 + E_2$, will be given as

$$E_1 + E_2 = P - A - E_{vib}$$

The conservation of momentum will require the two fragments to have equal momenta away from the center of mass of the original molecule. This will result in the fragment ion kinetic energies being partitioned between the fragments such that

$$E_1 = E_2(m_1/m_2)$$

with the result that

$$E_1 = (P - A - E_{vib})\left(\frac{m_2}{m_1 + m_2}\right)$$

Most of the kinetic energy will be imparted to the lighter fragment.

It has been well established[20] that the kinetic energy distribution of frag-ment ions can be determined from the width of mass peaks in the time-of-flight mass spectrum. Such an analysis provides the kinetic energy distribution of all ions of a particular mass, formed at all ionization energies. However, only in the case of a coincidence photoelectron time-of-flight mass spectrum can the kinetic energy of product ions resulting from a particular ionization be deduced. A photoelectron–photoion coincidence spectrum, in addition to elucidating fragmentation patterns of ions, will therefore be capable of providing (through a study of the shape of each mass peak) the kinetic energy of the ion products resulting from a particular ionization.

Analyses of coincidence time-of-flight peak widths have shown[21] that considerable kinetic energies are released in the fragmentation of CF_4, N_2O, and COS. As an illustration, consider the ionization from the second band of the photoelectron spectrum of N_2O, which produces NO^+ and N fragments with an average total kinetic energy release of more than 1 eV, of which about one-third is imparted to the NO^+ ion.

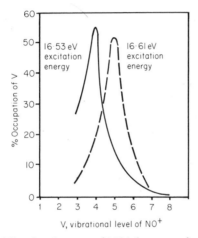

Fig. 4. Vibrational energy of NO^+ fragments from N_2O.

At an ionization from N_2O requiring 16·53 eV, an excess energy of 2·30 eV will be present which will appear as vibrational excitation of the NO^+ ion and kinetic energy of both fragments. Analysis of the shape of the NO^+ mass peak formed in coincidence with a 16·53 eV excitation energy reveals that the most probable kinetic energy imparted to the fragments is 1·18 eV. The remaining 1·12 eV of excitation energy will appear as integral energy of the NO^+ fragment. If it is assumed that the fragment has undergone no rotational excitation, vibrational internal energy will be present requiring the ion to be in the $v = 4$ vibrational level. Of course, a distribution of vibrational energy will be present among NO^+ ion fragments with $v = 4$ being the most probable excitation, as shown in Fig. 4.

One might expect that an increase in the energy of excitation will generally result in increased vibrational excitation of the fragments. Preliminary work[6,7,21] has indicated that this is usually the case, although not all of the additional excitation energy will always appear as vibrational energy of the fragments. For example, it has been found that in the dissociation of SO_2, only about 50% of the energy available in excess of that required for the appearance of the SO^+ ion product will actually appear as vibrational energy of the SO^+ fragment. In the case of formation of NO^+ from N_2O, an increase in excitation energy to 16·61 eV will result in higher vibrational energy of the NO^+ fragment, with most fragments being formed in the $v = 5$ vibrational level.

The vibrational energy distributions of fragment ions is of considerable interest, as internal energy of ions may play a large role in determining the outcome of reactions in ion–molecule collisions. In fact, in addition to studying the vibrational energy of fragment ions, a photoelectron–photoion coincidence method can be used as a technique to study the reaction of ions in known vibrational states.

III. COINCIDENCE STUDIES OF METASTABLE IONS

Upon formation, an excited ion state may exist for some considerable period of time before undergoing relaxation to the ground state. For such metastable ions, several alternate relaxation mechanisms will compete to dissipate the excitation energy of the ion. Possible mechanisms are the emission of radiation, unimolecular dissociation of the ion into fragments, or collision with a neutral molecule resulting in energy transfer or reaction. The extent to which each of these relaxation processes occur is determined by the corresponding lifetime of the excited ion.

A. Fluorescence

The existence of metastable positive ions capable of fluorescence as a means of dispelling excess energy has been known for some time. Detailed studies of fluorescence in the visible and ultraviolet regions of the spectrum have been performed on N_2,[22] CO,[23] COS,[24] CO_2,[25] HBr,[26a] N_2O,[26a] and CS_2[26a,26b] and, in addition, emission has been confirmed for ion states of O_2 and C_6F_6 (while found to be apparently absent in C_6H_6 and CCl_4). Each metastable ion state of a molecule, in undergoing a transition to the ion ground state, will emit a series of photon energies corresponding to the vibrational energies of the ground and metastable state. In the absence of competing relaxation processes, the population of ions formed in a metastable state will undergo an exponential decay with a characteristic time constant, which is referred to as the fluorescence lifetime.

For example, fluorescence in CO has been shown to occur in the ultraviolet region resulting from radiative relaxation of the $B^2\Sigma$ state, and in the visible region by relaxation of the $A^2\Pi$ state. The visible emission involves transitions from the lowest nine vibrational levels of the metastable $A^2\Pi$ state to the $v = 0$ vibrational level of the ground $X^2\Sigma$ state. Figure 5 compares the photoelectron spectrum of CO with the fluorescence emission spectrum resulting from ionization.

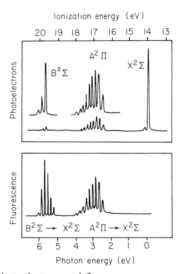

Fig. 5. Photoelectron and fluorescence spectrum of CO.

In order to determine the radiative lifetime of a metastable ion state, it is necessary to know when the ion is originally formed, and the time that the resulting photon is observed. This may be accomplished by detection of the photon in coincidence with a photoelectron ejected from the molecular orbital of interest. The detection of an electron of the proper energy, indicating that ionization into the desired metastable state has occurred, denotes the precise time of the ionization event. Thereafter, the arrival of the resulting photon is anticipated in order to determine the time delay between the ionization and emission of the resulting photon.

Experiments have recently been performed[27] in which light emitted by excited ions is detected in coincidence with photoejection of an electron of preselected energy. The technique, which is very similar to photoelectron–photoion coincidence spectroscopy, has allowed the lifetimes of ionic excited states to be determined for N_2, CO, N_2O, and CO_2. Since most ionic state lifetimes appear to be quite short, less than 250 ns, the detector need only be responsive to a light photon for a short time period following the detection of a photoelectron. Consequently, "false coincidences" due to the detection of randomly emitted light photons are minimized, thereby enhancing the attainable signal/noise ratio.

Another coincidence experiment[28] has utilized high-energy electron impact to ionize selectively CO molecules into a predetermined state. The radiative lifetime of the $B\,^2\Sigma$ state of CO^+ was determined to be 50 ± 5 ns, in good agreement with the photoelectron coincidence experiments.

B. Dissociation

As a means of dispelling surplus energy, unimolecular fragmentation may be the preferred relaxation process of a metastable ion. The initial electronic excitation energy imparted to the parent ion may be converted into vibrational energy of the ground state, whereby a quasi-equilibrium situation is established as the vibrational energy becomes redistributed throughout the ion. Eventually, sufficient vibrational energy becomes localized within one region of the molecular ion to exceed the bond dissociation energy. If enough vibrational energy is present, any of several possible bond cleavages may occur, each resulting in different ion fragmentation products.

Consider as an illustration the case of butadiene, a simple organic molecule, in which two fragmentation processes compete when sufficient excitation energy is present to result in the respective bond cleavages.

$$(C_4H_6^+)^* \begin{cases} \longrightarrow C_3H_3^+ + CH_3 \\ \longrightarrow C_4H_5^+ + H \end{cases}$$

The relative probability of formation of each product depends upon the amount of excitation energy available. It is reasonable to expect that if a minimal amount of vibrational energy is redistributed in a quasi-equilibrium condition, the fragmentation process requiring the least energy will be strongly favored. On the other hand, with increasing amounts of excitation energy, the dissociation would become progressively less specific.

The excitation energy imparted to the parent ion determines not only the relative probability of formation of product ions, but also the time required before dissociation of the parent ion occurs. As higher excitation energies are imparted, less time is required for energy to become sufficiently localized to fragment the ion. A knowledge of ion fragmentation lifetimes therefore provides additional insight into the fragmentation process.

Measurement of fragmentation lifetimes of metastable ions can be carried out in coincidence experiments very similar to those previously described. The detection of a photoelectron would not only denote that the formation of an ion with a predetermined excitation energy has occurred, but also would provide the exact time that the metastable ion originated. The ion could then drift for a selected period of time before mass analysis of the product ion is undertaken.

Such a coincidence study[29] was carried out on the fragmentation of butadiene into $C_3H_3^+$ fragment ions. The lifetime of $C_4H_6^+$ parent ions toward fragmentation varied considerably with the ionization potential of the electron removed by photoionization, as shown in Table II.

TABLE II

Dissociative Lifetime of $C_4H_6^+$

$(C_4H_6^+)^* \longrightarrow C_3H_3^+ + CH_3$	
Ionization energy (eV)	Lifetime of $(C_4H_6^+)^*$ (μs)
11·43	3·7
11·59	3·0
11·69	1·7
11·81	1·1
12·40	<0·1

C. Ion–Molecule Interactions

Should an excited metastable ion collide with a neutral molecule before undergoing radiative or dissociative relaxation, an ion–molecule reaction may occur. The reaction may be categorized as occurring by either a "direct" or "intimate" mechanism.

A "direct" mechanism involves the direct transfer of charge between the two interacting particles. The reaction is a simple affair, with no long-term relationship developing between the two particles. This mechanism is apparently responsible for the collision-induced fragmentation of $CH_2Br_2^+$ yielding CH_2Br^+ and Br atoms, studied in a recent coincidence experiment.[30]

An alternate mechanism consists of the combination of the excited ion and the neutral molecule to form an "intimate" reaction complex which exists for a relatively long period of time (several rotational periods). During the duration of the complex, the vibrational energy of the ion is redistributed among the internal vibrational degrees of freedom, and a quasi-equilibrium condition is present. Finally, when sufficient energy is localized within one bond, dissociation of the complex into the reaction products ensues.

The formation of an "intimate" complex is thought to be responsible for condensation reactions of small unsaturated hydrocarbons, as suggested by the extensive hydrogen rearrangement characteristic of these reactions. For example, a detailed analysis[31] of the ion–molecule reactions occurring in ethylene considers the reaction to occur via the formation of a $C_4H_8^+$ "intimate" complex.

$$C_2H_4^+ + C_2H_4 \longrightarrow [C_4H_8^+]^* \begin{cases} \nearrow C_3H_5^+ + CH_3 \\ \searrow C_4H_7^+ + H \end{cases}$$

The reaction proceeds in two distinct stages: the formation of the complex, and its subsequent dissociation. The amount of vibrational energy of the complex is determined by the initial excitation energy of the reactant ion and the relative kinetic energies of the particles at impact. Once formed, the original excitation energy imparted to the complex is equilibrated among the vibrational modes, and the $C_4H_8^+$ complex loses all "memory" of its origin. In fact, the complex will behave exactly like an excited $C_4H_8^+$ parent ion formed by ionization of a neutral molecule.

As in all ion dissociations, the fragmentation products of the "intimate" complex will depend upon the internal energy of the complex. This in turn depends primarily upon the internal energy of the original reactant ion and its kinetic energy. While the translational kinetic energy of an ion can be readily controlled through the application of an electric field, the internal energy of the reactant ion has often been beyond the control of the experimenter. In a few previous studies[32-34] (of reactions of H_2^+ and NH_3^+) experimenters have been able to investigate the effect of reactant ion excitation upon the reaction products formed.

The application of coincidence methods however offers a general technique of preparing ions in known vibrational states. Detection of photoelectrons of the proper energy indicates that a parent ion has been formed with a predetermined internal energy, and would allow the ion to be selectively transmitted in a beam of pure vibrationally excited ions. This beam could then be intersected with a gas containing neutral molecules to study the resulting products as a function of vibrational energy, or kinetic energy, of the reactant ion.

Similarly, ion fragments with known internal energy could be prepared by imparting a chosen excitation energy to the parent ion, and then selectively transmitting ions of the proper kinetic energy. Since energy in excess of that required for dissociation will be partitioned between the kinetic energy and internal energy of the fragments, selection of ions of known kinetic energy could effectively define the vibrational energy of the ion fragment.

IV. ELECTRON–ELECTRON COINCIDENCE EXPERIMENTS

Although rapid development of electron spectroscopy has ensued to a large extent from interest in photoelectron spectroscopy, the use of photon sources as a means of excitation is extremely limited. While gaseous discharge lamps can provide radiation in the ultraviolet and vacuum-uv regions, few usable photon sources are available with energies between 25 eV and x-rays (above 300 eV). The one available source of continuous radiation within this energy range is synchrotron radiation from high-energy particle accelerators.[35] Unfortunately, due to the prohibitive cost and large size, synchrotrons could not be considered as potentially useful to many chemists and physicists, who have no access to a synchrotron (or for that matter have even seen one). It is clear then that the limited available energies of photon excitation imposes severe restrictions, particularly for the determination of ionization cross-sections as a function of excitation energy.

These limitations, however, can be circumvented by the simulation of photon excitation with high energy electron impact in an electron–electron coincidence experiment.

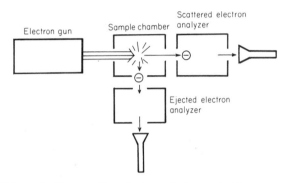

Fig. 6. Schematic diagram of an electron–electron coincidence experiment.

When a sufficiently energetic beam of electrons of energy E_{inc} undergoes collision with sample gas molecules, some electrons will be scattered in the original beam direction with a reduced energy E_{scat}. Effectively, the electron has acted as an excitation source of energy $(E_{inc} - E_{scat})$, and the detection of the scattered electron with some preselected energy will therefore indicate that an ionization has occurred with energy $(E_{inc} - E_{scat})$ imparted to the molecule. This energy will be able to ionize the molecule, the excess appearing as kinetic energy of the ejected electron. Measurement of the kinetic energy spectrum of ejected electrons, formed in coincidence with scattered electrons which have undergone a predetermined energy loss, allows the "photo-electron" spectrum of a molecule to be obtained, at any desired excitation energy.

The basic experimental apparatus for performing electron–electron coincidence experiments is depicted in Fig. 6. A high-energy electron beam enters the sample chamber where electrons undergo collision with molecules and are scattered. Those electrons scattered in the forward direction, which have undergone a loss of kinetic energy corresponding to a desired excitation energy imparted to the molecule, are transmitted by an energy analyzer. Upon detection of the forward scattered electron, indicating that an ionization has occurred, the arrival of the ejected electron is anticipated within a short period of time, if its energy is within the range transmitted by the ejected electron analyzer. In this manner, a spectrum of the ejected electron energy may be obtained for a predetermined excitation energy imparted by the incident beam.

Because such a coincidence technique allows photon sources of any desired energy to be simulated, the relative probability for any ionization process of the molecule can be investigated at various excitation energies. Relative ionization probabilities for the formation of various ion states have been determined for CO,[28, 36] NH$_3$,[37] and H$_2$O.[38] In addition, such coincidence experiments have allowed the angular distribution of ejected electrons from Ne,[39] Ar,[40] and Kr[40] to be derived as a function of excitation energy.

Consider as an illustration of the capabilities of the technique, the results of a study[36] of CO molecule using electron–electron coincidence spectroscopy. Carbon monoxide has three valence molecular orbitals giving rise to bands in the photoelectron spectrum: the $X^2\Sigma$ orbital resulting in one sharp peak at 14·0 eV, the $A^2\Pi$ vibrational series centered at about 17 eV, and the $B^2\Sigma$ band at about 19·7 eV. The appearance of the electron–electron coincidence spectrum of CO at an effective excitation energy of 21 eV is shown in Fig. 7. Also shown is how the probability of ionization from each of the molecular orbitals varies with the excitation energy.

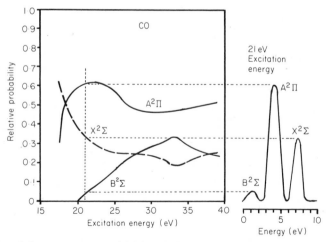

Fig. 7. Ejected electron spectrum of CO in coincidence with excitation by electron impact.

As is expected of any coincidence method, very low signal count rates are achieved in electron–electron coincidence experiments, requiring the sacrifice of resolution in order to achieve acceptable data collection rates. Typically, electron energy resolution in electron–electron coincidence experiments has only been within an energy range of 0·8 eV FWHM, less than one-tenth the resolution attained routinely in analogous one-parameter experiments (compare, for example, the coincidence ejected electron spectrum of CO in Fig. 7) with the photoelectron spectrum in Fig. 5.

In all fairness, it should be mentioned that the same variation of partial ionization cross-sections in CO has also recently been investigated[41] without the use of electron–electron coincidence detection. By selecting various resonance emission lines (from several gases) with a monochromator in the wavelength range 800–304 Å, a useful range of photon energies was available up to 40 eV.

Electron–electron coincidence spectra obtained so far have featured sufficiently poor resolution and signal count rates to discourage widespread interest for other applications. Nonetheless, as a general source of excitation, electron impact, followed by the coincidence detection of two electrons, appears in principle to be ideal. Remarkable versatility could be offered, where the experimenter could alter the energy of the source and its effective monochromaticity to suit his convenience. With additional refinement of the technique, considerable potential exists for the widespread adoption of electron–electron coincidence detection in the field of electron spectroscopy.

V. CONCLUSION

By now, the reader should be convinced that coincidence experiments, vital to nuclear research efforts for a long time, are for some applications the only unambiguous means of investigating chemical systems. Skeptics may argue that the characteristically poor data collection rates will require intolerably long periods of time to scan spectra, preventing the technique from progressing into general use.

The evolutionary development of a spectroscopic technique may be envisioned as occurring in two distinct stages. Initial work is done with research instruments, in order to acquire specific fundamental information about molecules or reactions unobtainable through other methods. After substantial development, the spectrometer may undergo a metamorphosis, losing superfluous cables and gauges while acquiring an exterior covering of panels, into a new species—the commercial spectrometer. Although almost identical in operation to the research instrument which preceded it, the commercial spectrometer fulfils an entirely different function, the routine analysis of samples.

It is entirely possible that electron spectrometers which detect two processes in coincidence may never become commercially available. To allow an expensive spectrometer to operate for days collecting data for a high-resolution spectrum of a routine sample would be economically unfeasible. Research instruments, on the other hand, are often operated only a small percentage of the time, and therefore complete utilization of the instrument (including operation during evenings and weekends) would virtually eliminate the restrictions of low data collection rates. Furthermore, as the solution of some fundamental problems may only be accessible through the use of coincidence experiments, it is reasonable to expect that coincidence techniques will play a widening role in the application of electron spectroscopy to basic chemical research.

Acknowledgement

We wish to thank Professor Tomas Baer for a most helpful communication explaining the coincidence experiments done by his group at the University of North Carolina.

REFERENCES

1. M. E. Gellender and A. D. Baker, *J. Electron Spectrosc.* **4**, 249 (1974).
2. N. W. Reid, *Int. J. Mass Spectrom. Ion Phys.* **6**, 1 (1971).
3. A. Maccoll (ed.), "MTP Int. Rev. of Sci." (Phys. Chem. Series 1), Vol. 5.
4. H. M. Rosenstock and M. Krause, *Advances in Mass Spectrosc.* **2**, 251 (1963).
5. J. H. D. Eland, *Int. J. Mass Spectrom. Ion Phys.* **9**, 3917 (1972).
6. B. Brehm, J. H. D. Eland, R. Frey, and A. Küstler, *Int. J. Mass Spectrom. Ion Phys.* **12**, 197 (1973).
7. B. Brehm, J. H. D. Eland, R. Frey, and A. Küstler, *Int. J. Mass Spectrom. Ion Phys.* **12**, 213 (1973).
8. J. H. D. Eland, *Int. J. Mass. Spectrom Ion Phys.* **12**, 389 (1973).
9. R. Stockbauer, *J. Chem. Phys.* **58**, 3800 (1973).
10. C. J. Danby and J. H. D. Eland, *Int. J. Mass Spectrom. Ion Phys.* **8**, 153 (1972).
11. P. Tsai, T. Baer, A. S. Werner, and S. F. Lin, *J. Phys. Chem.* **79**, 570 (1975).
12. I. G. Simm, C. J. Danby, and J. H. D. Eland, *Int. J. Mass Spectrom. Ion Phys.* **14**, 285 (1974).
13. I. G. Simm, C. J. Danby, and J. H. D. Eland, *J.C.S. Chem. Comm.* 833 (1973).
14. J. H. D. Eland, *Int. J. Mass Spectrom. Ion Phys.* **13**, 457 (1974).
15. B. Brehm and E. von Puttkamer, *Z. Naturforsch.* **22a**, 8 (1967).
16. E. von Puttkamer, *Z. Naturforsch.* **25a**, 1062 (1970).
17. B. Brehm, V. Fuchs, and P. Kebarle, *Int. J. Mass Spectrom. Ion Phys.* **6**, 279 (1971).
18. M. E. Gellender and A. D. Baker, *Int. J. Mass Spectrom. Ion Phys.* **17**, 1 (1975).
19. K. E. McCulloh, T. E. Sharp, and H. M. Rosenstock, *J. Chem. Phys.* **42**, 3501 (1965).
20. J. L. Franklin, P. M. Hierl, and D. A. Whan, *J. Chem. Phys.* **47**, 3148 (1967).
21. B. Brehm, R. Frey, A. Küstler, and J. H. D. Eland, *Int. J. Mass Spectrom. Ion Phys.* **13**, 251 (1974).

22. D. L. Judge and G. L. Weissler, *J. Chem. Phys.* **48**, 4590 (1968).
23. D. L. Judge and L. C. Lee, *J. Chem. Phys.* **57**, 455 (1972).
24. D. L. Judge and M. Ogawa, *J. Chem. Phys.* **51**, 2035 (1969).
25. D. L. Judge and L. C. Lee, *J. Phys.* **B6**, 2150 (1973).
26a. B. S. Schneider and A. L. Smith, in: "Electron Spectroscopy" (Proc. of an Int. Conf., 1972, D. A. Shirley, ed.), p. 335.
26b. J. Daintith, J. P. Maier, D. A. Sweigart, and D. W. Turner, in: "Electron Spectroscopy" (Proc. of an Int. Conf., 1972, D. A. Shirley, ed.), p. 189.
27. M. Block and D. W. Turner, *Chem. Phys. Letters*, **30**, 344 (1975).
28. C. Backx, M. Klewer, and M. J. Van der Wiel, *Chem. Phys. Letters*, **20**, 100 (1973).
29. A. S. Werner and T. Baer, Paper presented at 22nd Annual Conf. on Mass Spectrom., Phil., USA (1974); A. S. Werner and T. Baer, *J. Chem. Phys.* **62**, 2900 (1975).
30. T. Baer, L. Squires, and A. S. Werner, *Chem. Phys.* **6**, 325 (1974).
31. S. E. Buttrill, Jr., *J. Chem. Phys.* **52**, 6174 (1970).
32. W. A. Chupka, M. E. Russell, and K. Rafaey, *J. Chem. Phys.* **48**, 1518 (1968).
33. W. A. Chupka and M. E. Russell, *J. Chem. Phys.* **48**, 1527 (1968).
34. W. A. Chupka and M. E. Russell, *J. Chem. Phys.* **49**, 5426 (1968).
35. M. L. Perlman, E. M. Rove, and R. E. Watson, *Physics Today*, **27**, 30 (1974).
36. M. J. Van der Wiel and C. E. Brion, *J. Electron Spectrosc.* **1**, 309 (1973).
37. M. J. Van der Wiel and C. E. Brion, *J. Electron Spectrosc.* **1**, 443 (1973).
38. G. R. Branton and C. E. Brion, *J. Electron Spectrosc.* **3**, 129 (1974).
39. M. J. Van der Wiel and C. E. Brion, *J. Electron Spectrosc.* **1**, 439 (1973).
40. G. R. Branton and C. E. Brion, *J. Electron Spectrosc.* **3**, 123 (1974).
41. J. A. R. Samson and J. L. Gardner, *J. Electron Spectrosc.* **8**, 35 (1976).

Subject Index